CLASSICAL MECHANICS
A Modern Perspective

McGRAW-HILL SERIES IN FUNDAMENTALS OF PHYSICS: AN UNDERGRADUATE TEXTBOOK PROGRAM

INTRODUCTORY TEXTS

Young · Fundamentals of Mechanics and Heat
Kip · Fundamentals of Electricity and Magnetism
Young · Fundamentals of Optics and Modern Physics
Beiser · Concepts of Modern Physics

UPPER-DIVISION TEXTS

Barger and Olsson · Classical Mechanics: A Modern Perspective
Beiser · Perspectives of Modern Physics
Cohen · Concepts of Nuclear Physics
Elmore and Heald · Physics of Waves
Kraut · Fundamentals of Mathematical Physics
Longo · Fundamentals of Elementary Particle Physics
Meyerhof · Elements of Nuclear Physics
Reif · Fundamentals of Statistical and Thermal Physics
Tralli and Pomilla · Atomic Theory: An Introduction to Wave Mechanic

CLASSICAL MECHANICS

A Modern Perspective

V. BARGER, Ph.D.
Professor of Physics
University of Wisconsin

M. OLSSON, Ph.D.
Associate Professor of Physics
University of Wisconsin

McGRAW-HILL BOOK COMPANY
New York St. Louis San Francisco Düsseldorf Johannesburg
Kuala Lumpur London Mexico Montreal New Delhi Panama
Rio de Janeiro Singapore Sydney Toronto

Library of Congress Cataloging in Publication Data

Barger, Vernon D 1938-
Classical mechanics.

(McGraw-Hill series in fundamentals of physics)
1. Mechanics. I. Olsson, Martin, 1938- joint
author. I. Title.
QA805.B287 531 72-5697
ISBN 0-07-003723-X

This book was set in Times New Roman.
The editors were Jack L. Farnsworth and Michael Gardner;
the designer was J. E. O'Connor;
and the production supervisor was John A. Sabella.
The drawings were done by Vantage Art, Inc.
The printer and binder was Kingsport Press, Inc.

CLASSICAL MECHANICS
A Modern Perspective

1234567890 KPKP 798765432

Contents

Preface

The study of classical mechanics offers an unequaled opportunity for physical insights into events of everyday life. For this reason it is desirable that a textbook for an intermediate-level course in mechanics be suitable for use both by physics majors and by students from other disciplines. This textbook attempts to meet this basic goal by presentation of topics of widespread popular interest. By repeated application of the principles of mechanics to such diverse topics as sports, seagulls, boomerangs, satellites, and tides, we try to develop physical intuition as well as proficiency with mathematical methods.

This text was designed for an intensive one-semester course in theoretical mechanics at the junior-senior level. A knowledge of general physics, integral calculus, and differential equations is assumed. The problems at the end of each chapter are designed to illustrate the methods developed in the text and to further stimulate the student's interest in mechanics. Since a mastery of problem-solving techniques is an essential requirement for a mechanics course, we have included a number of easy problems to permit the student to get a wide range of practice.

A major departure in our book from the conventional approach to the subject is the introduction of the Lagrange formulation of the equations of motion at an early stage (end of Chapter 2). In the conventional organization, Lagrange's equations are presented near the end of a one-semester course, and the student rarely develops a reasonable familiarity with lagrangian methods. Since our organization encourages the student to solve problems in later chapters by direct application of Newton's laws and Lagrange's equations, he can achieve a mastery of both techniques.

In our experience, about 85 percent of this text can be covered in a 15-week semester with three lecture hours per week. A majority of our students also attended an optional session each week for discussion of solutions to assigned homework problems. In the choice of material for lectures, any of the following sections can be selectively omitted without loss of continuity in the text: 2-9, 2-10, 2-12, 4-6, 5-11, 6-4, 6-11, 6-12, 6-13. The last three sections of Chapter 7 can be covered or omitted, as time permits. Since the number of different topics in mechanics which can be discussed in the course of a semester is necessarily limited, we do not include chapters on strength of materials, continuous media, or relativity.

The first chapter contains novel one-dimensional applications involving frictional, gravitational, and harmonic forces in the sports of drag racing, sky diving, and archery. The simple harmonic oscillator with damping and driving forces is given appropriate attention. Chapters 2, 3, and 4 are organized around the fundamental conservation laws of energy, momentum, and angular momentum. As an application of energy conservation in Chapter 2, we calculate the minimum velocity needed to escape the earth's gravitational attraction. In Chapter 3 the Apollo moon rocket is used as a concrete example in a section dealing with variable mass. Collisions of billiard balls are discussed in center-of-mass and laboratory coordinate systems to develop familiarity with momentum-conservation methods. The concept of a differential cross section is introduced in a calculation of the likelihood that BBs ricochet off a cylindrical pipe in a given direction. In Chapter 4 the trajectories for planetary motion are derived by two alternative methods. The orbit period for a proposed NASA weather satellite is determined from Kepler's law. The forthcoming Grand Tours of our solar system on gravity-assistance trajectories are discussed as examples of the central-force problem. The differential cross section for Rutherford scattering is derived in the concluding section of Chapter 4.

In the treatment of rigid-body motion in Chapter 5, the return of a boomerang is explained in terms of the gyroscope effect. "Draw" and "follow" shots in billiards and the action of "superballs" are presented as intriguing examples of rigid-body rotations. Chapter 6 is concerned with applications of the law of motion in moving coordinate systems. The relevance of the centrifugal and Coriolis forces in a variety of physical situations is indicated. Several sections are devoted to the motion of spinning tops, concluding with an analysis of the flipping motion of the amazing tippie-top.

Chapter 7 begins with a proof that the net gravitational attraction of a point mass on a spherically symmetrical body acts as though the mass of the body were concentrated at its center. We then proceed to calculate the tides on earth due to the moon and sun. As a useful application of lagrangian methods in gravitation, we discuss the technique for automatic attitude stabilization of a satellite orbiting the earth. In somewhat more advanced sections of Chapter 7, we calculate the gravity field and shape of the oblate earth.

It is a pleasure to acknowledge helpful conversations with numerous colleagues and friends. Suggestions by Professors L. Durand, III, C. Goebel, and R. March were of particular value in the development and refinement of the text. A critical and thorough review of the manuscript by Professor C. Goebel, for which we are especially grateful, led to substantial improvements in various sections. We benefited by the able assistance of Mr. Kevin Geer as teaching assistant in charge of problem-solving sessions. Many thanks go to Mrs. Laurel Hermanson for typing the several drafts of the manuscript. One of us (V.B.) is grateful for the kind hospitality extended by Professor San Fu Tuan and other members of the Physics Department at the University of Hawaii, where part of the manuscript was prepared.

V. Barger
M. Olsson

CHAPTER 1

The Beginnings

Classical mechanics is one of the most satisfying subjects of study in all of science. In order to understand and appreciate how both everyday and esoteric things in our world work, some knowledge of the principles of mechanics is essential. In this age everyman needs to know mechanics to fulfill and enrich his daily existence.

The formulation of classical mechanics represents a giant milestone in man's intellectual and technological history, as the first mathematical abstraction of physical theory from empirical observation. This crowning achievement is rightly accorded to Isaac Newton (1642–1720), who modestly acknowledged that if he had seen further than others, "it is by standing upon the shoulders of Giants." However, the great Laplace characterized Newton's work as the supreme exhibition of individual intellectual effort in the history of the human race.

Newton translated his interpretation of various physical observations into a compact mathematical theory. Three centuries of experience indicate that all mechanical behavior in the everyday domain can be understood from Newton's theory. His simple hypotheses are now elevated to the exalted status of laws, and these are our point of embarkation into the subject.

1-1 NEWTONIAN THEORY

The newtonian theory of mechanics is customarily stated in three laws. According to the first law, a particle continues in uniform motion unless a force acts on it. The first law is a fundamental observation that physics is simpler when viewed from a certain kind of coordinate system, called an *inertial frame*. One cannot define an inertial frame except by saying that it is a frame in which Newton's laws hold. However, once one finds (or imagines) one such frame, all other inertial frames are moving in straight lines at constant velocity (i.e., nonaccelerating) with respect to it. A coordinate system fixed on the surface of the earth is not an inertial frame because of the accelerations due to the rotation of the earth, and its motion around the sun. Nevertheless, for many purposes it is an adequate approximation to regard a coordinate frame fixed on the earth's surface as an inertial frame. Indeed, Newton himself discovered nature's true laws while riding on the earth!

The meat of Newton's theory is contained in the second law, which states that *the time rate of change of momentum of a particle is equal to the force acting on the particle*,

$$F = \frac{dp}{dt} \tag{1-1}$$

where the momentum p is given by the product of (mass) $\times$ (velocity) for the particle.

$$p = mv \tag{1-2}$$

The second law provides a definition of force. The physics content of the second law depends on empirical forms for the forces as functions of positions and velocities. The force in Eq. (1-1) can be a function of x, v, and t, and so

$$F(x, v, t) = \frac{dp}{dt} = m\frac{d^2x}{dt^2}$$

is a differential equation. While Newton's laws promise to apply to any situation in which one can specify the force at all times, very few interesting physical problems lead to force laws amenable to simple mathematical solution. To approximate the true force law by a sufficiently accurate approximate form is one of the arts that will be taught in this book. However, in this modern age of digital computers, one can handle incredibly complicated force laws by the brute-force method of numerical integration.

In the special case $F = 0$, integration of Eq. (1-1) gives $p =$ constant in accordance with the first law. A more familiar expression of the second law in Eq. (1-1) is

$$F = ma \tag{1-3}$$

where $a = dv/dt$ is the acceleration.

The third law states that if particle A experiences a force due to particle B, then B feels simultaneously a force of equal magnitude but in the opposite direction. This law is extremely useful, especially in the treatment of rigid-body motion, but its range of applicability is not as universal as the first two laws. The third law breaks down when the interaction between the particles is electromagnetic.

It is a remarkable fact that macroscopic phenomena can be explained by such a simple set of mathematical laws. As we shall see, the mathematical solutions to some problems can be complex; nevertheless, the physical basis is just Eq. (1-1). Of course, there is still a great deal of physics to put into Eq. (1-1), namely, the laws of force for specific kinds of interactions.

1-2 INTERACTIONS

The gravitational and electromagnetic forces determine our whole condition of life. Newton deduced the following force law for gravitation by studying data phenomenologically fitted by Kepler on the motion of planets and satellites in our solar system.

$$F = -\frac{GM_1M_2}{r^2} \tag{1-4}$$

The force between masses M_1 and M_2 is proportional to the masses and inversely proportional to the square of the distance between them. The negative sign in Eq. (1-4) denotes an attractive force between the masses. Newton proposed that this gravitational law was universal, the same force applying on the earth as between celestial bodies. The universality of the gravitational law can be verified, and the proportionality constant G determined, by delicate experimental measurements of the force between masses in the laboratory. The value of G is

$$G = 6.67 \times 10^{-11} m^3/(\text{kg})(\text{s}^2) \tag{1-5}$$

The dominant gravitational force on an object located on the surface of the earth is due to the attraction from the earth. The gravitational attraction on a point mass from a spherically symmetric body acts as if all the mass of the body were concentrated at its center, as Newton rigorously proved from his invention of calculus. We will give a proof of this assertion in Chap. 7. For an object of mass m on the surface of the earth, the force law of Eq. (1-4) becomes

$$F = -m \frac{M_e G}{R_e^2} = -mg \tag{1-6}$$

where g is the gravitational acceleration,

$$g = 9.8 \text{ m/s}^2$$

The values of mass and radius of the earth in Eq. (1-6) are

$$R_e = 6{,}371 \text{ km}$$
$$M_e = 5.97 \times 10^{24} \text{ kg}$$

Since the earth's radius is large, the gravitational force on an object anywhere between the surface of the earth and the top of the atmosphere ($\approx$200 km up) is given with reasonable accuracy by Eq. (1-6). Consequently, in many applications on earth, we can neglect the variation of the gravitational force with position.

The static Coulomb force between two charges e_1 and e_2 is similar in form to the gravitational-force law of Eq. (1-4).

$$F = \frac{e_1 e_2}{r^2} \tag{1-7}$$

This force is attractive if the charges are opposite in sign and repulsive if the charges are of the same sign.

Another force with a wide range of application is the spring force or Hooke's law, which is expressed as

$$F = -kx \tag{1-8}$$

Here k is a spring constant which is dependent on the properties of the spring and x is the extension of the spring from its relaxed position. This particular force law is a very good approximation in many physical situations, such as the stretching or bending of materials which are initially in equilibrium.

Frictional forces play a crucial role in damping or retarding motion initiated by other forces. The static frictional force between two solid surfaces is

$$|f| \leq \mu_s N \tag{1-9}$$

The force f acts to prevent sliding motion. N is the perpendicular force (normal force) holding the surfaces together, and μ_s is a material-dependent coefficient. Equation (1-9) is an *approximate* formula for frictional forces which has been deduced from empirical observations. The frictional force which retards the motion of sliding objects is given by

$$f = \mu_k N$$

It is observed that this force is nearly independent of the velocity of the motion, for velocities which are neither too small or too large. For a given set of surfaces, the coefficient of kinetic friction μ_k is less than the coefficient of static friction μ_s.

Frictional laws to describe the motion of a solid through a fluid or a gas are often complicated by such effects as turbulence. However, for sufficiently small velocities, the approximate form

$$f = -bv \tag{1-10}$$

where b is a constant, holds. At higher, but still subsonic velocities, the frictional-force law

$$f = -cv^2 \tag{1-11}$$

is approximately true. The drag force on a propeller airplane is remarkably well represented by a constant times the square of the velocity.

Externally imposed forces can take on a variety of forms. Of those depending explicitly on time, sinusoidal oscillating applied forces like

$$F = F_0 \cos \omega t \tag{1-12}$$

are frequently encountered in physical situations.

In a general case the forces can be position-, velocity-, and time-dependent.

$$F = F(x, v, t) \tag{1-13}$$

Among the most interesting and easily solved examples are those in which the forces depend on only one of the above three variables, as illustrated by the examples in the following three sections.

1-3 THE DRAG RACER: FRICTIONAL FORCE

A number of interesting engineering-type problems can be solved from straightforward application of Newton's laws. As an illustration, suppose we want to design a drag racer which will achieve maximum possible acceleration when starting from rest. We assume that the engine of the racer can apply an arbitrary torque to the rear wheels, and our problem is to determine the optimal weight distribution of the racer. The external forces on the racer which must be taken into account are (1) gravity, (2) the normal forces supporting the racer at the wheels, and (3) the frictional forces which oppose the rotation of the powered rear wheels. A sketch indicating the various external forces is given in Fig. 1-1. The gravity force Mg acts as if all the weight were concentrated at the center of mass. This is a familiar

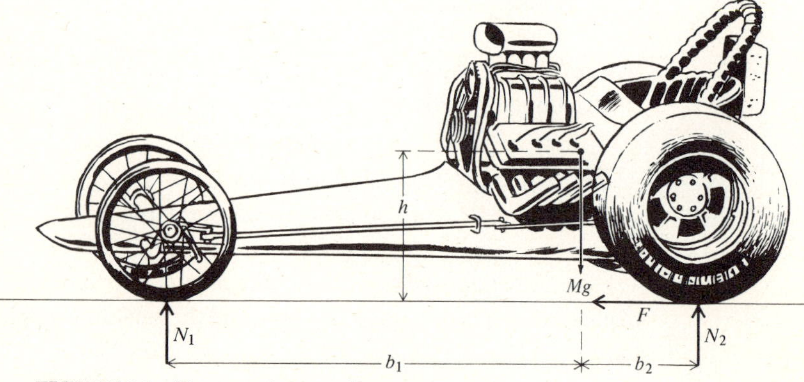

FIGURE 1-1 Forces acting on a drag racer.

fact that we shall prove in Chap. 5. Since the racer is in vertical equilibrium, the sum of the external vertical forces must vanish.

$$N_1 + N_2 - Mg = 0 \tag{1-14}$$

Both N_1 and N_2 must be positive. A further equilibrium requirement is imposed by the absence of any turning motion about the center of mass. The expression of Newton's law for angular equilibrium is that the sum of torques (or moments) about the center of mass must vanish,

$$N_2 b_2 - N_1 b_1 - Fh = 0 \tag{1-15}$$

For the horizontal motion we apply Newton's second law,

$$F = Ma \tag{1-16}$$

Bearing in mind that the frictional force F is bounded by

$$F \leq \mu N_2 \tag{1-17}$$

the maximum friction force occurs just as the racer tires begin to slip relative to the drag strip. All the information necessary for the racer design is contained in the preceding four equations. Our design criteria of maximal initial acceleration require in Eq. (1-16) the maximum friction force $F = \mu N_2$, with N_2 at its maximum value. Referring back to Eq. (1-14), a maximal $N_2 = Mg$ is obtained when $N_1 = 0$; that is, the back wheel completely supports the racer. The greatest possible acceleration is then

$$a_{\text{max}} = \frac{\mu (N_2)_{\text{max}}}{M} = \mu g \tag{1-18}$$

To achieve this, the design requirements found from Eq. (1-15) are

$$Mgb_2 = Ma_{\text{max}} h = M\mu g h$$

or

$$b_2 = \mu h \tag{1-19}$$

Hence the mass of the racer is not an important factor in its design. Under normal track conditions the coefficient of friction μ is about 1. Thus a racer can achieve an acceleration of about 9.8 m/s^2, provided that the racer is designed so that the distance of the center of mass from the race track is about the same as the distance forward of the rear axle. In actual design a small normal force N_1 on the front wheels is allowed for steering purposes.

The standard drag strip is 1/4 mi ($\approx$400 m) in length. If we assume that the racer can maintain the maximum acceleration for the duration of a race and that the coefficient of friction is constant, we can calculate the final velocity and the elapsed time. The differential form of the second law is

$$F = Ma = M\frac{dv}{dt} = M\frac{d^2x}{dt^2} \tag{1-20}$$

When the acceleration a is constant, a single integration

$$\int_{v_0}^{v} dv = a\int_0^t dt$$

gives

$$v - v_0 = at \tag{1-21}$$

Using $dx = v dt$, a second integration

$$\int_{x_0}^{x} dx = \int_0^t (v_0 + at)\,dt$$

yields

$$x - x_0 = v_0 t + \tfrac{1}{2}at^2 \tag{1-22}$$

With our initial conditions $v_0 = 0$, $x_0 = 0$, we can eliminate t from Eqs. (1-21) and (1-22) to obtain

$$v = \sqrt{2ax} \tag{1-23}$$

Substituting $a = 9.8$ m/s^2 and $x = 0.40$ km, we find $v = 89$ m/s, or 200 mi/h! The time elapsed, $t = v/a$, is about 9 s. The world drag-racing record held in 1970 by Don (Big Daddy) Garlits is $v = 107$ m/s (240 mi/h), with elapsed time of 6.5 s. With very wide tires, coefficients of friction considerably greater than $\mu = 1$ are realized in drag racing.

1-4 SPORT PARACHUTING: VISCOUS FORCE

The sport of skydiving visually illustrates the effect of the viscous frictional force of Eq. (1-11). Immediately upon leaving the aircraft, the jumper accelerates downward due to the gravity force. As his velocity increases, the air resistance exerts a greater retarding force, and eventually approximately balances the pull of gravity. From this time onward the descent of the diver is at a uniform rate, called the *terminal velocity*. The terminal velocity in a spread-eagle position is, roughly, 120 mi/h. By assuming a vertical head-down position, the diver can decrease his exposed surface area, thereby lowering the air resistance [smaller value of c in Eq. (1-11)], and increase his terminal velocity of descent. Eventually, of course, the diver opens his parachute. This dramatically increases the air resistance and correspondingly reduces his terminal velocity, to allow a soft impact with the ground.

To analyze the physics of skydiving, we shall assume that the motion is vertically downward and choose a coordinate system with $x = 0$ at the earth's surface and positive upward. In this coordinate frame, downward forces are negative. We approximate the external force on the diver as

$$F = -mg + cv^2 \tag{1-24}$$

The frictional force is positive, as required for an upward force. The terminal velocity is reached when the opposing gravity and frictional forces balance, giving $F = 0$. Under this condition, the terminal velocity is

$$v_t = \sqrt{\frac{mg}{c}} \tag{1-25}$$

To solve the differential equation of motion,

$$F = m\frac{dv}{dt} = -mg + cv^2 \tag{1-26}$$

we rearrange the factors and integrate.

$$\int_0^v \frac{dv}{v_t^2 - v^2} = -\frac{g}{v_t^2}\int_0^t dt \tag{1-27}$$

In Eq. (1-27) the frictional coefficient c has been replaced by v_t from Eq. (1-25). We obtain

$$\frac{1}{2v_t}\ln\left(\frac{v_t + v}{v_t - v}\right) = -\frac{g}{v_t^2}t \tag{1-28}$$

which can be inverted to express v in terms of t.

$$v = -v_t\frac{1 - \exp(-2gt/v_t)}{1 + \exp(-2gt/v_t)} \tag{1-29}$$

At large times the decreasing exponentials go rapidly to zero and v approaches the terminal velocity.

$$v \to -v_t$$

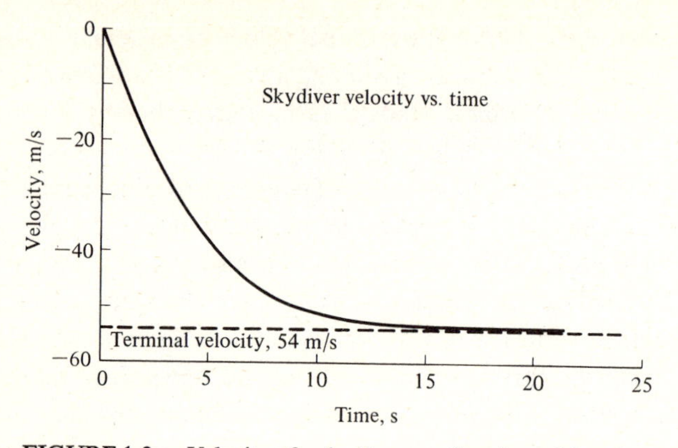

FIGURE 1-2*a* **Velocity of a skydiver as a function of time for a terminal velocity of 54 m/s.**

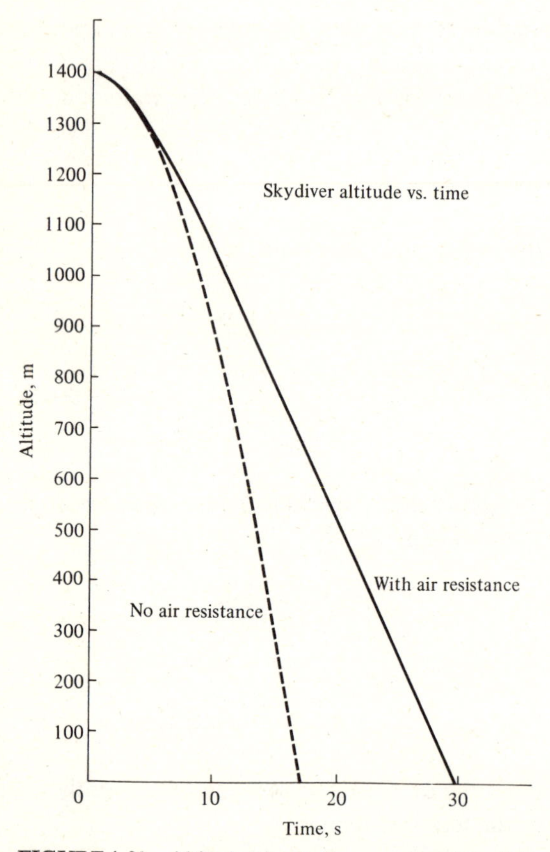

FIGURE 1-2*b* **Altitude of a skydiver as a function of time for a terminal velocity of 54 m/s.**

Although the limiting velocity is exactly reached only at infinite time, it is approximately reached for times $t \gg v_t/2g$. A typical value for v_t on a warm summer day is 54 m/s (120 mi/h) for a 70-kg diver in a spread-eagle position. After a time

$$t = \frac{2v_t}{g} = \frac{2(54)}{9.8} = 11 \text{ s} \tag{1-30}$$

the skydiver would be traveling about 52 m/s with his parachute unopened! The velocity of the diver as a function of time is plotted in Fig. 1-2. In actual practice skydivers reach terminal velocity in 11 to 12 s after falling about 380 m.

To calculate the distance the diver has fallen after a specific elapsed time, we integrate $dx = v\,dt$ from Eq. (1-29).

$$\int_h^x dx = -v_t \int_0^t \left(1 - \frac{2\exp(-2gt/v_t)}{1 + \exp(-2gt/v_t)}\right) dt \tag{1-31}$$

The result of the integration is

$$h - x = v_t\left[t - \frac{v_t}{g} \ln\left(\frac{2}{1 + \exp(-2gt/v_t)}\right)\right] \tag{1-32}$$

At time $t = 2v_t/g$, the diver has fallen a distance $(h - x)$, given by

$$h - x = \frac{v_t^2}{g}\left[2 - \ln\left(\frac{2}{1 + e^{-4}}\right)\right] = \frac{(54)^2}{9.8}(2 - 0.7) = 385 \text{ m}$$

Sky divers normally free-fall about 1,400 m (in 30 s) so that much of the descent is at terminal velocity. (See Fig. 1-3 on front endpaper.)

1-5 ARCHERY: SPRING FORCE

The action of the archer's bow in shooting an arrow is generated by the spring force of Eq. (1-8). A 30-lb (134-N) bow with a 0.72-m draw d has a spring constant k given by

$$k = \frac{|F|}{d} = \frac{134}{0.72} = 186 \text{ kg/s}^2 \tag{1-33}$$

Upon release of the bowstring, the motion of the arrow is described by the second law,

$$m\frac{dv}{dt} = -kx \tag{1-34}$$

To solve this differential equation for the velocity as the arrow leaves the bow, we use the chain rule of differentiation,

$$\frac{dv}{dt} = \frac{dv}{dx}\frac{dx}{dt} = v\frac{dv}{dx} \tag{1-35}$$

Substituting into Eq. (1-34) and rearranging factors, we obtain

$$m\int_0^v v\,dv = -k\int_{-d}^0 x\,dx$$

or

$$\tfrac{1}{2}mv^2 = \tfrac{1}{2}kd^2 \tag{1-36}$$

The velocity of the arrow as it leaves the bowstring is given by

$$v = d\sqrt{\frac{k}{m}} \tag{1-37}$$

Thus the longer the draw and the stronger the bow, the faster the arrow velocity. For a typical target arrow, with weight $m = 23$ g, the velocity is

$$v = 0.72\sqrt{\frac{186}{23 \times 10^{-3}}} = 65 \text{ m/s}$$

This is almost double the maximum speed of a fastball thrown by a professional baseball player!

1-6 METHODS OF SOLUTION

For the general motion of a particle in one dimension, the equation of motion is

$$m\frac{d^2x}{dt^2} = F\left(x, \frac{dx}{dt}, t\right) \tag{1-38}$$

Since this is a second-order differential equation, the solution for x as a function of t involves two arbitrary constants. These constants can be fixed from the physical boundary conditions, such as the position and velocity at the initial time. In the examples of Secs. 1-3 to 1-5, we have introduced several techniques for solving Eq. (1-38). In case F depends at most on one of the variables x, dx/dt, and t, the formal solution of Eq. (1-38) is straightforward. We now run through the methods of solution to the differential equations of motion for this limited class of force laws.

For a force that depends only on x, we may use the chain rule of Eq. (1-35), and integrate Eq. (1-38) to obtain

$$m\int v\,dv = \int F(x)\,dx + C_1 \tag{1-39}$$

or

$$v = \sqrt{\frac{2}{m}}\sqrt{\int^x F(x)\,dx + C_1} \tag{1-40}$$

where C_1 is a constant of integration. This method of solution was employed in the archery discussion of Sec. 1-5. The solution for $x(t)$ is found by substituting $v = dx/dt$ in Eq. (1-40), rearranging factors, and integrating.

$$\int^x \frac{dx}{\sqrt{\int^x F(x')\,dx' + C_1}} = \sqrt{\frac{2}{m}}\int^t dt + C_2 \tag{1-41}$$

The integration constants C_1, C_2 can be fixed from the boundary conditions, such as the initial velocity and position.

With a velocity-dependent force we can integrate Eq. (1-38) as follows:

$$m\int^v \frac{dv}{F(v)} = \int^t dt + C_1 \tag{1-42}$$

We used this technique in the sky-diving analysis of Sec. 1-4. The result of the integration gives $v(t)$, which can then be integrated over t to find $x(t)$.

The solution of Eq. (1-38) for a time-dependent force can be obtained from direct integration.

$$m\int^v dv = \int^t F(t)\,dt + C_1 \tag{1-43}$$

A second integration leads to the solution for $x(t)$.

$$m\int^x dx = \int^t \left[\int^t F(t')\,dt' + C_1\right] dt + C_2 \tag{1-44}$$

If the force law depends on more than one variable, the techniques for finding analytical solutions when they exist are more complicated. In general, solving an ordinary differential equation means reducing it, in effect, to an algebraic expression for the constants of integration, which are determined by the initial or other conditions.

For the forces involved in many physical problems, Eq. (1-38) cannot be solved in closed analytical form. However, we can then resort to numerical methods which can be evaluated with the aid of high-speed electronic computers. To illustrate the numerical approach, we assume that the position x_0 and velocity v_0 are known at the initial time t_0. The acceleration a_0 at this instant is given by Eq. (1-38) as

$$a_0 = \frac{F(x_0, v_0, t_0)}{m} \tag{1-45}$$

After a short time interval Δt, we have

$$t_1 = t_0 + \Delta t$$
$$x_1 = x_0 + v_0 \, \Delta t$$
$$v_1 = v_0 + a_0 \, \Delta t$$

From these new values of the variables, we can calculate the new acceleration, using Eq. (1-38).

$$a_1 = \frac{F(x_1, v_1, t_1)}{m} \tag{1-46}$$

By repetition of this procedure n times, we can calculate x and v at time $t_n = t_0 + n \, \Delta t$.

$$\begin{aligned} x_n &= x_{n-1} + v_{n-1} \, \Delta t \\ v_n &= v_{n-1} + a_{n-1} \, \Delta t \end{aligned} \tag{1-47}$$

We thereby obtain a complete numerical solution to the equation of motion. This illustrates that a unique solution to the differential equation of motion is always possible for any reasonable force law.

1-7 SIMPLE HARMONIC OSCILLATOR

The spring force in the archer's bow initiated the unbounded linear motion of the arrow. More common physical applications of the spring force involve oscillatory motion, such as vibrations of a mass attached to a spring. A system undergoing steady-state motion under the action of a spring is called a *harmonic oscillator*. The motion is called *simple harmonic* when the restoring force is proportional to the extension or compression of the spring. Any situation that has a linear restoring tendency (such as ac circuits, certain servomechanisms, etc.) exhibits simple harmonic oscillations.

The equation of motion for a simple harmonic oscillator,

$$m\frac{d^2x}{dt^2} = -kx \tag{1-48}$$

can be solved by Eqs. (1-40) and (1-41). However, we can cleverly write the solution by inspection. The functions cos ωt and sin ωt satisfy Eq. (1-48) if the angular frequency ω is given by

$$\omega = \sqrt{\frac{k}{m}} \tag{1-49}$$

The significance of the trigonometric functions as solutions to Eq. (1-48) is that two differentiations give back the original function, sign reversed. The general solution to Eq. (1-48) is

$$x(t) = A\cos \omega t + B\sin \omega t \tag{1-50}$$

where A and B are arbitrary constants. An equivalent form of the solution is

$$x(t) = a\cos(\omega t + \alpha) \tag{1-51}$$

with constants related by

$$A = a\cos\alpha \qquad B = -a\sin\alpha$$

The constant a is called the *amplitude* of the motion, and α is called the *initial phase*. The initial conditions can be used to specify the arbitrary constants a and α. From Eq. (1-51) the velocity of the oscillator is

$$v(t) = -a\omega\sin(\omega t + \alpha) \tag{1-52}$$

The period τ of the motion is the time required for the system to undergo a complete oscillation and return to the initial values of x and v. The period for the oscillator is

$$\tau = \frac{2\pi}{\omega} \tag{1-53}$$

The frequency of the oscillator (number of oscillations per unit time) is

$$\nu = \frac{1}{\tau} = \frac{\omega}{2\pi}$$

We can illustrate our harmonic-oscillator solution with the bow-and-arrow example of Sec. 1-5. At $t = 0$ the bow is at full draw, $x = -d$, and the arrow velocity is zero. From Eq. (1-52) we find

$$\alpha = 0$$

and from Eq. (1-51) we obtain

$$a = -d$$

The solution with proper boundary conditions is

$$x(t) = -d \cos \omega t$$
$$v(t) = \omega d \sin \omega t \qquad (1\text{-}54)$$

with $\omega = \sqrt{k/m}$. As time increases from $t = 0$, x increases to zero at

$$t = \frac{1}{2}\left(\frac{\pi}{\omega}\right) \qquad (1\text{-}55)$$

At this instant the arrow leaves the bowstring with velocity

$$v = \omega d = \sqrt{\frac{k}{m}}\, d \qquad (1\text{-}56)$$

which agrees with Eq. (1-37). For the bow described in Sec. 1-5 the arrow-propulsion time from Eq. (1-55) is

$$t = \frac{\pi}{2}\sqrt{\frac{m}{k}} = \frac{\pi}{2}\sqrt{\frac{23 \times 10^{-3}}{186}} \approx \frac{1}{60}\text{ s}$$

In our archery example the simple-harmonic-force law does not apply beyond this time (one-fourth of the period τ), as illustrated in Fig. 1-4.

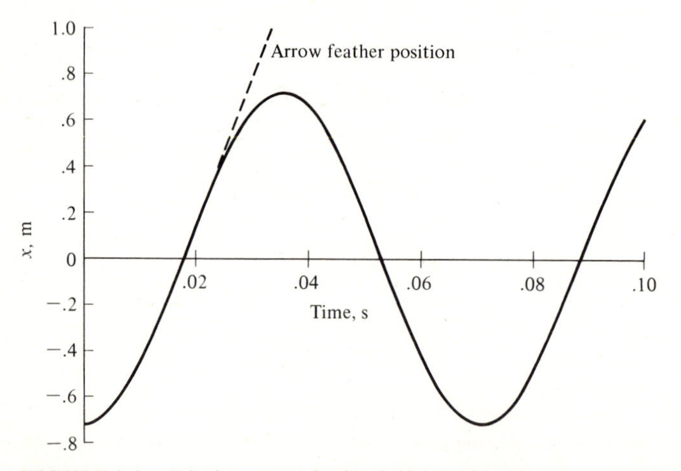

FIGURE 1-4*a* Displacement of a simple harmonic oscillator vs. time. The position of the feather end of the archer's arrow as a function of time is indicated by the dashed line after the arrow leaves the bow.

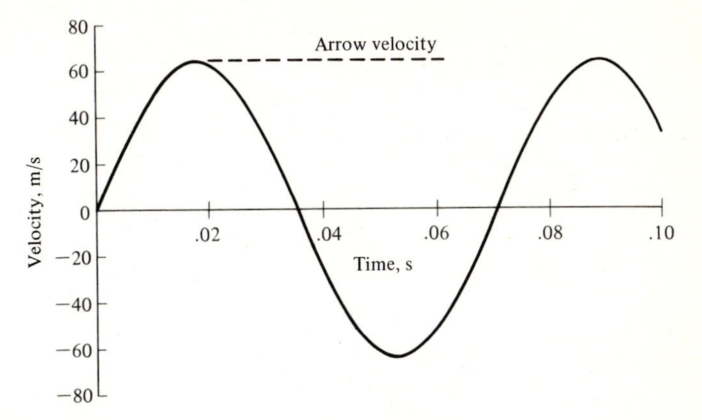

FIGURE 1-4*b* **Velocity of a simple harmonic oscillator vs. time. The velocity of the arrow after it leaves the bow is indicated by the dashed line.**

1-8 DAMPED HARMONIC MOTION

In almost all physical problems frictional forces play a role. For example, a harmonic oscillator that is subject to a damping force has an amplitude that continuously decreases with time. Since the simple harmonic oscillator applies to such a broad range of physical phenomena, we treat its solution in the presence of the frictional force of Eq. (1-10) at some length. (A further reason for our choice is that it is one of the few examples that can be completely solved analytically.)

For the damped harmonic oscillator the equation of motion is

$$m\frac{d^2x}{dt^2} = -kx - b\frac{dx}{dt} \tag{1-57}$$

This equation can be written in operator form as

$$\left(\frac{d^2}{dt^2} + 2\gamma\frac{d}{dt} + \omega_0{}^2\right)x = 0 \tag{1-58}$$

where $\gamma = \frac{1}{2}(b/m)$ and $\omega_0{}^2 = k/m$. Since the differentials have constant coefficients, we can write Eq. (1-58) as the product of two differential factors which operate on x:

$$\left\{\frac{d}{dt} - [-\gamma + (\gamma^2 - \omega_0{}^2)^{1/2}]\right\}\left\{\frac{d}{dt} - [-\gamma - (\gamma^2 - \omega_0{}^2)^{1/2}]\right\}x(t) = 0 \tag{1-59}$$

A solution of this equation is obtained if the operation of either factor on $x(t)$ gives zero.

$$\left\{\frac{d}{dt} - [-\gamma + (\gamma^2 - \omega_0^2)^{1/2}]\right\} x(t) = 0 \tag{1-60}$$

$$\left\{\frac{d}{dt} - [-\gamma - (\gamma^2 - \omega_0^2)^{1/2}]\right\} x(t) = 0 \tag{1-61}$$

By inspection, the solution of Eq. (1-60) is

$$x(t) = C_1 \exp\{[-\gamma + (\gamma^2 - \omega_0^2)^{1/2} t]\} \tag{1-62}$$

and the solution of Eq. (1-61) is

$$x(t) = C_2 \exp\{[-\gamma - (\gamma^2 - \omega_0^2)^{1/2} t]\} \tag{1-63}$$

The general solution is a sum of Eqs. (1-62) and (1-63), which we can rewrite,

$$x(t) = e^{-\gamma t}\left(\frac{C\{\exp[(\gamma^2 - \omega_0^2)^{1/2} t] + \exp[-(\gamma^2 - \omega_0^2)^{1/2} t]\}}{2} + \frac{C'\{\exp[(\gamma^2 - \omega_0^2)^{1/2} t] - \exp[-(\gamma^2 - \omega_0^2)^{1/2} t]\}}{2}\right) \tag{1-64}$$

where

$$C = C_1 + C_2$$
$$C' = C_1 - C_2$$

The exponential terms may be expressed in terms of the hyperbolic functions

$$\sinh x = \frac{e^x - e^{-x}}{2}$$

$$\cosh x = \frac{e^x + e^{-x}}{2}$$

leading to

$$x(t) = e^{-\gamma t}[C \cosh (\gamma^2 - \omega_0^2)^{1/2} t + C' \sinh (\gamma^2 - \omega_0^2)^{1/2} t] \tag{1-65}$$

The general solution must have two independent constants to satisfy the boundary conditions. We notice that in the case $\gamma = \omega_0$, the second term in Eq. (1-65) vanishes unless the constant coefficient C' is infinite. To guarantee the appropriate behavior at $\gamma = \omega_0$, it is convenient to set $C' = D(\gamma^2 - \omega_0^2)^{-1/2}$. The final form of the solution becomes

$$x(t) = e^{-\gamma t}\left[C \cosh (\gamma^2 - \omega_0^2)^{1/2} t + \frac{D}{\sqrt{\gamma^2 - \omega_0^2}} \sinh (\gamma^2 - \omega_0^2)^{1/2} t\right] \tag{1-66}$$

The boundary conditions at $t = 0$ can be readily imposed on Eq. (1-66). We find

$$C = x_0 \qquad D = v_0 + \gamma x_0 \tag{1-67}$$

The properties of the solution in Eq. (1-66) are dependent on the relative magnitude of the viscous coefficient γ and the natural frequency ω_0. We distinguish the three cases (I) $\gamma = \omega_0$, (II) $\gamma > \omega_0$, and (III) $\gamma < \omega_0$.

For Case I, the solution reduces to

$$\text{(I)} \qquad x(t) = e^{-\gamma t}[x_0 + (v_0 + \gamma x_0)t] \tag{1-68}$$

In deriving this from Eqs. (1-66) and (1-67), we have used the power series expansion

$$\sinh(\gamma^2 - \omega_0^2)^{1/2}t = (\gamma^2 - \omega_0^2)^{1/2}t + (\gamma^2 - \omega_0^2)^{3/2}t^3/6 + \cdots$$

for the limit to $\gamma = \omega_0$.

For Case II, $\gamma > \omega_0$, we simply rewrite Eqs. (1-66) and (1-67) in exponential form.

$$\text{(II)} \quad x(t) = \tfrac{1}{2}\exp[(-\gamma + \sqrt{\gamma^2 - \omega_0^2})t]\left(x_0 + \frac{v_0 + \gamma x_0}{\sqrt{\gamma^2 - \omega_0^2}}\right)$$

$$+ \tfrac{1}{2}\exp[(-\gamma - \sqrt{\gamma^2 - \omega_0^2})t]\left(x_0 - \frac{v_0 + \gamma x_0}{\sqrt{\gamma^2 - \omega_0^2}}\right) \tag{1-69}$$

In Case III, $\gamma < \omega_0$, the square roots in Eq. (1-66) become

$$\sqrt{\gamma^2 - \omega_0^2} = i\sqrt{\omega_0^2 - \gamma^2}$$

We can then use the identities

$$\sinh(ix) = i \sin x$$
$$\cosh(ix) = \cos x$$

to express the $\gamma < \omega_0$ solution as

$$x(t) = e^{-\gamma t}\left[x_0 \cos(\omega_0^2 - \gamma^2)^{1/2}t + \frac{v_0 + \gamma x_0}{\sqrt{\omega_0^2 - \gamma^2}} \sin(\omega_0^2 - \gamma^2)^{1/2}t\right] \tag{1-70}$$

By reference to Eqs. (1-50) and (1-51), the result in Eq. (1-70) can be written in the form

$$\text{(III)} \qquad x(t) = e^{-\gamma t}a\cos[(\omega_0^2 - \gamma^2)^{1/2}t + \alpha] \tag{1-71}$$

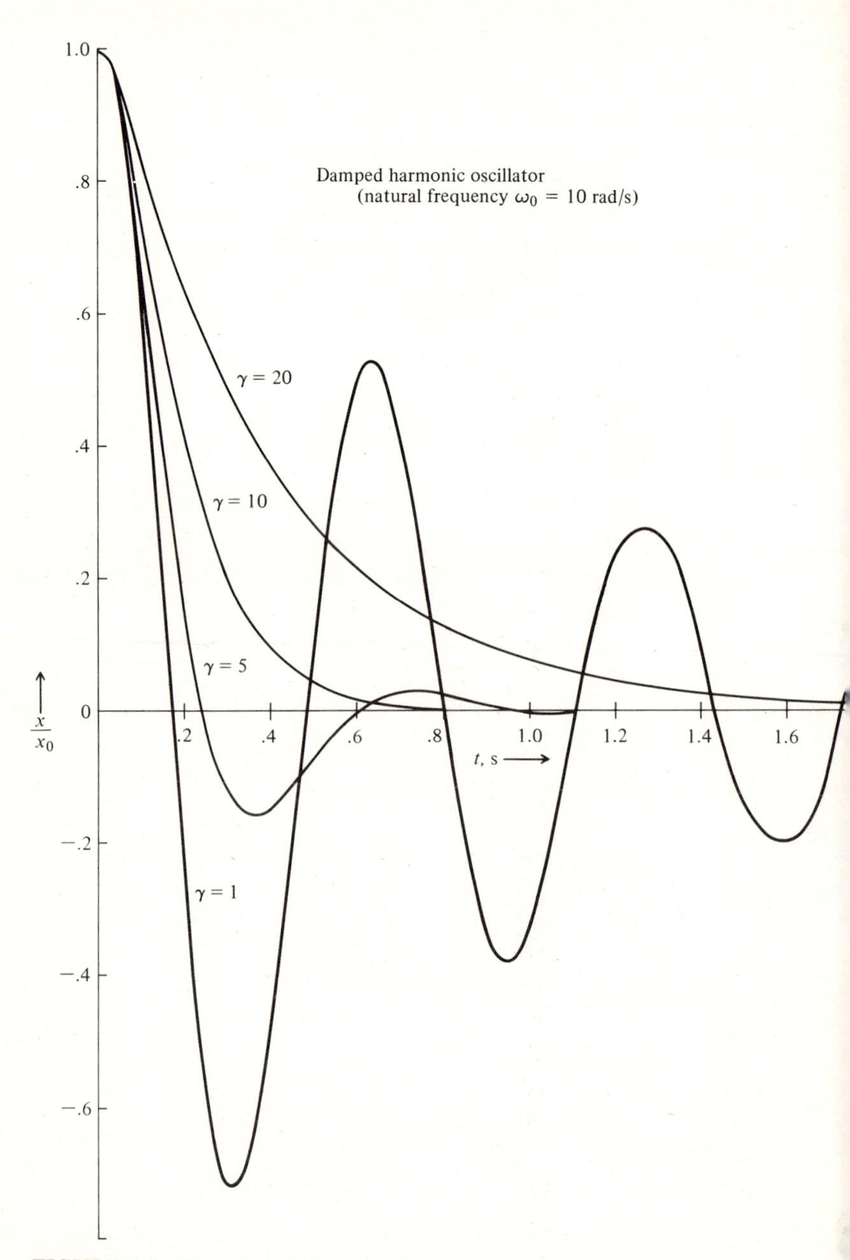

FIGURE 1-5 Time dependence of the displacement of a damped harmonic oscillator for the initial conditions $x = x_0$, $v = 0$. The natural frequency of the oscillator is $\omega_0 = 10$ rad/s. Results for various strengths of the damping constant γ are illustrated.

with

$$a = \sqrt{\frac{\omega_0^2 x_0^2 + 2\gamma v_0 x_0 + v_0^2}{\omega_0^2 - \gamma^2}}$$

$$\tan \alpha = -\frac{v_0 + \gamma x_0}{x_0 \sqrt{\omega_0^2 - \gamma^2}}$$

In all three cases the amplitude of the displacement decays exponentially with time, although in Case I the exponential factor is multiplied by a linear function of t. At large times the rates of falloff are characterized by the exponentials:

$$\begin{array}{lll} \text{(I)} & e^{-\gamma t} * \text{(linear function of } t) & \gamma = \omega_0 \text{ (critically damped)} \\ \text{(II)} & \exp\{[-\gamma + (\gamma^2 - \omega_0^2)^{1/2}]t\} & \gamma > \omega_0 \text{ (overdamped)} \\ \text{(III)} & e^{-\gamma t} * \text{(sinusoidal function of } t) & \gamma < \omega_0 \text{ (underdamped)} \end{array} \tag{1-72}$$

Illustrations of the time dependences for the three cases are given in Fig. 1-5 for the initial conditions $x = x_0$, $v_0 = 0$. The only exception to the above rates of decrease occurs when the initial conditions are such that the coefficient of the $\exp\{[-\gamma + (\gamma^2 - \omega_0^2)^{1/2}]t\}$ term of solution II vanishes. In that unique circumstance, the spring returns to rest like $\exp\{-[\gamma + (\gamma^2 - \omega_0^2)^{1/2}]t\}$.

There are an endless number of applications of damped harmonic oscillators. The pneumatic-spring return on a door represents an everyday situation where solution I is the ideal. Upon releasing the door with no initial velocity, we want it to close as rapidly as possible without slamming. Equation (1-72) indicates that solution I should be selected; the spring-tube system should be designed with $\gamma = \omega_0$. Solution III might bring the door to a close faster, due to the vanishing of the cosine factor in Eq. (1-71), but this would let the door slam! On the other hand, solution III is the relevant choice for physical systems that undergo damped periodic oscillations.

The behavior of simple electric circuits is determined by a differential equation which has the same mathematical form as the damped harmonic oscillator. As an example we consider the circuit of Fig. 1-6a with an inductor L, resistor R, and charged capacitor C in series. The sum of the voltage drops across the elements of the circuit must add up to zero after the switch is closed. This leads to the differential equation

$$L\frac{di}{dt} + Ri + \frac{q}{C} = 0 \tag{1-73}$$

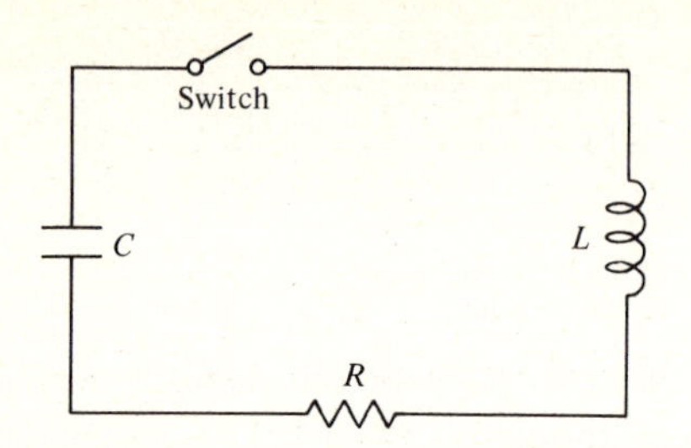

FIGURE 1-6*a* Simple L, R, C electric series circuit.

where $i(t)$ is the current flowing in the circuit and $q(t)$ is the charge on one of the capacitor plates. Since $i = dq/dt = \dot{q}$, the circuit equation can be written

$$L\ddot{q} + R\dot{q} + \frac{q}{C} = 0 \tag{1-74}$$

This equation is analogous to the damped-harmonic-oscillator equation (1-57) with the following correspondence:

$$\begin{aligned} x &\to q & \gamma &= \frac{b}{2m} \to \frac{R}{2L} \\ m &\to L \\ b &\to R & \omega_0 &= \sqrt{\frac{k}{m}} \to \sqrt{\frac{1}{LC}} \\ k &\to \frac{1}{C} \end{aligned} \tag{1-75}$$

Since it is often far easier to connect circuit elements than to build and test a mechanical system, this analogy is of considerable practical importance. Electrical-analog models of complex mechanisms like airplane wings can be tested and modified to optimize the design at a small fraction of the cost of building actual prototypes.

If the switch on the circuit in Fig. 1-6*a* is closed at time $t = 0$, the initial conditions for a capacitor charged to a voltage V_0 are

$$\begin{aligned} q(t=0) &= q_0 \\ \dot{q}(t=0) &= 0 \end{aligned} \tag{1-76}$$

where $q_0 = CV_0$. By reference to Eqs. (1-66) and (1-67), the solution for the charge as a function of time is

$$q(t) = q_0\, e^{-\gamma t}\left[\cosh\,(\gamma^2 - \omega_0{}^2)^{1/2}t + \frac{\gamma}{\sqrt{\gamma^2 - \omega_0{}^2}}\sinh\,(\gamma^2 - \omega_0{}^2)^{1/2}t\right] \tag{1-77}$$

The condition for critical damping is

$$\gamma = \omega_0 \tag{1-78}$$

which, from Eq. (1-75), can be expressed as

$$R = 2\sqrt{\frac{L}{C}} \tag{1-79}$$

If this condition is met, the charge on the capacitor damps exponentially to zero according to

$$q(t) = q_0(1 + \gamma t)e^{-\gamma t} \tag{1-80}$$

as the resistance converts the electric energy to heat. For a smaller resistance

$$R < 2\sqrt{\frac{L}{C}}$$

the circuit is underdamped ($\gamma < \omega_0$), and the charge in the circuit undergoes oscillations that damp out in time.

For a circuit with a voltage source, as in Fig. 1-6*b*, the sum of the voltage drops across the elements must match the impressed voltage. Thus the differential equation for the circuit in Fig. 1-6*b* is

$$L\frac{di}{dt} + Ri + \frac{q}{C} = V(t) \tag{1-81}$$

where $V(t)$ is the voltage of the generator. This differential equation is completely analogous to the equation of motion for a damped harmonic oscillator subjected to an external force, a topic which we take up in the following section.

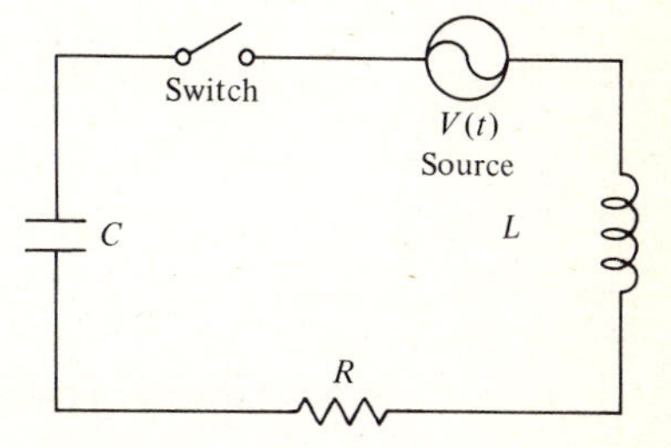

FIGURE 1-6*b* Series *L*, *R*, *C* circuit with a voltage generator *V*(*t*).

1-9 FORCED OSCILLATOR WITH DAMPING: RESONANCE

Numerous physical systems undergo steady-state motion that can be described in terms of a damped harmonic oscillator driven by an external force that oscillates sinusoidally with time as

$$F(t) = F_0 \cos \omega t \tag{1-82}$$

The equation of motion from Eq. (1-58) gets modified to

$$\frac{d^2x}{dt^2} + 2\gamma \frac{dx}{dt} + \omega_0{}^2 x = \frac{F_0}{m} \cos \omega t \tag{1-83}$$

A solution to this equation is most readily obtained by using complex numbers. For this purpose we introduce

$$\begin{aligned} z &= x + iy \\ e^{i\omega t} &= \cos \omega t + i \sin \omega t \end{aligned} \tag{1-84}$$

and observe that the real part of

$$\frac{d^2z}{dt^2} + 2\gamma \frac{dz}{dt} + \omega_0{}^2 z = \frac{F_0}{m} e^{i\omega t} \tag{1-85}$$

is identical with Eq. (1-83). This latter form is more convenient to solve. Once we find the solution for z, the physical displacement x is obtained from $x = \text{Re } z$. The complex arithmetic method can be used only for linear differential equations in which the variable x appears in first or zero power. Since the derivative $e^{i\omega t}$ is $i\omega e^{i\omega t}$, it is clear that the time dependence of the solution for z will be $e^{i\omega t}$. Thus, as a possible solution to Eq. (1-85), we try

$$z = RF_0 e^{i\omega t} \tag{1-86}$$

where R is a time-independent factor. The differential equation is satisfied by this choice if

$$[(i\omega)^2 + 2\gamma(i\omega) + \omega_0{}^2]\, RF = \frac{F_0}{m}$$

or

$$R = \frac{1}{m(\omega_0{}^2 - \omega^2 + 2i\gamma\omega)} \tag{1-87}$$

The complex factor R can be written in polar form

$$R = re^{-i\theta} \tag{1-88}$$

where

$$r^2 = |R|^2 = \frac{1}{m^2[(\omega_0{}^2 - \omega^2)^2 + 4\gamma^2\omega^2]}$$

and

$$\tan\theta = -\left(\frac{\text{Im } R}{\text{Re } R}\right) = \frac{2\gamma\omega}{\omega_0{}^2 - \omega^2}$$

Using Eqs. (1-84), (1-86), and (1-88), we arrive at the desired solution to Eq. (1-83)

$$x = \text{Re } z = \text{Re}(rF_0 e^{i(\omega t - \theta)})$$

or

$$x(t) = rF_0 \cos(\omega t - \theta) \tag{1-89}$$

The response $x(t)$ to the force $F_0 \cos \omega t$ is thus proportional to the magnification factor r. The response oscillates with a phase $(\omega t - \theta)$ that lags the oscillations of the force by a phase angle θ.

The magnification r and phase lag θ depend critically on the relative size of the driving frequency ω and natural frequency ω_0. For small damping $\gamma \ll \omega_0$ and values of ω near to ω_0, we can make the following approximations in Eq. (1-88):

$$r^2 = \frac{1}{m^2[(\omega_0 - \omega)^2(\omega_0 + \omega)^2 + 4\gamma^2\omega^2]} \approx \frac{1}{4\omega_0{}^2 m^2[(\omega_0 - \omega)^2 + \gamma^2]} \tag{1-90}$$

$$\tan\theta = \frac{2\gamma\omega}{(\omega_0 - \omega)(\omega_0 + \omega)} \approx \frac{\gamma}{\omega_0 - \omega} \tag{1-91}$$

From these approximate expressions, we see that r^2 has a maximum when the driving force is at the natural frequency of the oscillator, $\omega = \omega_0$. The large response $x(t)$ produced by a driving frequency in the vicinity $\omega = \omega_0$ is called a *resonance*. The magnitude of r^2 at the resonance frequency $\omega = \omega_0$ is limited by the size of the frictional coefficient γ.

$$r_0{}^2 = \frac{1}{4\omega_0{}^2 m^2 \gamma^2} \qquad \text{at } \omega = \omega_0 \tag{1-92}$$

The half-width of the resonance is defined as the value of $(\omega - \omega_0)$ at which r^2 has fallen to one-half of its maximum value $r_0{}^2$. From Eqs. (1-90) and (1-92) we find that the half-width is γ. The resonance becomes narrower and higher as frictional effects are made smaller.

A plot of $r^2(\omega)$ and $\theta(\omega)$ is shown in Fig. 1-7. The phase lag θ is 90° on resonance. At small frequencies ω, the phase lag tends to zero, and far above resonance it approaches 180°, as is evident from Eq. (1-88). Resonance phenomena analogous to that discussed here play an extremely important role in all branches of physics and engineering.

The general solution to the forced-oscillator differential equation is obtained by adding the particular solution in Eq. (1-89) to the solution in Eq. (1-66) of the oscillator problem with no driving force. The result is

$$x(t) = e^{-\gamma t}\left[C\cosh(\gamma^2 - \omega_0{}^2)^{1/2}t + \frac{D}{\sqrt{\gamma^2 - \omega_0{}^2}}\sinh(\gamma^2 - \omega_0{}^2)^{1/2}t\right]$$
$$+ \frac{F_0}{m[(\omega_0{}^2 - \omega^2)^2 + 4\gamma^2\omega^2]^{1/2}}\cos\left(\omega t - \arctan\frac{2\gamma\omega}{\omega_0{}^2 - \omega^2}\right) \tag{1-93}$$

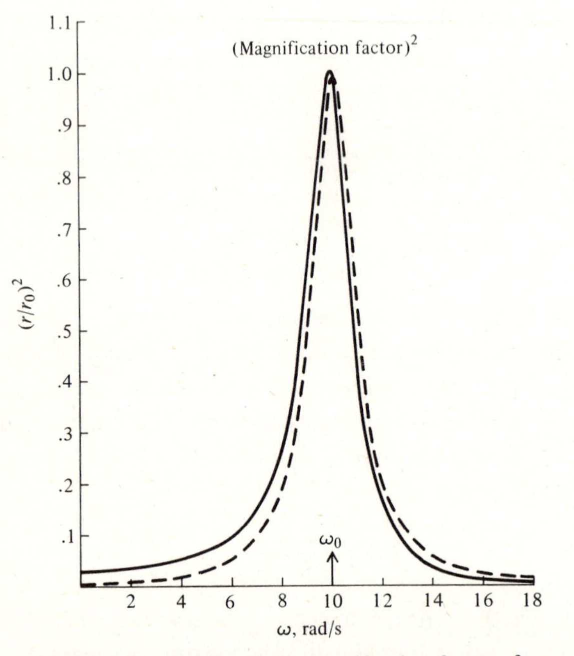

FIGURE 1-7a Square of the magnification factor, r^2, as a function of driving frequency ω for a forced oscillator of natural frequency $\omega_0 = 10$ and damping constant $\gamma = 1$. The solid curve is the exact result, and the dashed curve represents the resonance approximation of Eq. (1-90).

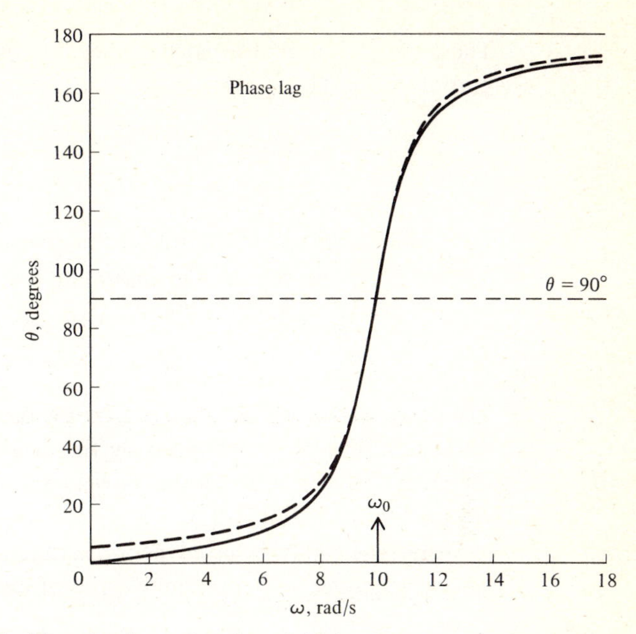

FIGURE 1-7*b* **Phase lag θ as a function of driving frequency ω for the forced oscillator of Fig. 1-7*a*.**

The initial conditions determine the constants C and D. The term with the decaying exponential is a transient which goes away at large times. The force-dependent term provides the steady-state oscillatory motion of the harmonic system.

The differential of the energy dissipated by the oscillator is

$$dW = F_{fr}\, dx \tag{1-94}$$

where F_{fr} is the frictional damping force

$$F_{fr} = -b\frac{dx}{dt} = -2m\gamma\frac{dx}{dt} \tag{1-95}$$

The power dissipated by friction is

$$P = \frac{dW}{dt} = F_{fr}\frac{dx}{dt} = -2m\gamma\left(\frac{dx}{dt}\right)^2 \tag{1-96}$$

To calculate the power loss in steady-state motion, we substitute the solution in Eq. (1-89) into Eq. (1-96). We find

$$P = -2m\gamma\omega^2 r^2 F_0^2 \sin^2(\omega t - \theta) \tag{1-97}$$

The average power dissipated per cycle can be calculated by integration of Eq. (1-97) over a period of the motion

$$\langle P \rangle = -2m\gamma\omega^2 r^2 F_0^2 \int_t^{t+2\pi/\omega} \sin^2(\omega t - \theta)\, dt = -m\gamma\omega^2 r^2 F_0^2 \tag{1-98}$$

For small damping and for driving frequencies close to the resonance frequency ω_0, we can use the approximation in Eq. (1-90) to obtain

$$\langle P \rangle \approx -\frac{\gamma F_0^2}{4m[(\omega_0^2 - \omega^2)^2 + \gamma^2]} \tag{1-99}$$

The maximum power is dissipated by the oscillator at the resonance frequency. The power supplied by the driving force $P = F\dot{x}$ is just the negative of the dissipated power, as can be verified by direct calculation using Eq. (1-82).

Any reasonably behaved periodic time-dependent force can be Fourier-analyzed into an infinite series of $\cos(\omega_n t + \phi_n)$ terms.

$$F(t) = \sum_n F_n \cos(\omega_n t + \phi_n) \tag{1-100}$$

where F_n and ϕ_n are constants. The solution, Eq. (1-93), for a driving force $F_0 \cos \omega t$ can be used for a force $F_n \cos(\omega_n t + \phi_n)$. Then the solution for a superposition of driving frequencies in Eq. (1-100) can be obtained as a summation over solutions with pure driving frequencies ω_n.

PROBLEMS

1. A car traveling at 100 km/h smashes into a concrete abutment along a superhighway. Show that the disaster is equivalent to driving off the top of a ten-story building. Allow 4 m per story and neglect air resistance. Next find the equivalent vertical height to represent the effect of a two-car head-on collision, each car traveling 100 km/h.

2. A vehicle has brakes on all four wheels. Find the deceleration which corresponds to maximum possible braking. Indicate why disk brakes are usually put on only the front wheels. (*Hint:* Calculate the normal forces on front and back wheels.)

3. A drag racer experiences a retarding force due to wind resistance that is proportional to the square of the racer's velocity. Assuming

that the racer is designed for optimum acceleration, set up the equation of motion and solve for the velocity as a function of time $v(t)$ and for the velocity as a function of distance $v(x)$. Solve the resulting transcendental equations numerically for the coefficient of friction and the terminal velocity which can reproduce the world record of $v = 107$ m/s, $t = 6.5$ s for $x = 0.4$ km.

4. An archer using the equipment described in Sec. 1-5 aims horizontally at a target 50 m distant. How far below the aiming point will the arrow strike? (Neglect air resistance.) At what angle should the arrow be released so as to hit the target?

5. An athlete can throw a javelin 60 m from a standing position. If he can run 100 m at constant velocity in 10 s, how far could he hope to throw the javelin from a running start? (Neglect air resistance.) Compare your answer with the world record of about 95 m for the javelin throw. (*Hint:* At what angle to the horizontal should he throw it for best results?)

6. Integrate the equation of motion in Eq. (1-26) to find the velocity as a function of height for a sky diver in free fall. At what free-fall distance does the velocity reach two-thirds of the terminal velocity?

7. A ball of mass m is thrown vertically upward with initial velocity v_i. If the air resistance is proportional to v^2 and the terminal velocity is v_t, show that the ball returns to its initial position with velocity v_f given by

$$\frac{1}{{v_f}^2} = \frac{1}{{v_i}^2} + \frac{1}{{v_t}^2}$$

8. A perfectly flexible cable has length l. Initially, a length l_0 of the cable hangs at rest over the edge of a table. Neglecting friction, compute the length hanging over the edge after a time t.

9. A boat is slowed by a frictional force $F(v)$. Its velocity decreases according to the formula

$$v = c^2(t - t_1)^2$$

where c is a constant and t_1 is the time at which it stops. Find the force $F(v)$ as a function of v.

10. A particle of mass m, initially at rest, moves on a horizontal line subject to a force $F(t) = ae^{-bt}$. Find its position and velocity as a function of time.

11. A massless spring of length l and spring constant k hangs freely from a support. If a mass m is attached to the free end of the spring and then released at the relaxed length of the spring, find the subsequent motion.

12. Find the initial conditions such that a damped harmonic oscillator will immediately begin steady-state motion under the time-dependent force $F = F_0 \cos \omega t$.

13. Find the steady-state solution for a damped harmonic oscillator driven by the force

$$F(t) = F_0 \sin \omega t$$

14. An electric motor weighing 100 kg is suspended by vertical springs which stretch 10 cm when the motor is attached. If the flywheel on the motor is not properly balanced, for what r/min would resonance be expected?

15. A tuning fork with frequency of 440 cps (i.e., A above middle C) is observed to damp to one-tenth in amplitude in 10 s. If this damping is primarily due to sound production, what would be the frequency of oscillation of the fork in a vacuum?

16. The tuning fork described in Prob. 15 sets up a sympathetic vibration in a second fork across the room that is also tuned to A above middle C. If the second fork is retuned to A#, by what ratio does its induced vibration amplitude decrease? With what frequency does the A# fork vibrate? (A tempered semitone interval such as the interval between A# and A is a frequency ratio of the twelfth root of 2.)

17. For steady-state forced harmonic motion, find the frequency ω at which the maximum in the exact resonance amplitude occurs.

18. An undamped harmonic oscillator with natural frequency ω_0 is subjected to a driving force $F(t) = ae^{-bt}$. The oscillator starts from rest at the origin ($x = 0$) at time $t = 0$. Find the solution of the equation of motion which satisfies the specified boundary condition.

CHAPTER 2

Energy Conservation

There are three great immutable conservation laws of mechanics: energy, momentum, and angular momentum. The three laws can be derived from newtonian theory. However, their range of validity is much broader, extending even to the domain of relativistic elementary particles, where Newton's laws are replaced by the laws of relativistic quantum mechanics. In its ramifications in all branches of science, energy conservation has the most far-reaching consequences.

2-1 POTENTIAL ENERGY

We begin our discussion of the energy-conservation law with the case of one-dimensional motion. For a particle of constant mass m, the second law of motion

$$\frac{d}{dt}(mv) = F(x, v, t)$$

can be written in the form

$$\frac{d}{dt}(\tfrac{1}{2}mv^2) = F(x, v, t)v \tag{2-1}$$

The differentiated quantity on the left-hand side is the familiar expression for the kinetic energy

$$T = \tfrac{1}{2}mv^2 \tag{2-2}$$

Many forces depend only on the position rather than explicitly on the velocity or time. In such cases it is valuable to express the equation of motion in a form that does not explicitly mention time. If the force is a function of x only, Eq. (2-1) can be written

$$\frac{d}{dt}(T) - F(x)\frac{dx}{dt} = \frac{d}{dt}\left[T - \int_{x_s}^{x} F(x)\,dx\right] = 0 \tag{2-3}$$

where x_s is an arbitrary but fixed reference point. We identify the quantity

$$E = T - \int_{x_s}^{x} F(x)\,dx \tag{2-4}$$

with the total energy of the particle. Since $dE/dt = 0$, the energy is constant in time; i.e., the energy is conserved. If F has an explicit dependence on either v or t, then the energy of the particle is not conserved in general. However, the change in energy of the particle is accompanied by a corresponding change in energy of the surroundings, so that in an enlarged sense energy conservation is not violated. As an example, kinetic frictional forces always oppose the direction of motion, and hence are velocity-dependent. The change in energy of a particle caused by frictional dissipation is absorbed by the surroundings as thermal energy.

The expression for the energy in Eq. (2-4) is often written

$$E = T + V(x) = \tfrac{1}{2}mv^2 + V(x) \tag{2-5}$$

The quantity

$$V(x) = -\int_{x_s}^{x} F(x)\,dx \tag{2-6}$$

which is the negative of the *work* done on the particle by the force during the motion of the particle from the reference point x_s to point x, is called the *potential energy*. This name was given to signify that $E = T + V =$ constant means that V is a form of energy which potentially may appear as kinetic energy.

By differentiating Eq. (2-6), we can solve for the force in terms of the potential energy.

$$F(x) = -\frac{dV(x)}{dx} \tag{2-7}$$

A force that is derivable in this way from a potential energy is called a *conservative force*. In one-dimensional motion any force which is a function only of position is conservative. The effect of changing from a reference point x_s to a new reference point x_s' is just to change $V(x)$ in Eq. (2-6) by a constant, independent of x. The force is unaffected by a change in reference point since it is calculated from the derivative of the potential energy. Because the motion of the particle is determined by the force, all measurable quantities are independent of x_s; x_s can be chosen to make the expression for the potential energy as simple as possible.

Since the energy is a constant of the motion for a conservative force, we can use

$$E = \tfrac{1}{2}mv_0^2 + V(x_0) = \tfrac{1}{2}mv_1^2 + V(x_1) \tag{2-8}$$

to relate velocities at two different points on the path. If the energy is known for a conservative system, it can be used as a boundary condition instead of the initial velocity or the initial coordinate.

The conservative nature of spring and gravitational forces allows the use of energy-conservation methods in solving related problems. The potential energy of the spring force is

$$V(x) = -\int_0^x (-kx)\,dx = \tfrac{1}{2}kx^2 \tag{2-9}$$

where we have chosen $x_s = 0$. Equation (1-36) of the archery example is a special case of Eq. (2-8) with this potential energy. The total

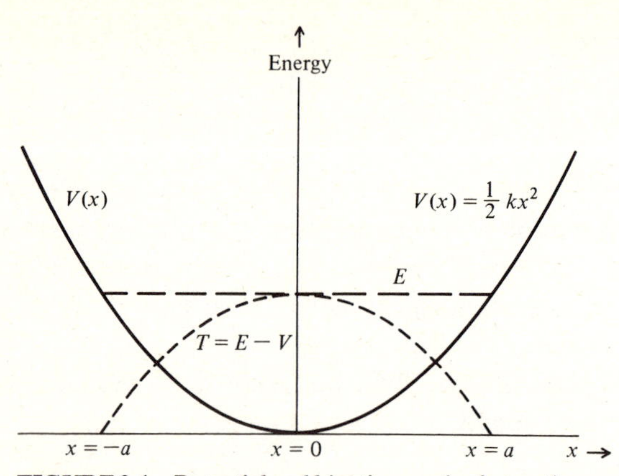

FIGURE 2-1 Potential and kinetic energies for motion under the spring force.

energy of an oscillator can be calculated from Eq. (2-8) using the solutions for x and v from Eqs. (1-51) and (1-52). We find

$$E = \tfrac{1}{2}m[-\omega a \sin(\omega t + \alpha)]^2 + \tfrac{1}{2}k[a \cos(\omega t + \alpha)]^2$$

which reduces to

$$E = \tfrac{1}{2}ka^2 \tag{2-10}$$

where we have used $\omega^2 = k/m$. The energy is proportional to the square of the maximum displacement a. At the turning points of the motion, $x = \pm a$, the energy is entirely potential energy. At $x = 0$, the energy is all kinetic.

2-2 GRAVITATIONAL ESCAPE

The gravitational potential energy due to the earth's attraction on a mass m at a distance $x \geq R_e$ from the center of the earth is

$$V(x) = -\int_{\infty}^{x} \left(-\frac{GmM_e}{x^2}\right) dx = -\frac{GmM_e}{x} \tag{2-11}$$

We have chosen x_s so that the potential energy vanishes at infinite distance. We may express the gravitational constant G in terms of the gravitational acceleration on earth using Eq. (1-6),

$$GM_e = gR_e^2$$

to obtain

$$V(x) = -\frac{mgR_e^2}{x} \tag{2-12}$$

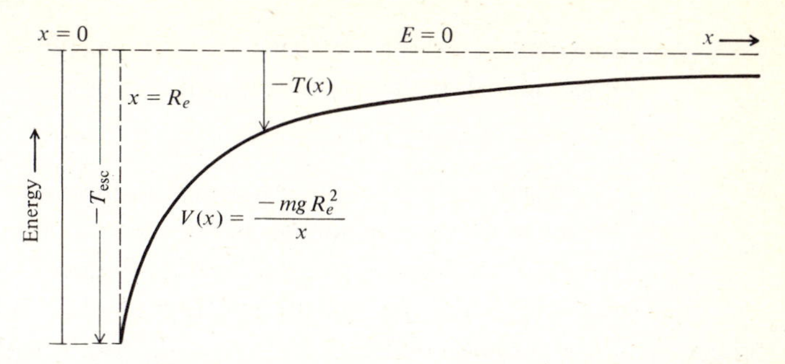

FIGURE 2-2 Gravitational potential energy due to the earth and the minimum kinetic energy T_{esc} needed for escape from the earth's gravitational attraction.

We can use Eq. (2-12) in Eq. (2-5) to calculate the minimum velocity needed by a rocket at the earth's surface to go to $x = \infty$, that is, to "escape from the earth's gravitational attraction." From Eq. (2-5) the velocity at some position x is

$$v(x) = \sqrt{\frac{2}{m}\left(E + \frac{mgR_e^2}{x}\right)} \qquad (2\text{-}13)$$

The argument under the radical sign must be positive for all x in order for the velocity to be real. In particular, this requires $E \geq 0$ for $x = \infty$. The minimum escape velocity from the earth's surface is consequently obtained by putting $E = 0$, $x = R_e$ in Eq. (2-13), giving

$$\begin{aligned} v_{esc} &= \sqrt{2gR_e} \\ &= \sqrt{2(9.8 \times 10^{-3})(6{,}371)} \\ &= 11.2 \text{ km/s} \qquad (25{,}000 \text{ mi/h}) \end{aligned} \qquad (2\text{-}14)$$

The escape velocity is independent of the mass of the rocket. To get to the moon, a spacecraft needs a velocity nearly equal to the escape velocity.

2-3 SMALL OSCILLATIONS

For a general potential energy the velocity can be calculated from Eq. (2-5) to be

$$v(x) = \pm\sqrt{\frac{2}{m}[E - V(x)]} \qquad (2\text{-}15)$$

This expression determines only the magnitude of the velocity. The sign depends on the previous history of the motion. Since the velocity must be real, the classically accessible region is

$$V(x) \leq E \tag{2-16}$$

The positions at which $V(x) = E$ are turning points. The velocity goes to zero at the turning points and reverses its direction. The qualitative nature of the motion of a particle can frequently be described using Eq. (2-15). For the potential energy sketched in Fig. 2-3 there are three turning points, x_1, x_2, x_3, at the total energy indicated by the dashed horizontal line. The regions $0 \leq x < x_1$ and $x_2 < x < x_3$ are classically forbidden by Eq. (2-16). The motion of a particle in the region $x_1 \leq x \leq x_2$ will be oscillatory since the particle cannot escape. The sign of the velocity in this region changes at the turning points as in Eq. (2-15). Finally, a particle approaching x_3 from infinity will slow down, reverse its motion at $x = x_3$, and go back out toward infinity in unbounded motion.

The motion of a particle in the potential valley $x_1 \leq x \leq x_2$ is particularly simple if the maximum displacements from the minimum potential energy at $x = x_e$ are small. In such a case, we can approximate the potential by a few terms of a series expansion about $x = x_e$.

$$V(x) = V(x_e) + (x - x_e)\left[\frac{dV(x)}{dx}\right]_{x=x_e} + \tfrac{1}{2}(x - x_e)^2\left[\frac{d^2V(x)}{dx^2}\right]_{x=x_e} + \cdots \tag{2-17}$$

The derivative dV/dx must vanish at an extremum, and hence $F = 0$ at $x = x_e$. Since the second derivative of $V(x)$ is positive at a minimum of $V(x)$, a particle at $x = x_e$ is in stable equilibrium.

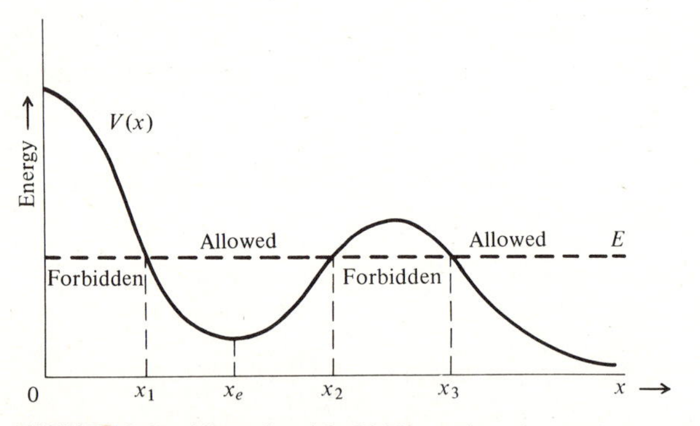

FIGURE 2-3 Allowed and forbidden regions for motion of a particle with energy E for a potential energy $V(x)$.

For small displacements the potential energy can be approximated by

$$V(x) = \tfrac{1}{2}k(x - x_e)^2 + V(x_e) \tag{2-18}$$

where

$$k \equiv \left[\frac{d^2V(x)}{dx^2}\right]_{x=x_e} \geqq 0$$

The constant term $V(x_e)$ can be dropped since it has no consequences for the physical motion. If we make a change of variable to the displacement from equilibrium,

$$x' = x - x_e$$

the potential energy can be written in the form of a simple harmonic oscillator,

$$V(x') = \tfrac{1}{2}kx'^2 \tag{2-19}$$

Small oscillations in complicated systems can often be approximately treated in terms of simple harmonic motion. The expansion about a potential-energy minimum as in Eq. (2-18) also provides justification for Hooke's law on the springlike elastic deformation in solids.

As an illustration, we find an approximate solution for the motion of a particle of mass m in the attractive potential well.

$$V(x) = \frac{-g^2}{x} + \frac{h^2}{x^2} \tag{2-20}$$

At the equilibrium position,

$$\left[\frac{dV(x)}{dx}\right]_{x=x_e} = \frac{g^2}{x_e^2} - \frac{2h^2}{x_e^3} = 0$$

which gives

$$x_e = \frac{2h^2}{g^2} \tag{2-21}$$

From Eq. (2-18) the effective spring constant for small oscillations about x_e is

$$k = \left[\frac{d^2V(x)}{dx^2}\right]_{x=x_e} = \frac{-2g^2}{x_e^3} + \frac{6h^2}{x_e^4}$$

or upon substitution from Eq. (2-21),

$$k = \frac{g^8}{8h^6} \tag{2-22}$$

The approximate equation of motion from Eqs. (2-22) and (1-51) is

$$x(t) - \frac{2h^2}{g^2} = a\cos\left(\frac{g^4}{2h^3}\sqrt{\frac{1}{2m}}\,t - \alpha\right) \tag{2-23}$$

where a and α are arbitary constants.

2-4 THREE-DIMENSIONAL MOTION: VECTOR NOTATION

In three dimensions the position of a particle of mass m can be specified by its cartesian coordinates (x, y, z). Newton's second law can then be stated as the three equations

$$\begin{aligned} m\ddot{x} &= F_x \\ m\ddot{y} &= F_y \\ m\ddot{z} &= F_z \end{aligned} \tag{2-24a}$$

where (F_x, F_y, F_z) are called the x, y, z components of the force on the particle. If one chooses to use a different cartesian coordinate system, which is translated and rotated with respect to the original system, the equations of motion have the same mathematical form; in the new coordinate system the equations of motion are

$$\begin{aligned} m\ddot{x}' &= F_{x'} \\ m\ddot{y}' &= F_{y'} \\ m\ddot{z}' &= F_{z'} \end{aligned} \tag{2-24b}$$

where (x', y', z') are the coordinates of the particle in the new frame. Each of Eqs. (2-24*b*) is a linear combination of Eqs. (2-24*a*). As an example, suppose the new coordinate system has the same origin as the original system but is rotated by an angle ϕ around the z axis, as illustrated in Fig. 2-4. The coordinates of the particle in the two frames are related by

$$\begin{aligned} x' &= x\cos\phi + y\sin\phi \\ y' &= -x\sin\phi + y\cos\phi \\ z' &= z \end{aligned} \tag{2-25}$$

By time-differentiating, we see that analogous relations hold for velocities and accelerations; e.g.,

$$\begin{aligned} \ddot{x}' &= \ddot{x}\cos\phi + \ddot{y}\sin\phi \\ \ddot{y}' &= -\ddot{x}\sin\phi + \ddot{y}\cos\phi \\ \ddot{z}' &= \ddot{z} \end{aligned} \tag{2-26}$$

Substituting Eqs. (2-24*a*) into Eqs. (2-26), we obtain

$$m\ddot{x}' = F_x \cos\phi + F_y \sin\phi$$
$$m\ddot{y}' = -F_x \sin\phi + F_y \cos\phi$$
$$m\ddot{z}' = F_z$$

Thus the equation of motion $m\ddot{x}' = F_{x'}$ is not the same equation as $m\ddot{x} = F_x$. Nonetheless, when we identify

$$\begin{aligned} F_{x'} &= F_x \cos\phi + F_y \sin\phi \\ F_{y'} &= -F_x \sin\phi + F_y \cos\phi \\ F_{z'} &= F_z \end{aligned} \tag{2-27}$$

the set of all three new equations is equivalent to the old set; the new equations are just linear combinations of the old equations. For instance, $m\ddot{x}' = F_{x'}$ is just $\cos\phi$ times the equation $m\ddot{x} = F_x$ plus $\sin\phi$ times the equation $m\ddot{y} = F_y$.

To symbolize the above state of affairs and at the same time realize a great simplification in notation, we introduce the concept of a vector. A *vector* is a set of three quantities whose components in differently oriented (i.e., rotated) coordinate systems are related in the same way as the set of coordinates (x, y, z). Symbolically, we

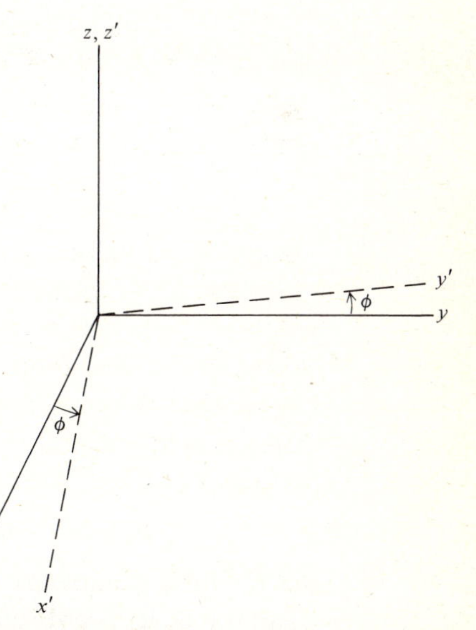

FIGURE 2-4 **Two coordinate systems related by a rotation by an angle ϕ about the z axis.**

denote a vector with components (a_x, a_y, a_z) by $\mathbf{a}$. Examples of vectors which we have already encountered are the position vector $\mathbf{r} \equiv (x, y, z)$, the acceleration vector $\ddot{\mathbf{r}} = (\ddot{x}, \ddot{y}, \ddot{z})$, and the force vector $\mathbf{F} \equiv (F_x, F_y, F_z)$. The basic idea of a vector is that it is a definite quantity, despite the changes of its components when the coordinate system is changed [e.g., Eq. (2-25)].

In vector notation Newton's second law can be written as a single equation.

$$m\ddot{\mathbf{r}} = \mathbf{F} \tag{2-28}$$

This is shorthand for the set of Eqs. (2-24*a*) or (2-24*b*). An advantage of vector notation is that no specific reference frame is necessary in this statement of the laws of motion. Vector notation is also often useful in manipulating and solving the equations of motion.

The distance of a point (x, y, z) from the origin of the coordinate system is $\sqrt{x^2 + y^2 + z^2}$. This distance is independent of the orientation of the coordinate system,

$$\sqrt{x'^2 + y'^2 + z'^2} = \sqrt{x^2 + y^2 + z^2}$$

as can easily be checked for the transformation in Eqs. (2-25). The above quantity is called the length, or magnitude, of $\mathbf{r}$ and is denoted by

$$|\mathbf{r}| \equiv r \equiv \sqrt{x^2 + y^2 + z^2}$$

For a general vector $\mathbf{a} = (a_x, a_y, a_z)$, the length, or magnitude, is similarly defined as

$$|\mathbf{a}| \equiv a \equiv \sqrt{a_x^2 + a_y^2 + a_z^2} \tag{2-29}$$

Since by the definition of a vector given above the components of $\mathbf{a}$ transform under rotations of the coordinate system in the same way as the components of $\mathbf{r}$, the length of $\mathbf{a}$ is independent of the orientation of the coordinate frame. A quantity, such as $|\mathbf{a}|$, that is independent of frame orientation is called a *scalar*, to distinguish it from a quantity such as F_x, which is the component of a vector.

If vectors are multiplied by numbers or added together by the rule

$$\alpha\mathbf{a} + \beta\mathbf{b} = (\alpha a_x + \beta b_x, \alpha a_y + \beta b_y, \alpha a_z + \beta b_z)$$

the resulting quantity is again a vector, because its components transform under coordinate-system rotations according to the definition of a vector. Since any linear combination of vectors is a vector, many new vectors can be generated from the position vector $\mathbf{r}$. For instance, the relative coordinate of two particles $\mathbf{r}_2 - \mathbf{r}_1 = (x_2 - x_1, y_2 - y_1, z_2 - z_1)$ is a vector, as is the change of coordinate of a particle between two times.

$$\Delta\mathbf{r} = \mathbf{r}(t + \Delta t) - \mathbf{r}(t) = [x(t + \Delta t) - x(t), y(t + \Delta t) - y(t), z(t + \Delta t) - z(t)]$$

It follows that the velocity

$$\mathbf{v} = \dot{\mathbf{r}} = \lim_{\Delta t \to 0} \frac{\Delta\mathbf{r}}{\Delta t}$$

and the acceleration

$$\mathbf{a} = \ddot{\mathbf{r}} = \lim_{\Delta t \to 0} \frac{\mathbf{v}(t + \Delta t) - \mathbf{v}(t)}{\Delta t}$$

are vectors. We note that all vectors which are constructed from the difference of two position vectors (such as $\mathbf{r}_2 - \mathbf{r}_1$, $\Delta\mathbf{r}$, $\dot{\mathbf{r}}$, $\ddot{\mathbf{r}}$) are unchanged by a shift in origin of the coordinate frame. Under a change in origin, all position vectors $\mathbf{r}$ are replaced by $\mathbf{r}' = \mathbf{r} + \mathbf{s}$, where $\mathbf{s}$ is a constant vector. It follows that the vectors formed from differences

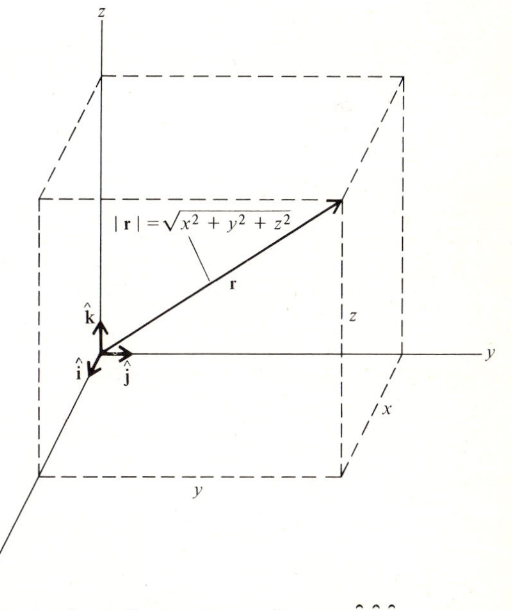

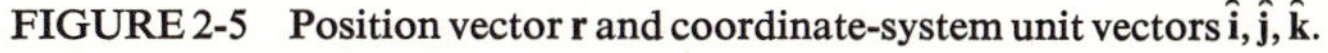
FIGURE 2-5 Position vector r and coordinate-system unit vectors $\hat{\mathbf{i}}, \hat{\mathbf{j}}, \hat{\mathbf{k}}$.

of two position vectors are independent of **s**. Since the acceleration is unchanged by a shift in origin, the force vector must also share this property in order for Eq. (2-28) to hold in translated frames. The position vector **r** is the only vector which depends upon the origin of the coordinate system.

The geometrical representation of a vector as a directed line segment, or "arrow," is a powerful intuitive tool. We represent the position vector $\mathbf{r} = (x, y, z)$ by an arrow drawn from the origin to the point (x, y, z), as illustrated in Fig. 2-5. The length of **r** is then just the length of the arrow. The components of **r** are the coordinates of the orthogonal projections of the arrow's point onto the coordinate axes. We can also represent an arbitrary vector **a** by an arrow, since under rotations of the coordinate frame the components of **a** transform the same way as the components of **r**. The length of the arrow is proportional to the magnitude of the vector, and the projections of the arrow on the coordinate axes are proportional to the components of the vector. The location of the arrow is arbitrary (so long as the arrow represents a "true" vector, not the position vector) and

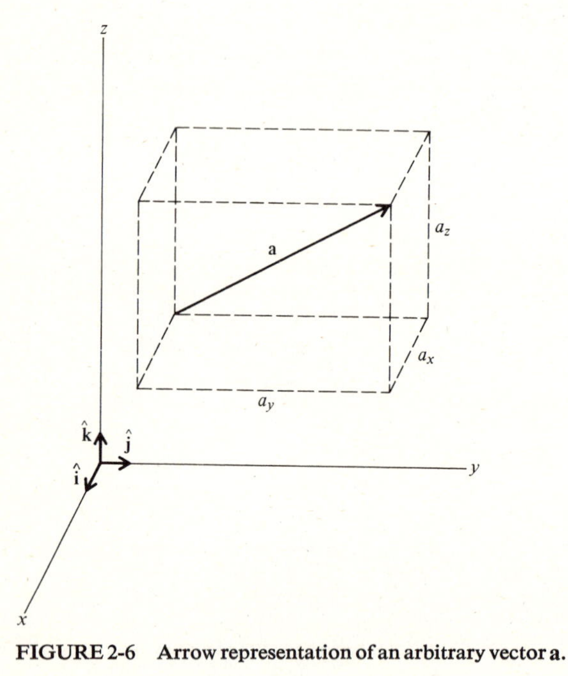

FIGURE 2-6 Arrow representation of an arbitrary vector a.

may be chosen for convenience. For instance, arrows representing the velocity, acceleration, or force on a particle may be attached to the point representing the position of the particle. The addition of vectors is represented by the head-to-tail construction illustrated in Fig. 2-7.

The *dot product* of two vectors is defined as

$$\mathbf{a} \cdot \mathbf{b} \equiv a_x b_x + a_y b_y + a_z b_z \tag{2-30}$$

The dot product is a scalar (i.e., independent of the frame orientation), as we can readily demonstrate from the identity

$$\begin{aligned} a_x b_x + a_y b_y + a_z b_z &= \tfrac{1}{2}[(a_x + b_x)^2 - a_x^2 - b_x^2 + (a_y + b_y)^2 - a_y^2 - b_y^2 \\ &\quad + (a_z + b_z)^2 - a_z^2 - b_z^2] \\ &= \tfrac{1}{2}(|\mathbf{a} + \mathbf{b}|^2 - |\mathbf{a}|^2 - |\mathbf{b}|^2) \end{aligned}$$

Since the vector magnitudes $|\mathbf{a}|$, $|\mathbf{b}|$, $|\mathbf{a} + \mathbf{b}|$ are scalars, it follows that the dot product is a scalar. From the defining equation (2-30), we further observe that

$$\begin{aligned} \mathbf{a} \cdot \mathbf{a} &= |\mathbf{a}|^2 \equiv \mathbf{a}^2 \\ \mathbf{a} \cdot \mathbf{b} &= \mathbf{b} \cdot \mathbf{a} \\ (\mathbf{a} + \mathbf{b}) \cdot \mathbf{c} &= \mathbf{a} \cdot \mathbf{c} + \mathbf{b} \cdot \mathbf{c} \end{aligned} \tag{2-31}$$

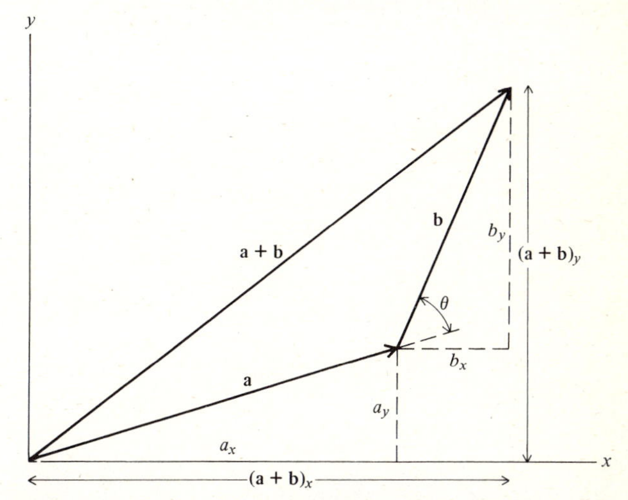

FIGURE 2-7 Head-to-tail construction of the addition of two vectors a and b. (For convenience of illustration the *x*, *y*-coordinate axes are taken to lie in the plane defined by a and b.)

The magnitude of the vector $\mathbf{a} + \mathbf{b}$ is given in terms of the dot product $\mathbf{a} \cdot \mathbf{b}$ by

$$|\mathbf{a} + \mathbf{b}|^2 = (\mathbf{a} + \mathbf{b}) \cdot (\mathbf{a} + \mathbf{b}) = |\mathbf{a}|^2 + |\mathbf{b}|^2 + 2\mathbf{a} \cdot \mathbf{b}$$

Furthermore, inasmuch as the vectors $\mathbf{a}$, $\mathbf{b}$, and $\mathbf{a} + \mathbf{b}$ form a triangle as illustrated in Fig. 2-7, the opposite side $|\mathbf{a} + \mathbf{b}|$ of the triangle is related by trigonometry to the adjacent sides $|\mathbf{a}|$ and $|\mathbf{b}|$ by

$$|\mathbf{a} + \mathbf{b}|^2 = |\mathbf{a}|^2 + |\mathbf{b}|^2 + 2|\mathbf{a}|\,|\mathbf{b}| \cos \theta$$

where θ is the angle between the arrows representing $\mathbf{a}$ and $\mathbf{b}$. Equating the above two formulas for $|\mathbf{a} + \mathbf{b}|^2$, we deduce the following result for the dot product:

$$\mathbf{a} \cdot \mathbf{b} = |\mathbf{a}|\,|\mathbf{b}| \cos \theta \tag{2-32}$$

Thus the dot product represents the product of the length of one vector times the orthogonal projection of the other vector on it, as indicated in Fig. 2-8. If $\mathbf{a} \cdot \mathbf{b} = 0$, even though $|\mathbf{a}| \neq 0$, $|\mathbf{b}| \neq 0$, the angle between the vectors is 90° and the vectors are said to be orthogonal.

It is useful to define a set of coordinate-axis vectors $\hat{\mathbf{i}}$, $\hat{\mathbf{j}}$, $\hat{\mathbf{k}}$ of unit length $|\hat{\mathbf{i}}| = |\hat{\mathbf{j}}| = |\hat{\mathbf{k}}| = 1$ which are directed along the x, y, z axes of the coordinate system, as in Fig. 2-6. The components of these orthogonal unit vectors are

$$\begin{aligned} \hat{\mathbf{i}} &= (1, 0, 0) \\ \hat{\mathbf{j}} &= (0, 1, 0) \\ \hat{\mathbf{k}} &= (0, 0, 1) \end{aligned} \tag{2-33}$$

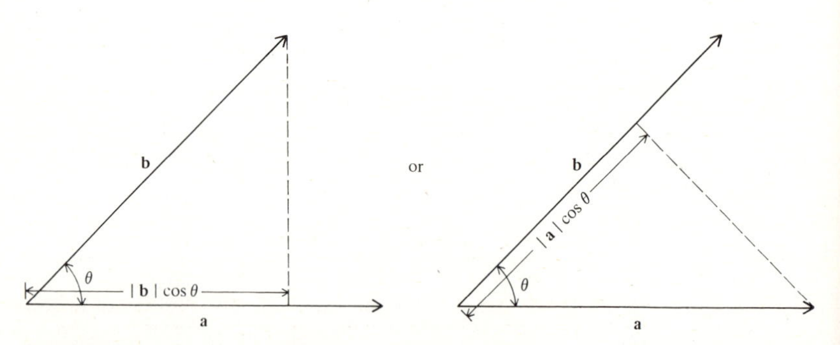

FIGURE 2-8 Geometrical illustration of the dot product.

From this definition the dot products of unit vectors with each other are

$$\mathbf{\hat{j}} \cdot \mathbf{\hat{j}} = \mathbf{\hat{i}} \cdot \mathbf{\hat{i}} = \mathbf{\hat{k}} \cdot \mathbf{\hat{k}} = 1$$
$$\mathbf{\hat{j}} \cdot \mathbf{\hat{i}} = \mathbf{\hat{i}} \cdot \mathbf{\hat{k}} = \mathbf{\hat{k}} \cdot \mathbf{\hat{j}} = 0 \qquad (2\text{-}34)$$

In a given frame a general vector **a** can be represented in terms of the unit vectors of the frame as

$$\mathbf{a} = a_x \mathbf{\hat{i}} + a_y \mathbf{\hat{j}} + a_z \mathbf{\hat{k}}$$

The sum of two vectors can be expressed as

$$\mathbf{a} + \mathbf{b} = (a_x + b_x)\mathbf{\hat{i}} + (a_y + b_y)\mathbf{\hat{j}} + (a_z + b_z)\mathbf{\hat{k}}$$

Another type of product of two vectors of considerable importance is the *cross product*, written $\mathbf{a} \times \mathbf{b}$. The cross product has three components, defined by

$$(\mathbf{a} \times \mathbf{b})_x = a_y b_z - a_z b_y$$
$$(\mathbf{a} \times \mathbf{b})_y = a_z b_x - a_x b_z \qquad (2\text{-}35)$$
$$(\mathbf{a} \times \mathbf{b})_z = a_x b_y - a_y b_x$$

Thus, in terms of the unit vectors of the coordinate system, the cross product is

$$\mathbf{a} \times \mathbf{b} = (a_y b_z - a_z b_y)\mathbf{\hat{j}} + (a_z b_x - a_x b_z)\mathbf{\hat{i}} + (a_x b_y - a_y b_x)\mathbf{\hat{k}} \qquad (2\text{-}36)$$

Alternatively, the definition can be written as the determinant

$$\mathbf{a} \times \mathbf{b} = \det \begin{vmatrix} \mathbf{\hat{j}} & \mathbf{\hat{i}} & \mathbf{\hat{k}} \\ a_x & a_y & a_z \\ b_x & b_y & b_z \end{vmatrix} \qquad (2\text{-}37)$$

From the symmetry properties of the determinant or from Eq. (2-36), we note that

$$\mathbf{a} \times \mathbf{b} = -\mathbf{b} \times \mathbf{a} \qquad (2\text{-}38)$$

The cross product of a vector with itself vanishes.

$$\mathbf{a} \times \mathbf{a} = 0$$

The cross product transforms like an ordinary vector under rotations of the coordinate system. For instance, consider the transformation equation (2-25). The vectors **a** and **b** transform in the same way as the position vector **r**; that is,

$$a_{x'} = a_x \cos\phi + a_y \sin\phi \qquad b_{x'} = b_x \cos\phi + b_y \sin\phi$$
$$a_{y'} = -a_x \sin\phi + a_y \cos\phi \qquad b_{y'} = -b_x \sin\phi + b_y \cos\phi$$
$$a_{z'} = a_z \qquad b_{z'} = b_z$$

The components of $\mathbf{a} \times \mathbf{b}$ in the rotated frame are then found to be

$$\begin{aligned}
(\mathbf{a} \times \mathbf{b})_{x'} &= (a_{y'} b_{z'} - a_{z'} b_{y'}) \\
&= (a_y b_z - a_z b_y) \cos \phi + (a_z b_x - a_x b_z) \sin \phi \\
&= (\mathbf{a} \times \mathbf{b})_x \cos \phi + (\mathbf{a} \times \mathbf{b})_y \sin \phi \\
(\mathbf{a} \times \mathbf{b})_{y'} &= -(\mathbf{a} \times \mathbf{b})_x \sin \phi + (\mathbf{a} \times \mathbf{b})_y \cos \phi \\
(\mathbf{a} \times \mathbf{b})_{z'} &= (\mathbf{a} \times \mathbf{b})_z
\end{aligned}$$

which correspond to the transformation of (x, y, z) in Eq. (2-25). For this reason the cross product is sometimes called the vector product. However, the cross product behaves differently from ordinary vectors under *inversion* of the coordinate axes (that is, $x' = -x, y' = -y, z' = -z$). We have

$$\mathbf{r}' = -\mathbf{r} \qquad \mathbf{a}' = -\mathbf{a} \qquad (\mathbf{a} \times \mathbf{b})' = (\mathbf{a} \times \mathbf{b}) \tag{2-39}$$

A three-component quantity such as $(\mathbf{a} \times \mathbf{b})$ which behaves like a vector under rotation of the coordinate axes but does not change sign under inversion is called an *axial vector*.

The dot product of the vector $\mathbf{a}$ with $\mathbf{a} \times \mathbf{b}$ is zero, as we show by use of Eqs. (2-30) and (2-35).

$$\mathbf{a} \cdot (\mathbf{a} \times \mathbf{b}) = a_x(a_y b_z - a_z b_y) + a_y(a_z b_x - a_x b_z) + a_z(a_x b_y - a_y b_x) \tag{2-40a}$$

In a similar manner we find that

$$\mathbf{b} \cdot (\mathbf{a} \times \mathbf{b}) = 0 \tag{2-40b}$$

Thus the cross-product vector $\mathbf{a} \times \mathbf{b}$ is orthogonal to the vectors $\mathbf{a}$ and $\mathbf{b}$. The arrow representing $\mathbf{a} \times \mathbf{b}$ must be perpendicular to the plane defined by the arrows of $\mathbf{a}$ and $\mathbf{b}$. By the definition in Eq. (2-36), the direction of $\mathbf{a} \times \mathbf{b}$ is the direction in which a right-hand screw moves when it turns from $\mathbf{a}$ toward $\mathbf{b}$, as indicated in Fig. 2-9. The square of the magnitude of the cross product,

$$|\mathbf{a} \times \mathbf{b}|^2 = (a_y b_z - a_z b_y)^2 + (a_z b_x - a_x b_z)^2 + (a_x b_y - a_y b_x)^2$$

can be rewritten

$$\begin{aligned}
|\mathbf{a} \times \mathbf{b}|^2 &= (a_x^2 + a_y^2 + a_z^2)(b_x^2 + b_y^2 + b_z^2) - (a_x b_x + a_y b_y + a_z b_z)^2 \\
&= |\mathbf{a}|^2 |\mathbf{b}|^2 - |\mathbf{a} \cdot \mathbf{b}|^2
\end{aligned}$$

By substitution of Eq. (2-32), we obtain

$$|\mathbf{a} \times \mathbf{b}|^2 = |\mathbf{a}|^2 |\mathbf{b}|^2 (1 - \cos^2 \theta)$$

and so

$$|\mathbf{a} \times \mathbf{b}| = |\mathbf{a}| \, |\mathbf{b}| \sin \theta \tag{2-41}$$

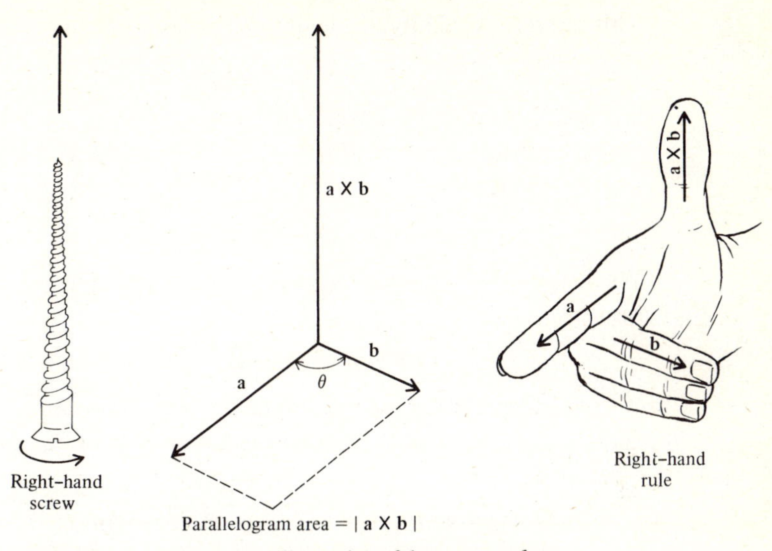

FIGURE 2-9 Geometrical illustration of the cross product.

where θ is the angle between the arrows representing **a** and **b**. The length of **a** × **b** is just the area of the parallelogram, with the arrows **a** and **b** as sides.

The cross products of the unit vectors $\hat{\imath}, \hat{\jmath}, \hat{k}$ of Eqs. (2-33) are found from Eq. (2-36) to be

$$\begin{array}{lll} \hat{\imath} \times \hat{\jmath} = \hat{k} & \hat{\jmath} \times \hat{\imath} = -\hat{k} & \hat{\imath} \times \hat{\imath} = 0 \\ \hat{\jmath} \times \hat{k} = \hat{\imath} & \hat{k} \times \hat{\jmath} = -\hat{\imath} & \hat{\jmath} \times \hat{\jmath} = 0 \\ \hat{k} \times \hat{\imath} = \hat{\jmath} & \hat{\imath} \times \hat{k} = -\hat{\jmath} & \hat{k} \times \hat{k} = 0 \end{array} \tag{2-42}$$

A new kind of scalar can be formed by taking the dot product of a vector **a** with an axial vector (**b** × **c**). This scalar, **a** · (**b** × **c**), is called the *triple product*. From Eqs. (2-30) and (2-31) the triple product can be written as a determinant of the vector components.

$$\mathbf{a} \cdot (\mathbf{b} \times \mathbf{c}) = \det \begin{vmatrix} a_x & a_y & a_z \\ b_x & b_y & b_z \\ c_x & c_y & c_z \end{vmatrix}$$

The symmetry properties of the determinant under interchange of rows imply that

$$\mathbf{a} \cdot (\mathbf{b} \times \mathbf{c}) = \mathbf{c} \cdot (\mathbf{a} \times \mathbf{b}) = \mathbf{b} \cdot (\mathbf{c} \times \mathbf{a})$$

This interchangeability of the dot and cross products,

$$\mathbf{a} \cdot (\mathbf{b} \times \mathbf{c}) = (\mathbf{a} \times \mathbf{b}) \cdot \mathbf{c} \tag{2-43}$$

is a useful property of vector algebra.

The repeated cross product of three vectors $\mathbf{a} \times (\mathbf{b} \times \mathbf{c})$ can be worked out to

$$\mathbf{a} \times (\mathbf{b} \times \mathbf{c}) = \mathbf{b}(\mathbf{a} \cdot \mathbf{c}) - \mathbf{c}(\mathbf{a} \cdot \mathbf{b}) \tag{2-44a}$$

When the cross products are carried out in a different order, the result is

$$(\mathbf{a} \times \mathbf{b}) \times \mathbf{c} = \mathbf{b}(\mathbf{a} \cdot \mathbf{c}) - \mathbf{a}(\mathbf{b} \cdot \mathbf{c}) \tag{2-44b}$$

A useful formula for the dot product of two cross products can be derived from Eq. (2-44*b*). We take the dot product of Eq. (2-44*b*) with a vector **d**,

$$(\mathbf{a} \times \mathbf{b}) \times \mathbf{c} \cdot \mathbf{d} = (\mathbf{a} \cdot \mathbf{c})(\mathbf{b} \cdot \mathbf{d}) - (\mathbf{a} \cdot \mathbf{d})(\mathbf{b} \cdot \mathbf{c})$$

and then interchange the dot and cross products on the left-hand side to obtain

$$(\mathbf{a} \times \mathbf{b}) \cdot (\mathbf{c} \times \mathbf{d}) = (\mathbf{a} \cdot \mathbf{c})(\mathbf{b} \cdot \mathbf{d}) - (\mathbf{a} \cdot \mathbf{d})(\mathbf{b} \cdot \mathbf{c}) \tag{2-45}$$

The components of a vector are often labeled $\mathbf{a} = (a_1, a_2, a_3)$, the subscripts 1, 2, 3 denoting the x, y, z components, respectively. In this notation the dot product of two vectors can be written

$$\mathbf{a} \cdot \mathbf{b} = \sum_i a_i b_i$$

where the summation is over $i = 1, 2, 3$. As a convenient shorthand notation, we shall often omit the $\sum_i$ symbol and simply write

$$\mathbf{a} \cdot \mathbf{b} = a_i b_i$$

where a summation over the repeated vector component index i is implied. This is known as the *summation convention.*

From the components a_i and b_i of two vectors **a** and **b**, we can form $3 \times 3 = 9$ products $a_i b_j$. We denote these nine components (also called *elements*) by the symbol T_{ij}.

$$\mathsf{T}_{ij} = a_i b_j$$

In vector notation we regard the nine quantities as components of

$$\mathbf{T} = \mathbf{ab} \tag{2-46}$$

with no dot or cross between the vectors **a** and **b**. This is sometimes called the *outer product* of the vectors **a** and **b**. Any such quantity whose nine elements in one coordinate system transform to those in a rotated coordinate system in the same way as a product of vector components transform is called a *tensor* (more precisely, a *tensor of second rank*). Any linear combination of tensors is also a tensor. The sum of the "diagonal" elements ($i = j$) of the tensor $\mathbf{T} = \mathbf{ab}$,

$$\mathsf{T}_{11} + \mathsf{T}_{22} + \mathsf{T}_{33} = a_1 b_1 + a_2 b_2 + a_3 b_3$$

is just the dot product $\mathbf{a} \cdot \mathbf{b}$. The components of the cross product $\mathbf{a} \times \mathbf{b}$ are constructed from the off-diagonal elements ($i \neq j$) of this tensor.

If we make a dot product of the tensor (**ab**) with a vector **c**, we get a vector

$$(\mathbf{ab}) \cdot \mathbf{c} = \mathbf{a}(\mathbf{b} \cdot \mathbf{c})$$

$$\mathbf{c} \cdot (\mathbf{ab}) = \mathbf{b}(\mathbf{c} \cdot \mathbf{a})$$

because ($\mathbf{b} \cdot \mathbf{c}$) and ($\mathbf{c} \cdot \mathbf{a}$) are scalars, and a vector multiplied by a scalar is a vector. Hence, for a general tensor **T**, the dot products $\mathbf{T} \cdot \mathbf{c}$ and $\mathbf{c} \cdot \mathbf{T}$ are vectors. In terms of components,

$$(\mathbf{T} \cdot \mathbf{c})_i = \mathsf{T}_{ij} c_j$$

$$(\mathbf{c} \cdot \mathbf{T})_i = c_j \mathsf{T}_{ji}$$

with a summation over the index j implied. The most important use of a tensor is to relate two vectors in this way. The unit tensor **I** with the property that

$$\mathbf{a} \cdot \mathbf{I} = \mathbf{I} \cdot \mathbf{a} = \mathbf{a}$$

for any vector **a**, is given in terms of unit vectors by

$$\mathbf{I} = \hat{\mathbf{i}}\hat{\mathbf{i}} + \hat{\mathbf{j}}\hat{\mathbf{j}} + \hat{\mathbf{k}}\hat{\mathbf{k}}$$

The components of a second-rank tensor are often written in a 3×3 matrix array as

$$\mathbf{T} = \begin{pmatrix} \mathsf{T}_{11} & \mathsf{T}_{12} & \mathsf{T}_{13} \\ \mathsf{T}_{21} & \mathsf{T}_{22} & \mathsf{T}_{23} \\ \mathsf{T}_{31} & \mathsf{T}_{32} & \mathsf{T}_{33} \end{pmatrix} \tag{2-47a}$$

and a vector **c** is represented by a column vector:

$$\mathbf{T} \cdot \mathbf{c} = \begin{pmatrix} c_1 \\ c_2 \\ c_3 \end{pmatrix} \tag{2-47b}$$

The dot product $\mathbf{T} \cdot \mathbf{c}$ can then be worked out by matrix multiplication.

$$\mathbf{T} \cdot \mathbf{c} = \begin{pmatrix} T_{11} & T_{12} & T_{13} \\ T_{21} & T_{22} & T_{23} \\ T_{31} & T_{32} & T_{33} \end{pmatrix} \begin{pmatrix} c_1 \\ c_2 \\ c_3 \end{pmatrix} = \begin{pmatrix} T_{11}c_1 + T_{12}c_2 + T_{13}c_3 \\ T_{21}c_1 + T_{22}c_2 + T_{23}c_3 \\ T_{31}c_1 + T_{32}c_2 + T_{33}c_3 \end{pmatrix}$$

2-5 CONSERVATIVE FORCES IN THREE DIMENSIONS

We want to find the conditions on the force **F** for which energy-conservation methods apply in three dimensions. With vector notation, Newton's laws of motion can compactly be expressed as

$$m \frac{d\mathbf{v}}{dt} = \mathbf{F}(\mathbf{r}, \mathbf{v}, \mathbf{t}) \tag{2-48}$$

The appearance of the vectors **r** and **v** in the argument of **F** is schematic only. It indicates that each component of **F** can depend on all the components of **r** and **v** [for example, $F_x(x, y, z, v_x, v_y, v_z, t)$].

Following our derivation in Eqs. (2-1) to (2-6) of energy conservation in one-dimensional motion, we take the dot product with **v** of both sides of Eq. (2-48) to obtain

$$\mathbf{v} \cdot \frac{d}{dt}(m\mathbf{v}) = \mathbf{F}(\mathbf{r}, \mathbf{v}, \mathbf{t}) \cdot \mathbf{v}$$

or equivalently,

$$\frac{d}{dt}(\tfrac{1}{2}m\mathbf{v} \cdot \mathbf{v}) = \mathbf{F}(\mathbf{r}, \mathbf{v}, \mathbf{t}) \cdot \frac{d\mathbf{r}}{dt} \tag{2-49}$$

From Eq. (2-30), the dot product $\mathbf{v} \cdot \mathbf{v}$ gives

$$\mathbf{v} \cdot \mathbf{v} = v^2 = v_x{}^2 + v_y{}^2 + v_z{}^2$$

Thus the differentiated quantity on the left-hand side of Eq. (2-49) is the kinetic energy for three-dimensional motion. In order to be able to write

$$\mathbf{F}(\mathbf{r}, \mathbf{v}, \mathbf{t}) \cdot \frac{d\mathbf{r}}{dt}$$

in an integral form analogous to Eq. (2-3),

$$\frac{d}{dt} \int_{\mathbf{r}_s}^{\mathbf{r}} \mathbf{F} \cdot d\mathbf{r} \tag{2-50}$$

the force must be independent of **v** and t. Furthermore, the integral in Eq. (2-50) must be unique, independent of the integration path between $\mathbf{r}_s$ and $\mathbf{r}$. If these conditions are met, we can demonstrate energy conservation from Eq. (2-49), namely,

$$\frac{dE}{dt} = \frac{d}{dt}\left[T - \int_{\mathbf{r}_s}^{\mathbf{r}} \mathbf{F}(\mathbf{r}) \cdot d\mathbf{r}\right] = 0 \qquad (2\text{-}51)$$

As in the one-dimensional case, we define a potential energy $V(\mathbf{r})$ by the line integral

$$V(\mathbf{r}) = -\int_{\mathbf{r}_s}^{\mathbf{r}} \mathbf{F}(\mathbf{r}) \cdot d\mathbf{r} \qquad (2\text{-}52)$$

Before proceeding further, we investigate the necessary condition on the force for the above line integral to be path-independent.

To find this condition on **F**, we first consider integration paths around a differential square in the yz plane, as shown in Fig. 2-11. We locate a corner of the square at the point (y_s, z_s) and calculate $V(\mathbf{r})$ at the diagonal corner $(y_s + dy, z_s + dz)$ by two different paths.

Path I: $(y_s, z_s) \to (y_s, z_s + dz) \to (y_s + dy, z_s + dz)$

Path II: $(y_s, z_s) \to (y_s + dy, z_s) \to (y_s + dy, z_s + dz)$

The value of $V(\mathbf{r})$ calculated from Path I is

$$V(\mathbf{r}) = -F_z(x_s, y_s, z_s)dz - F_y(x_s, y_s, z_s + dz)dy$$

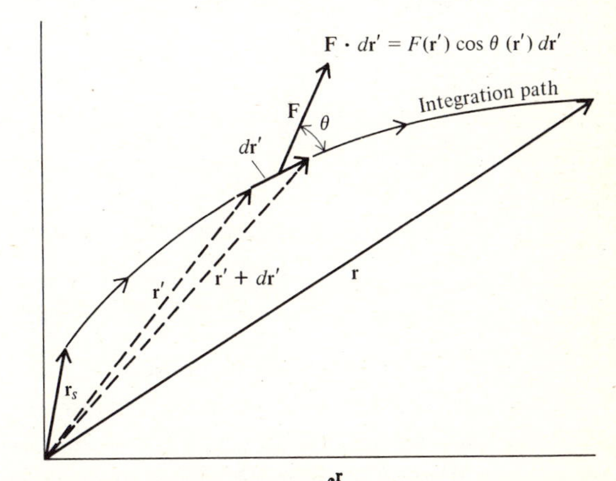

FIGURE 2-10 Interpretation of the line integral $\int_{\mathbf{r}_s}^{\mathbf{r}} \mathbf{F}(\mathbf{r}') \cdot d\mathbf{r}'$, where $\mathbf{r}'$ is the integration variable and $\mathbf{r}_s$, $\mathbf{r}$ are the limits of integration. The projection angle θ depends on the value of $\mathbf{r}'$ along the path.

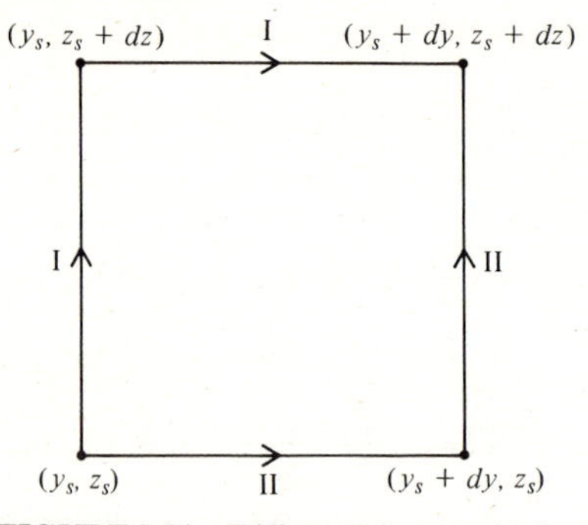

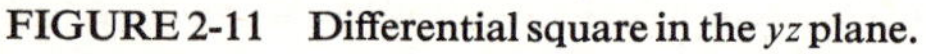
FIGURE 2-11 **Differential square in the *yz* plane.**

The corresponding result from Path II is

$$V(\mathbf{r}) = -F_y(x_s, y_s, z_s)\,dy - F_z(x_s, y_s + dy, z_s)\,dz$$

Demanding the same $V(\mathbf{r})$ from both integration paths yields by subtraction

$$[F_y(x_s, y_s, z_s + dz) - F_y(x_s, y_s, z_s)]\,dy \\ -[F_z(x_s, y_s + dy, z_s) - F_z(x_s, y_s, z_s)]\,dz = 0 \tag{2-53}$$

The quantities in parentheses are immediately recognizable as

$$\frac{\partial F_y}{\partial z}(x_s, y_s, z_s)\,dz$$

and

$$\frac{\partial F_z}{\partial y}(x_s, y_s, z_s)\,dy$$

Canceling the differentials in Eq. (2-53) gives the condition

$$\frac{\partial F_y}{\partial z} - \frac{\partial F_z}{\partial y} = 0 \tag{2-54}$$

for the force **F** to be conservative. Since the physics is independent of the choice of reference point (x_s, y_s, z_s), this condition must hold for any coordinates (x, y, z). At this point it is convenient to introduce the vector differentiation operator $\boldsymbol{\nabla}$, defined as

$$\boldsymbol{\nabla} = \hat{\mathbf{i}}\frac{\partial}{\partial x} + \hat{\mathbf{j}}\frac{\partial}{\partial y} + \hat{\mathbf{k}}\frac{\partial}{\partial z} \tag{2-55}$$

This is called the *del operator*. In vector notation the requirement in Eq. (2-54) can then be written

$$(\nabla \times \mathbf{F})_x = 0$$

To verify this assertion we recall the cross-product definition from Eqs. (2-35):

$$(\nabla \times \mathbf{F})_x = \nabla_y F_z - \nabla_z F_y = \frac{\partial F_z}{\partial y} - \frac{\partial F_y}{\partial z}$$

To derive the above condition on an energy-conserving force, we have used an integration path in the yz plane. If instead we integrate along a differential square in the xy plane, we get

$$(\nabla \times \mathbf{F})_z = 0$$

In general, the requirement for a conservative force to be derivable from a unique (path-independent) potential is

$$curl\, \mathbf{F} \equiv \nabla \times \mathbf{F} = 0 \tag{2-56}$$

The conservation of energy

$$E = \tfrac{1}{2}mv^2 + V(\mathbf{r}) \tag{2-57}$$

of a particle moving in a conservative force field $\mathbf{F}(\mathbf{r})$ is exceedingly useful in discussing the motion. We will make special use of it in our later treatment of planetary motion.

To compute a conservative vector force from the potential energy, we use Eq. (2-52) to deduce that

$$dV(\mathbf{r}) = -\mathbf{F}(\mathbf{r}) \cdot d\mathbf{r} \tag{2-58}$$

The expression for the total differential dV is

$$dV = \frac{\partial V}{\partial x}\, dx + \frac{\partial V}{\partial y}\, dy + \frac{\partial V}{\partial z}\, dz = (\nabla V) \cdot d\mathbf{r} \tag{2-59}$$

Combining the results of Eqs. (2-58) and (2-59), we find

$$\mathbf{F}(\mathbf{r}) = -\nabla V(\mathbf{r}) \equiv -grad[V(\mathbf{r})] \tag{2-60}$$

In component form we can write

$$F_k(\mathbf{r}) = -\nabla_k V(\mathbf{r}) = -\frac{\partial V(\mathbf{r})}{\partial x_k} \tag{2-61}$$

We have shown that if $\nabla \times \mathbf{F} = 0$, there exists a potential energy function. The converse is also easily shown to be true. We compute the *curl* of $\mathbf{F}$ from Eq. (2-60). For the x component of $\nabla \times \mathbf{F}$ we obtain

$$(\nabla \times \mathbf{F})_x = \nabla_y \nabla_z V(\mathbf{r}) - \nabla_z \nabla_y V(\mathbf{r}) \tag{2-62}$$

Since $\nabla_y \nabla_z = \nabla_z \nabla_y$, we get

$$(\nabla \times \mathbf{F})_x = 0$$

Among the most important physical examples of conservative forces are central forces. The magnitude of a central force is a function only of $|\mathbf{r}|$, and the direction is along $\mathbf{r}$. The gravitation and Coulomb forces are both of this type. We can write a general central force as

$$\mathbf{F}(\mathbf{r}) = F(r)\frac{\mathbf{r}}{r} \tag{2-63}$$

The force is attractive for $F(r) < 0$ and repulsive for $F(r) > 0$.

To prove that all central forces are conservative, we use the cartesian components of Eq. (2-63),

$$F_x = \frac{x}{r}F(r) \qquad F_y = \frac{y}{r}F(r) \qquad F_z = \frac{z}{r}F(r) \tag{2-64}$$

in the expression for dV from Eq. (2-59).

$$\begin{aligned} dV &= -(F_x\,dx + F_y\,dy + F_z\,dz) \\ &= -\frac{F(r)}{r}(x\,dx + y\,dy + z\,dz) = -F(r)\,dr \end{aligned} \tag{2-65}$$

In the last step we have used the differential of

$$r^2 = x^2 + y^2 + z^2$$

Since the result for dV in Eq. (2-65) depends only on the radial coordinate r (not on θ or ϕ), the integral of dV is path-independent. This establishes the conservative nature of a central force. We can also use Eq. (2-65) to derive the central potential energy from the force law as

$$V(r) = -\int_{r_S}^{r} F(r)\,dr \tag{2-66}$$

This is of the same form as the one-dimensional integral which we used in Eq. (2-11) to calculate the gravitational potential energy.

2-6 MOTION IN A PLANE

All physical input needed for the solution of any problem in mechanics is contained in Newton's laws with the appropriate initial conditions. For the analysis of the mechanical motion of a system, it is not always convenient to choose cartesian coordinates. For example, some kinds of motion in a plane can frequently be described more simply in terms of polar coordinates (r, θ) than (x, y). Unfortunately, the structure of Newton's equations of motion does not have the same form in all coordinate systems. It is convenient to use a reformulation of the second law of motion, which has a universal structure in all coordinate frames.

To elucidate on these remarks, we contrast the cartesian and polar equations of motion of a particle in a plane. In cartesian coordinates we have

$$m\ddot{x} = F_x \qquad m\ddot{y} = F_y$$

which is written in vector form as

$$m\ddot{\mathbf{r}} = \mathbf{F} \tag{2-67}$$

with

$$\mathbf{r} = \hat{\mathbf{i}}x + \hat{\mathbf{j}}y \qquad \text{and} \qquad \mathbf{F} = \hat{\mathbf{i}}F_x + \hat{\mathbf{j}}F_y$$

In polar coordinates the vector $\mathbf{r}$ is given by

$$\mathbf{r} = \hat{\mathbf{i}}r\cos\theta + \hat{\mathbf{j}}r\sin\theta \tag{2-68}$$

The unit vectors $\hat{\mathbf{i}}$ and $\hat{\mathbf{j}}$ in the cartesian system do not change with time. The velocity and acceleration in terms of polar variables are

$$\begin{aligned} \mathbf{v} &= \frac{d\mathbf{r}}{dt} = \dot{r}(\hat{\mathbf{i}}\cos\theta + \hat{\mathbf{j}}\sin\theta) + r\dot{\theta}(-\hat{\mathbf{i}}\sin\theta + \hat{\mathbf{j}}\cos\theta) \\ \mathbf{a} &= \frac{d^2\mathbf{r}}{dt^2} = (\ddot{r} - r\dot{\theta}^2)(\hat{\mathbf{i}}\cos\theta + \hat{\mathbf{j}}\sin\theta) \\ &\quad + (r\ddot{\theta} + 2\dot{r}\dot{\theta})(-\hat{\mathbf{i}}\sin\theta + \hat{\mathbf{j}}\cos\theta) \end{aligned} \tag{2-69}$$

From the geometrical picture in Fig. 2-12, we identify

$$\begin{aligned} \hat{\mathbf{r}} &= \hat{\mathbf{i}}\cos\theta + \hat{\mathbf{j}}\sin\theta \\ \hat{\mathbf{l}} &= -\hat{\mathbf{i}}\sin\theta + \hat{\mathbf{j}}\cos\theta \end{aligned} \tag{2-70}$$

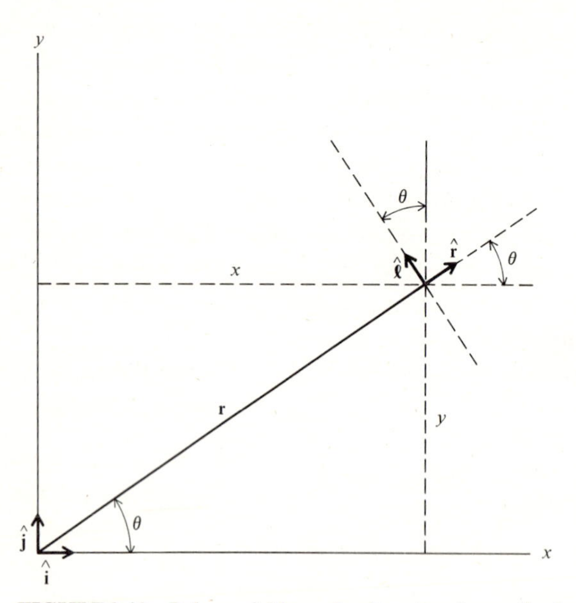

FIGURE 2-12 Polar variables and unit vectors for motion in a plane.

with unit vectors $\hat{\mathbf{r}}$ along $\mathbf{r}$ and $\hat{\boldsymbol{\ell}}$ perpendicular to $\mathbf{r}$ in a counterclockwise sense. Thus we have

$$\begin{aligned} \mathbf{v} &= \hat{\mathbf{r}}\dot{r} + \hat{\boldsymbol{\ell}}r\dot{\theta} \\ \mathbf{a} &= \hat{\mathbf{r}}(\ddot{r} - r\dot{\theta}^2) + \hat{\boldsymbol{\ell}}(r\ddot{\theta} + 2\dot{r}\dot{\theta}) \\ \mathbf{F} &= \hat{\mathbf{r}}F_r + \hat{\boldsymbol{\ell}}F_\theta \end{aligned} \tag{2-71}$$

The equations of motion in the polar system are then

$$\begin{aligned} m(\ddot{r} - r\dot{\theta}^2) &= F_r \\ m(r\ddot{\theta} + 2\dot{r}\dot{\theta}) &= F_\theta \end{aligned} \tag{2-72}$$

Notice the difference between the left-hand side of these equations and the cartesian equations (2-67). In other coordinate systems the structure of the equations in motion can be even more complicated, and the derivation of results similar to Eqs. (2-72) correspondingly more difficult. Therefore it is understandable why a formulation of the equations of motion with the same structure in any coordinate frame could be advantageous. The Lagrange equations of motion fulfill this promise.

2-7 LAGRANGE'S EQUATIONS

We will first derive the Lagrange equation for particle motion in one dimension. With a little extra effort we will then extend the derivation to a system with an arbitrary number of degrees of freedom. The derivation is purely mathematical and involves formal manipulations with partial and total derivatives.

The momentum $p = m\dot{x}$ of a particle in one-dimensional motion can be written in terms of the kinetic energy $T(\dot{x}) = \frac{1}{2}m\dot{x}^2$ as

$$p = \frac{dT}{d\dot{x}} \tag{2-73}$$

We consider a new coordinate $q(t)$ for which q and x are related by the transformation

$$q(t) = q[x(t), t]$$

or inversely,

$$x(t) = x[q(t), t] \tag{2-74}$$

An explicit dependence on t in the transformation allows for the possibility that the q- and x-coordinate axes are in relative translational motion. The derivative $\dot{x} = dx/dt$ can be related to the new coordinate by chain differentiation.

$$\dot{x} = \frac{\partial x}{\partial q}\dot{q} + \frac{\partial x}{\partial t} \tag{2-75}$$

Thus, in general, the functional dependence is

$$\begin{aligned} \dot{x} &= \dot{x}(\dot{q}, q, t) \\ T(\dot{x}) &= T(\dot{q}, q, t) \end{aligned} \tag{2-76}$$

We next introduce a new momentum $\mathfrak{p}(t)$ in correspondence with Eq. (2-73).

$$\mathfrak{p}(t) = \frac{\partial T}{\partial \dot{q}}(\dot{q}, q, t) \tag{2-77}$$

The new variables q and $\mathfrak{p}$ are usually called the *generalized coordinate* and the *generalized momentum*, respectively. The generalized momentum $\mathfrak{p}$ is related to the ordinary momentum p by

$$\mathfrak{p} = \frac{dT}{d\dot{x}}\frac{\partial \dot{x}}{\partial \dot{q}} = p\frac{\partial \dot{x}}{\partial \dot{q}}$$

By use of Eqs. (2-74) and (2-75), the partial derivative $\partial \dot{x}/\partial \dot{q}$ simplifies to

$$\frac{\partial \dot{x}}{\partial \dot{q}} = \frac{\partial x}{\partial q}$$

Therefore we have

$$\mathfrak{p} = p \frac{\partial x}{\partial q} \tag{2-78}$$

In the x coordinate, the equation of motion is

$$\dot{p} = F(\dot{x}, x, t) \tag{2-79}$$

To find the corresponding equation of motion in the q coordinate, we differentiate both sides of Eq. (2-78) with respect to t.

$$\dot{\mathfrak{p}} = \dot{p} \frac{\partial x}{\partial q} + p \frac{d}{dt}\left(\frac{\partial x}{\partial q}\right) \tag{2-80}$$

To simplify the second term, we interchange the order of differentiation,

$$\frac{d}{dt}\left(\frac{\partial x}{\partial q}\right) = \frac{\partial \dot{x}}{\partial q}$$

as can be justified from Eq. (2-75), and we use Eq. (2-73) to obtain

$$p \frac{d}{dt}\left(\frac{\partial x}{\partial q}\right) = \frac{dT}{d\dot{x}} \frac{\partial \dot{x}}{\partial q} = \frac{\partial T}{\partial q} \tag{2-81}$$

From Eqs. (2-79) to (2-81), the equation of motion in the q-coordinate system is

$$\dot{\mathfrak{p}} = F \frac{\partial x}{\partial q} + \frac{\partial T}{\partial q} \tag{2-82}$$

The quantity

$$Q(\dot{q}, q, t) = F \frac{\partial x}{\partial q} \tag{2-83}$$

is called the *generalized force*. The equation of motion

$$\frac{d\mathfrak{p}}{dt} = Q + \frac{\partial T}{\partial q} \tag{2-84}$$

is of universal form for an arbitrary choice of coordinate q. The term $\partial T/\partial q$ in this equation represents a "fictitious" force which appears whenever $\partial q/\partial x \neq$ constant, i.e., when equal intervals of q do not correspond to equal intervals of distance.

If the force F is separated into a conservative part $-[dV(x)]/dx$ and a nonconservative part F', the generalized force can be separated into corresponding parts.

$$Q = -\frac{dV(x)}{dx}\frac{\partial x}{\partial q} + F'\frac{\partial x}{\partial q} = -\frac{\partial V(q)}{\partial q} + Q'$$

What we mean by $V(q)$ is the quantity at each point that is the same as $V(x)$, that is, $V[x(q)]$. For simplicity, we use the notation $V(q)$, although it is not the same *function* as $V(x)$. Notice that the conservative part of the generalized force has the same form, $-\partial V/\partial q$, as the conservative cartesian force $-\partial V/\partial x$. When Eq. (2-84) for Q is substituted into Eq. (2-83), we can write

$$\frac{d\mathfrak{p}}{dt} = \frac{\partial L}{\partial q} + Q' \tag{2-85}$$

where

$$L(\dot{q}, q, t) \equiv T(\dot{q}, q, t) - V(q) \tag{2-86}$$

is the *lagrangian* function. Since

$$\mathfrak{p} = \frac{\partial T}{\partial \dot{q}} = \frac{\partial L}{\partial \dot{q}} \tag{2-87}$$

which follows from Eq. (2-77) and $\partial V(q)/\partial \dot{q} = 0$, the Lagrange equation of motion (2-85) can be also written

$$\frac{d}{dt}\left(\frac{\partial L}{\partial \dot{q}}\right) - \frac{\partial L}{\partial q} = Q' \tag{2-88}$$

The generalized force Q' must include all forces F' on the particle which are not implicitly included in the potential energy.

The derivation of the Lagrange equation given above was for the motion of one particle in one dimension. We now generalize the derivation to apply to the motion of systems of particles with many degrees of freedom. If n coordinates are needed to fully specify the motion of a system possibly consisting of several particles, the transformation from generalized coordinates to cartesian coordinates is

$$\begin{aligned} x_1 &= x_1(q_1, q_2, \ldots, q_n; t) \\ &\vdots \\ x_i &= x_i(q_1, q_2, \ldots, q_n; t) \\ &\vdots \\ x_n &= x_n(q_1, q_2, \ldots, q_n; t) \end{aligned} \tag{2-89}$$

For example, if N particles move without constraint in three-dimensional space, each particle requires three coordinates to specify its motion. The coordinates x_1, x_2, and x_3 label the first particle; x_4, x_5, and x_6, the second; and so on, up to a total of $3N$ cartesian coordinates and corresponding $3N$ generalized coordinates. Thus, in this case, $n = 3N$. If the N particles are all constrained to slide on a surface, $n = 2N$, since only two coordinates are needed to fix each particle. In terms of cartesian coordinates the ith component of momentum is

$$p_i = m\dot{x}_i$$

where x_5 might denote, for example, the y coordinate of the second particle. The kinetic energy of the system is

$$T = \sum_i \tfrac{1}{2} m_i \dot{x}_i^2$$

In analogy to Eq. (2-73),

$$p_i = \frac{\partial T}{\partial \dot{x}_i} \tag{2-90}$$

From Eqs. (2-89) we find that the total time derivative of x_i is

$$\dot{x}_i = \frac{\partial x_i}{\partial q_j} \dot{q}_j + \frac{\partial x_i}{\partial t} \tag{2-91}$$

In this and subsequent equations a summation is implied over repeated subscript indices of vector quantities. From Eq. (2-91) it follows that

$$\frac{\partial \dot{x}_i}{\partial \dot{q}_j} = \frac{\partial x_i}{\partial q_j} \tag{2-92}$$

The generalized momenta $\mathfrak{p}_j$ is now introduced as

$$\mathfrak{p}_j = \frac{\partial T}{\partial \dot{q}_j} \tag{2-93}$$

and $\mathfrak{p}_j$ is related to p_i by

$$\mathfrak{p}_j = \frac{\partial T}{\partial \dot{q}_j} = \frac{\partial T}{\partial \dot{x}_i} \frac{\partial \dot{x}_i}{\partial \dot{q}_j}$$

Using Eqs. (2-90) and (2-92),

$$\mathfrak{p}_i = p_i \frac{\partial x_i}{\partial q_j} \tag{2-94}$$

The time derivative of $\mathfrak{p}_j$ is

$$\dot{\mathfrak{p}}_j = \dot{p}_i \frac{\partial x_i}{\partial q_j} + p_i \frac{d}{dt}\left(\frac{\partial x_i}{\partial q_j}\right) \tag{2-95}$$

Using Newton's law, the $\dot{p}_i$ are given by the components of the force

$$\dot{p}_i = F_i(x_1, \ldots, x_n; \dot{x}_1, \ldots, \dot{x}_n; t) \tag{2-96}$$

With the use of Eqs. (2-90) and (2-96), the result in Eq. (2-95) becomes

$$\dot{\mathfrak{p}}_j = F_i \frac{\partial x_i}{\partial q_j} + \frac{\partial T}{d\dot{x}_i} \frac{d}{dt}\left(\frac{\partial x_i}{\partial q_j}\right) \tag{2-97}$$

The first term on the right-hand side of Eq. (2-97) is called the generalized force

$$Q_j = F_i \frac{\partial x_i}{\partial q_j} \tag{2-98}$$

The last term in Eq. (2-97) can be simplified by the identity

$$\frac{d}{dt}\left(\frac{\partial x_i}{\partial q_j}\right) = \frac{\partial \dot{x}_i}{\partial q_j}$$

which follows from Eq. (2-91). With the definition in Eq. (2-98) and the above identity, we get

$$\dot{\mathfrak{p}}_j = Q_j + \frac{\partial T}{\partial \dot{x}_i} \frac{\partial \dot{x}_i}{\partial q_j} = Q_j + \frac{\partial T}{\partial q_j} \tag{2-99}$$

When we separate the generalized force into conservative and non-conservative parts as

$$Q_j = -\frac{\partial V}{\partial q_j} + Q'_j \tag{2-100}$$

Eq. (2-99) can be written

$$\dot{\mathfrak{p}}_j = \frac{\partial(T - V)}{\partial q_j} + Q'_j$$

Introducing the lagrangian function

$$L(q_1, q_2, \ldots, q_n; \dot{q}_1, \dot{q}_2, \ldots, \dot{q}_n; t) = T - V \tag{2-101}$$

the equation of motion for the jth generalized coordinate is

$$\dot{\mathfrak{p}}_j = \frac{\partial L}{\partial q_j} + Q'_j \tag{2-102}$$

Since V is assumed to be independent of the $\dot{q}_j$, we find from Eq. (2-93) that

$$\begin{aligned} \mathfrak{p}_j &= \frac{\partial T}{\partial \dot{q}_j} = \frac{\partial L}{\partial \dot{q}_j} \\ \dot{\mathfrak{p}}_j &= \frac{d}{dt}\left(\frac{\partial T}{\partial \dot{q}_j}\right) = \frac{d}{dt}\left(\frac{\partial L}{\partial \dot{q}_j}\right) \end{aligned} \tag{2-103}$$

The combination of Eqs. (102) and (103) gives

$$\frac{d}{dt}\left(\frac{\partial L}{\partial \dot{q}_j}\right) - \frac{\partial L}{\partial q_j} = Q'_j \tag{2-104}$$

This set of equations for $j = 1, 2, \ldots, n$ is known as the *Lagrange equations of motion.*

As an elementary application of lagrangian techniques, we calculate the r and θ equations of motion for a particle moving in a plane under the influence of a central potential energy $V(r)$. As our generalized coordinates we take

$$q_1 = r \qquad q_2 = \theta$$

Since a central potential energy gives rise to a conservative force, we have

$$Q'_2 = Q'_1 = 0$$

The kinetic energy

$$T = \tfrac{1}{2}m(\dot{x}^2 + \dot{y}^2)$$

is easily expressed in polar coordinates

$$\begin{aligned} x &= r\cos\theta \qquad & \dot{x} &= \dot{r}\cos\theta - r\dot{\theta}\sin\theta \\ y &= r\sin\theta \qquad & \dot{y} &= \dot{r}\sin\theta + r\dot{\theta}\cos\theta \end{aligned}$$

as

$$T = \tfrac{1}{2}m(\dot{r}^2 + r^2\dot{\theta}^2) \tag{2-105}$$

We note that T is a function of $\dot{q}_1$, q_1, and $\dot{q}_2$. The lagrangian is

$$L = T - V = \tfrac{1}{2}m(\dot{r}^2 + r^2\dot{\theta}^2) - V(r) \tag{2-106}$$

From Eq. (2-104), Lagrange's equations of motion are

$$\begin{aligned} \frac{d}{dt}\left(\frac{\partial L}{\partial \dot{r}}\right) - \frac{\partial L}{\partial r} &= 0 \\ \frac{d}{dt}\left(\frac{\partial L}{\partial \dot{\theta}}\right) - \frac{\partial L}{\partial \theta} &= 0 \end{aligned} \tag{2-107}$$

Differentiating L according to Eqs. (2-107) leads to

$$m\ddot{r} - mr\dot{\theta}^2 = -\frac{\partial V}{\partial r}$$

$$mr\ddot{\theta} + 2m\dot{r}\dot{\theta} = 0 \tag{2-108}$$

For $F_r = -\partial V/\partial r$ and $F_\theta = 0$, these correspond to Eqs. (2-72), obtained from direct application of Newton's laws. The generalized momentum corresponding to the variable θ,

$$p_\theta = \frac{\partial L}{\partial \dot{\theta}} = mr^2\dot{\theta}$$

is in fact the angular momentum.

The utility of the lagrangian scheme is that we are free to choose those coordinates in which the forces or the motions have their simplest form, without having to return to Newton's laws and transforming the equations of motion in cartesian coordinates to the desired coordinates.

2-8 SIMPLE PENDULUM

In physical situations that involve forces of constraint, we write Lagrange's equations for coordinates q_k along which motion actually occurs in order that the forces of constraint do not enter into the equations of motion. The simple pendulum is an example involving constraints that can be easily solved by lagrangian methods. A simple pendulum is a point mass m at the end of a weightless rod of length l which swings back and forth in a vertical plane. Since the force of constraint on m due to the rod tension T_r is perpendicular to the circular arc of motion, we choose a variable along the arc as the generalized coordinate. The position s on the arc is a possible choice. It is somewhat more convenient to use the angle θ of the rod from vertical, which is proportional to s,

$$\theta = \frac{s}{l}$$

The x (vertical) and y (horizontal) coordinates of the pendulum bob measured from the origin at the support are

$$x = l\cos\theta$$

$$y = l\sin\theta \tag{2-109}$$

as illustrated in Fig. 2-13. Thus the kinetic energy of the bob is

$$T = \tfrac{1}{2}m(\dot{x}^2 + \dot{y}^2) = \tfrac{1}{2}ml^2\dot{\theta}^2 \tag{2-110}$$

The potential energy due to gravity is

$$V = -mgx = -mgl\cos\theta \tag{2-111}$$

with the reference point at $x = 0$. The lagrangian for the pendulum is

$$L = T - V = \tfrac{1}{2}ml^2\dot{\theta}^2 + mgl\cos\theta \tag{2-112}$$

The Lagrange equation for the degree of freedom θ

$$\frac{d}{dt}\left(\frac{\partial L}{\partial \dot{\theta}}\right) - \frac{\partial L}{\partial \theta} = 0$$

leads to

$$ml^2\ddot{\theta} + mgl\sin\theta = 0 \tag{2-113}$$

For small oscillations $\theta \ll 1$, we can approximate $\sin\theta \approx \theta$ to obtain

$$\ddot{\theta} + \omega_0{}^2\theta = 0 \tag{2-114}$$

where $\omega_0 = \sqrt{g/l}$. This is readily recognizable as simple harmonic motion in θ or y with angular frequency ω_0 [see Eqs. (1-48) and (1-49)]. The solution of Eq. (2-114) is

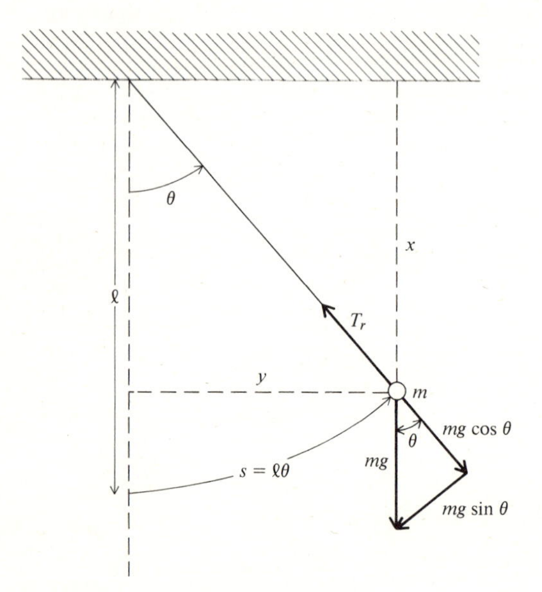

FIGURE 2-13 Simple pendulum.

$$\theta = a \cos(\omega_0 t + \alpha) \tag{2-115}$$

with arbitrary constants a and α. The period of small oscillations,

$$\tau = \frac{2\pi}{\omega_0} = 2\pi\sqrt{\frac{l}{g}} \tag{2-116}$$

is independent of a and α. This approximate amplitude-independent feature of the period of motion, called *isochronism*, is incorporated in escapement mechanisms that regulate the ticking rate of pendulum clocks.

It is instructive to compare the lagrangian method of solving the simple pendulum with direct application of Newton's laws. The forces on the pendulum bob can be resolved into components tangent and perpendicular to the arc of motion as follows:

$$F_\theta = -mg \sin\theta$$
$$F_r = mg\cos\theta - T_r$$

With $\dot{r} = 0$ and $r = l$ in the polar form of Newton's law in Eqs. (2-72), we have

$$F_\theta = ml\ddot{\theta} = -mg\sin\theta \tag{2-117}$$
$$F_r = -ml\dot{\theta}^2 = mg\cos\theta - T_r \tag{2-118}$$

The θ equation describes the motion, as is also found in Eq. (2-113) by the lagrangian method. Although $\dot{r} = 0$, the forces in the r direction are not in equilibrium, since the bob is undergoing radial acceleration to maintain a circular arc. The r equation can be solved for the tension T_r:

$$T_r = mg\cos\theta + ml\dot{\theta}^2 \tag{2-119}$$

To compute the magnitude of T_r, we first solve Eq. (2-117) for $\dot{\theta}$. Using the chain rule,

$$\ddot{\theta} = \frac{d\dot{\theta}}{d\theta}\frac{d\theta}{dt} = \frac{\dot{\theta}\, d\dot{\theta}}{d\theta} = \frac{d}{d\theta}\left(\frac{\dot{\theta}^2}{2}\right)$$

we integrate Eq. (2-117).

$$\int_0^{\dot{\theta}} d\left(\frac{\dot{\theta}^2}{2}\right) = -\frac{g}{l}\int_{\theta_0}^{\theta} \sin\theta\, d\theta \tag{2-120}$$

The boundary condition $\dot{\theta} = 0$ at the maximum angle θ_0 is imposed. The solution is

$$\dot{\theta}^2 = \frac{2g}{l}(\cos\theta - \cos\theta_0) \tag{2-121}$$

which is the statement of energy conservation. When we substitute Eq. (2-121) into Eq. (2-119), we find

$$T_r = 3mg \cos \theta - 2mg \cos \theta_0 \tag{2-122}$$

The $ml\dot{\theta}^2$ contribution to the tension is a centrifugal force due to the circular motion.

In the lagrangian approach to the pendulum problem, we did not solve for the tension since there was no physical motion in the r direction. But if it is desired, such a force of constraint can be found by lagrangian methods by writing down the equation of motion for the coordinate which is constrained, and then afterward imposing the constraint condition. The motion is called *virtual* since it is not actually made by the physical system. In the pendulum example, we write down Lagrange's equation for a virtual motion of the r coordinate,

$$\frac{d}{dt}\left(\frac{\partial L}{\partial \dot{r}}\right) - \frac{\partial L}{\partial r} = Q' \tag{2-123}$$

and afterward set $r = l$, $\dot{r} = 0$, $\ddot{r} = 0$. Allowing for radial motion, the lagrangian from Eqs. (2-106) and (2-111) is

$$L = \tfrac{1}{2}m(\dot{r}^2 + r^2\dot{\theta}^2) + mgr \cos \theta \tag{2-124}$$

The generalized force Q_r' in Eq. (2-123) is due to the force of constraint in the r direction. From Eq. (2-98)

$$Q_r' = F_x' \frac{\partial x}{\partial r} + F_y' \frac{\partial y}{\partial r} \tag{2-125}$$

The constraint force can be resolved into x and y components from the geometry in Fig. 2-13.

$$F_x' = -T_r \cos \theta$$

$$F_y' = -T_r \sin \theta$$

The derivatives $\partial x/\partial r$ and $\partial y/\partial r$ are found from Eqs. (2-109) with variable $r = l$.

$$\frac{\partial x}{\partial r} = \cos \theta \qquad \frac{\partial y}{\partial r} = \sin \theta$$

Thus, in terms of the tension, the generalized force in Eq. (2-125) is

$$Q_r' = -T_r \tag{2-126}$$

From Eqs. (2-123), (2-124), and (2-126) we obtain the equation governing the r motion.

$$m\ddot{r} - mr\dot{\theta}^2 - mg\cos\theta = -T_r$$

By application of the physical constraints $\ddot{r} = 0$, $r = l$, we get

$$T_r = ml\dot{\theta}^2 + mg\cos\theta \tag{2-127}$$

in agreement with Eq. (2-119).

2-9 PENDULUM WITH FINITE DISPLACEMENT

The equation of motion (2-113) for a simple pendulum leads to simple harmonic motion only for infinitesimal angles of oscillation. For finite displacements the angular velocity from Eq. (2-121) is

$$\dot{\theta}^2 = \frac{2g}{l}(\cos\theta - \cos\theta_0)$$

where θ_0 is the maximum angle. The exact solution to this equation for finite angles can be written in the integral form

$$\sqrt{\frac{2g}{l}}\,t = \int^{\theta} \frac{d\theta}{\sqrt{\cos\theta - \cos\theta_0}} \tag{2-128}$$

Unfortunately, this integration cannot be carried out in terms of simple functions. We can find an approximate solution for small, but not infinitesimal, θ_0 by making an expansion of the integrand in powers of θ_0. To do this, we substitute the identity

$$\cos\theta = 1 - 2\sin^2\frac{\theta}{2} \tag{2-129}$$

in Eq. (2-128).

$$2\sqrt{\frac{g}{l}}\,t = \int^{\theta} \frac{d\theta}{\sqrt{\sin^2(\theta_0/2) - \sin^2(\theta/2)}} \tag{2-130}$$

For small θ_0 the upper limit of integration θ is also small. Hence we introduce a new variable so that all small parameters occur in the integrand. An expansion in powers of θ_0 is then more easily carried out. We define

$$\sin\beta = \frac{\sin(\theta/2)}{\sin(\theta_0/2)} \tag{2-131}$$

The solution in Eq. (2-130) becomes

$$\sqrt{\frac{g}{l}}\, t = \int^{\beta} \frac{d\beta}{\sqrt{1 - \sin^2(\theta_0/2)\sin^2\beta}} \tag{2-132}$$

This integral is known as an elliptic integral. It cannot be evaluated in closed form, but numerical evaluations are available in tabular form or as subroutines on most large computers. To find an approximate solution for small angular displacements, the integrand in Eq. (2-132) can be expanded by power series and then integrated term by term. We find

$$\begin{aligned}\sqrt{\frac{g}{l}}\, t &= \int^{\beta} d\beta\left[1 + \tfrac{1}{2}\sin^2\left(\frac{\theta_0}{2}\right)\sin^2\beta + \cdots\right] \\ &= \beta + \tfrac{1}{4}\sin^2\left(\frac{\theta_0}{2}\right)\left(\beta - \frac{\sin 2\beta}{2}\right) + \cdots \end{aligned} \tag{2-133}$$

Over a half period of the motion ($\Delta t = \tau/2$), θ goes from $-\theta_0$ to θ_0 and β goes from $-\pi/2$ to $\pi/2$, as can be deduced from Eq. (2-131). Thus the period is given by

$$\tau = \tau_0\left[1 + \tfrac{1}{4}\sin^2\left(\frac{\theta_0}{2}\right) + \cdots\right]$$

where $\tau_0 = 2\pi\sqrt{l/g}$ is the simple harmonic period. Again using the fact that θ_0 is small, $\sin^2(\theta_0/2) \approx \theta_0{}^2/4$, we find

$$\tau = \tau_0\left(1 + \frac{\theta_0{}^2}{16} + \cdots\right) \tag{2-134}$$

The period is increased over the simple harmonic period. The fractional lengthening of the period is

$$\frac{\tau - \tau_0}{\tau_0} = \frac{\theta_0{}^2}{16}$$

For a 30° displacement, the fractional lengthening of the period

$$\frac{\tau - \tau_0}{\tau_0} = \frac{1}{16}\left(\frac{30°}{57.3°}\right)^2 = 0.017$$

is less than 2 percent.

2-10 PENDULUM WITH OSCILLATING SUPPORT

So far you may have wondered why the lagrangian method is so advantageous, since the examples we have solved are just as easy to do by more traditional methods. To illustrate the utility of the lagrangian approach, we shall treat the motion of a pendulum with an oscillating support. This example also provides a simple demonstration of the forced harmonic oscillator.

The point of suspension of a simple pendulum is moved horizontally as a function of time, as shown in Fig. 2-14. The vertical and horizontal coordinates of the bob are $(x, y + y_s)$ with respect to a fixed system, where y_s is the horizontal coordinate of the support and

$$x = l \cos \theta$$
$$y = l \sin \theta$$

The bob's kinetic energy is

$$T = \tfrac{1}{2}m[(\dot{y} + \dot{y}_s)^2 + \dot{x}^2] = \tfrac{1}{2}m(\dot{x}^2 + \dot{y}^2 + 2\dot{y}\dot{y}_s + \dot{y}_s^2) \qquad \text{(2-135)}$$

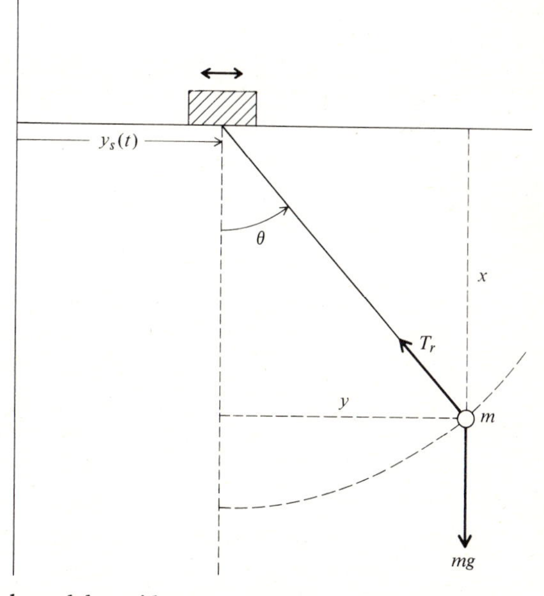

FIGURE 2-14 Simple pendulum with a support which oscillates horizontally.

and the potential energy is

$$V = -mgl\cos\theta \tag{2-136}$$

The lagrangian is

$$L = \tfrac{1}{2}m(l^2\dot{\theta}^2 + 2l\dot{\theta}\dot{y}_s\cos\theta + \dot{y}_s^2) + mgl\cos\theta \tag{2-137}$$

For the generalized coordinate θ, the derivatives appearing in the Lagrange equation are

$$\begin{aligned}
\frac{\partial L}{\partial \dot{\theta}} &= ml^2\dot{\theta} + ml\dot{y}_s\cos\theta \\
\frac{d}{dt}\left(\frac{\partial L}{\partial \dot{\theta}}\right) &= ml^2\ddot{\theta} + ml\ddot{y}_s\cos\theta - ml\dot{\theta}\dot{y}_s\sin\theta \\
\frac{\partial L}{\partial \theta} &= -ml\dot{\theta}\dot{y}_s\sin\theta - mgl\sin\theta
\end{aligned} \tag{2-138}$$

The equation of motion is

$$\ddot{\theta} + \frac{g}{l}\sin\theta = -\frac{\ddot{y}_s}{l}\cos\theta \tag{2-139}$$

For small angular displacements and a sinusoidal motion of the support,

$$y_s = y_0\cos\omega t \tag{2-140}$$

the equation of motion (2-139) becomes

$$\ddot{\theta} + \omega_0{}^2\theta = \frac{y_0}{l}\,\omega^2\cos\omega t \tag{2-141}$$

where $\omega_0 = \sqrt{g/l}$ is the natural frequency. This equation is mathematically identical with an undamped forced-harmonic oscillator [see Eqs. (1-76)].

The above result can be used to provide a simple demonstration of the main properties of the forced-harmonic oscillator. A pendulum consisting of a bob on a string is held in one hand, and the natural frequency is roughly noted. If the supporting hand is then made to move back and forth at a frequency smaller than the natural frequency, the bob will move in phase with the hand. As the hand is moved more rapidly, the bob starts to lag behind and the amplitude increases. Finally, for still more rapid hand motion, the bob moves in an opposite sense to the hand, and the amplitude decreases. All the observations are predicted by the mathematical solution in Sec. 1-9, illustrated in Fig. 1-7.

We stress that the advantage of using lagrangian methods is the methodical and straightforward procedure that is involved. Once a lagrangian function is constructed from the kinetic and potential energies, the task of obtaining the equations of motion is simply a matter of differentiation. In complex problems there is often less chance of error using this method.

2-11 COUPLED HARMONIC OSCILLATORS

The equations of motion for a system of interacting harmonic oscillators can be solved in closed form. The solution is of general interest since numerous physical systems are well approximated by coupled harmonic oscillators. As a specific example, we investigate the motion of two simple pendulums whose bobs are connected by a spring, as indicated in Fig. 2-15.

For small angular displacements the equation of motion of a single isolated pendulum is given by Eq. (2-114).

$$\ddot{\theta} + \omega_0^2 \theta = 0 \tag{2-142}$$

This equation can be alternatively expressed in terms of the x and y coordinates of the pendulum bob. For $\theta \ll 1$, we have

$$\begin{aligned} x &= l \cos \theta \approx l \\ y &= l \sin \theta \approx l\theta \end{aligned} \tag{2-143}$$

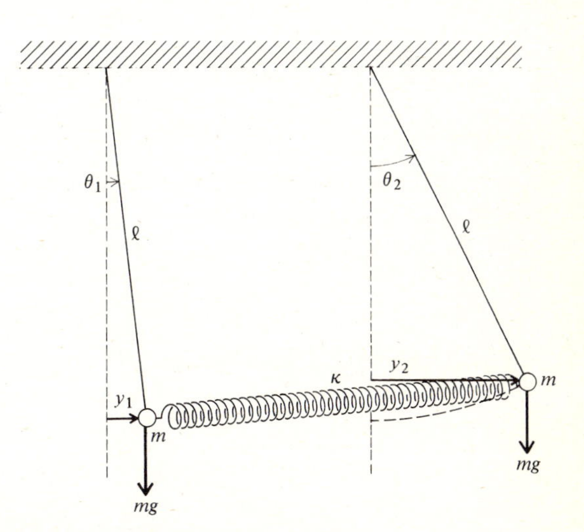

FIGURE 2-15 Two simple pendulums coupled by a spring.

and Eq. (2-142) becomes

$$\ddot{y} + \omega_0^2 y = 0 \tag{2-144}$$

In this approximation the pendulum executes simple harmonic motion in the horizontal direction with no vertical motion. The pendulum spring system of Fig. 2-15 is therefore equivalent to the three-spring system of Fig. 2-16, with spring constants $k = m\omega_0^2 = mg/l$ for the outer springs and κ for the inner spring.

The kinetic and potential energies for the spring system are given by

$$\begin{aligned} T &= \tfrac{1}{2}m(\dot{y}_1^2 + \dot{y}_2^2) \\ V &= \tfrac{1}{2}k(y_1^2 + y_2^2) + \tfrac{1}{2}\kappa(y_1 - y_2)^2 \end{aligned} \tag{2-145}$$

From the lagrangian function

$$L = \tfrac{1}{2}m(\dot{y}_1^2 + \dot{y}_2^2) - \tfrac{1}{2}k(y_1^2 + y_2^2) - \tfrac{1}{2}\kappa(y_1 - y_2)^2 \tag{2-146}$$

we determine the differential equations of motion to be

$$\begin{aligned} m\ddot{y}_1 &= -ky_1 - \kappa(y_1 - y_2) \\ m\ddot{y}_2 &= -ky_2 + \kappa(y_1 - y_2) \end{aligned} \tag{2-147}$$

To solve the differential equations, we look for linear combinations of y_1 and y_2 that are of simple harmonic form. If we add the equations, we find

$$m(\ddot{y}_1 + \ddot{y}_2) = -k(y_1 + y_2) \tag{2-148}$$

Upon subtraction of Eqs. (2-147), we get

$$m(\ddot{y}_1 - \ddot{y}_2) = -(k + 2\kappa)(y_1 - y_2) \tag{2-149}$$

The solutions of these two uncoupled equations are found directly from Eq. (1-51) as

$$\begin{aligned} y_1 + y_2 &= a\cos(\omega t + \alpha) \\ y_1 - y_2 &= b\cos(\tilde{\omega} t + \beta) \end{aligned} \tag{2-150}$$

where the angular frequencies are

$$\begin{aligned} \omega &= \sqrt{\frac{k}{m}} \\ \tilde{\omega} &= \sqrt{\frac{k + 2\kappa}{m}} \end{aligned} \tag{2-151}$$

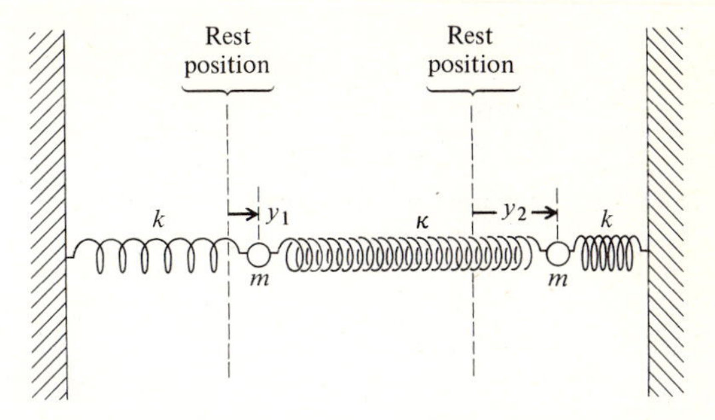

FIGURE 2-16 **Equivalent three-spring system for the coupled-pendulum system of Fig. 2-15.**

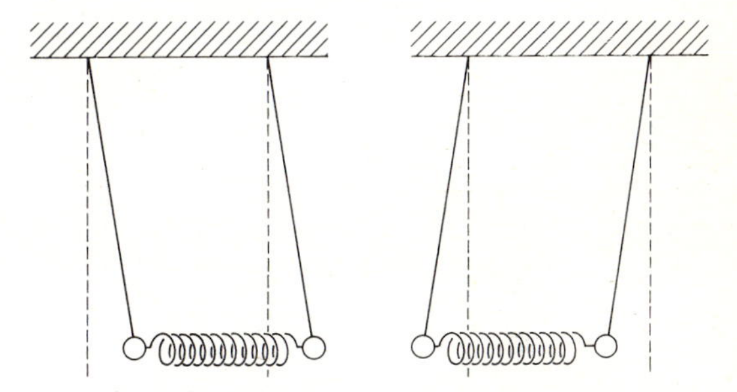

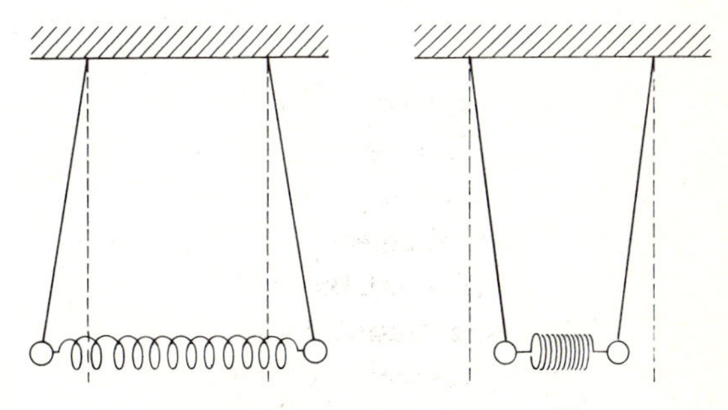

FIGURE 2-17 **Normal modes for the coupled-pendulum system.**

The combinations $(y_1 + y_2)$ and $(y_1 - y_2)$ oscillate independently with pure frequencies, and are called the *normal modes*. The motion of the pendulum bobs is in general a superposition of the two normal modes of vibration.

$$\begin{aligned} y_1 &= \tfrac{1}{2}a \cos(\omega t + \alpha) + \tfrac{1}{2}b \cos(\tilde{\omega} t + \beta) \\ y_2 &= \tfrac{1}{2}a \cos(\omega t + \alpha) - \tfrac{1}{2}b \cos(\tilde{\omega} t + \beta) \end{aligned} \tag{2-152}$$

The four constants a, b, α, β in Eqs. (2-152) are to be determined by the boundary conditions. To excite a single frequency (i.e., one normal mode) either a or b must be zero. These two possibilities correspond to the bobs swinging synchronously with frequency ω or antisynchronously with frequency $\tilde{\omega}$, as illustrated in Fig. 2-17.

In the weak coupling limit $\kappa \ll k$, the coupling between the two pendulums causes a very gradual interchange of energy between the two oscillators. To demonstrate this we suppose that both bobs are initially at rest and the motion of the system is started by displacing the first bob by a distance δ. When these boundary conditions are imposed on Eqs. (2-152), we obtain

$$\begin{aligned} y_1 &= \frac{\delta}{2}(\cos \omega t + \cos \tilde{\omega} t) \\ y_2 &= \frac{\delta}{2}(\cos \omega t - \cos \tilde{\omega} t) \end{aligned} \tag{2-153}$$

From trigonometric identities for the sum and difference of cosine functions, Eqs. (2-153) can be written

$$\begin{aligned} y_1 &= \delta \cos\left(\frac{\tilde{\omega} + \omega}{2} t\right) \cos\left(\frac{\tilde{\omega} - \omega}{2} t\right) \\ y_2 &= \delta \sin\left(\frac{\tilde{\omega} + \omega}{2} t\right) \sin\left(\frac{\tilde{\omega} - \omega}{2} t\right) \end{aligned} \tag{2-154}$$

For the weak coupling limit, $\tilde{\omega} - \omega \ll \omega$, the last factors in Eqs. (2-154) are slowly varying functions of time. These slowly varying factors constitute an envelope for the rapidly oscillating sinusoidal factors of argument $[(\tilde{\omega} + \omega)/2]t$, as illustrated in Fig. 2-18. At time $t = \pi/(\tilde{\omega} - \omega)$, the first pendulum has come to rest and all the energy has been transferred through the coupling to the second oscillator. This gives rise to the phenomenon known as *beats*. The beat frequency is $(\tilde{\omega} - \omega)/2$, and the frequency of the envelope of the amplitude is $(\tilde{\omega} - \omega)$.

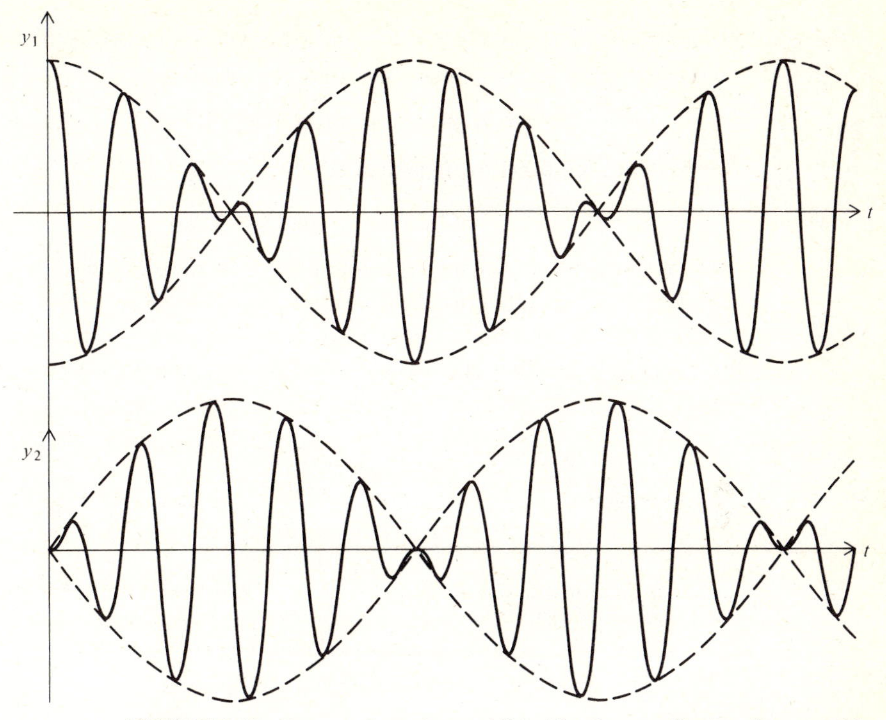

FIGURE 2-18 Beats and envelope exhibited by the coordinates of two weakly coupled oscillators.

2-12 HAMILTON'S EQUATIONS

The Lagrange equations of motion represent an equivalent formulation of Newton's second law in terms of a set of second-order differential equations. An alternative formulation of Newton's law consists of two sets of first-order differential equations known as *Hamilton's equations.* Although the hamiltonian formulation of classical mechanics offers no particular advantage in the solution of ordinary problems, this formalism is basic to the foundations of statistical and quantum mechanics.

In lagrangian mechanics the independent variables are q_j, $\dot{q}_j$, and t, and the generalized momentum $\mathfrak{p}_j$ is a derived variable given by Eqs. (2-103).

$$\mathfrak{p}_j = \frac{\partial T}{\partial \dot{q}_j} = \frac{\partial L}{\partial \dot{q}_j} \tag{2-155}$$

In hamiltonian mechanics, q_j, $\mathfrak{p}_j$, and t are chosen as independent variables and $\dot{q}_j$ is a dependent function.

$$\dot{q}_j = \dot{q}_j(q_1, q_2, \ldots, q_n; \mathfrak{p}_1, \mathfrak{p}_2, \ldots, \mathfrak{p}_n; t) \tag{2-156}$$

The hamiltonian function H is defined as

$$H(q_1, q_2, \ldots, q_n; \mathfrak{p}_1, \mathfrak{p}_2, \ldots, \mathfrak{p}_n; t) = \mathfrak{p}_k \dot{q}_k - L \tag{2-157}$$

where a summation over the repeated index k on the right-hand side is implied. In this definition the variables $\dot{q}_j$ are understood to be functions of the generalized coordinates, momenta, and time as in Eq. (2-156). To derive Hamilton's equations of motion, we differentiate H with respect to the independent variables $\mathfrak{p}_j$ and q_j.

$$\begin{aligned} \frac{\partial H}{\partial \mathfrak{p}_j} &= \dot{q}_j + \left(\mathfrak{p}_k - \frac{\partial L}{\partial \dot{q}_k}\right)\frac{\partial \dot{q}_k}{\partial \mathfrak{p}_j} \\ \frac{\partial H}{\partial q_j} &= \left(\mathfrak{p}_k - \frac{\partial L}{\partial \dot{q}_k}\right)\frac{\partial \dot{q}_k}{\partial q_j} - \frac{\partial L}{\partial q_j} \end{aligned} \tag{2-158}$$

From Eq. (2-155) the factors in parentheses in Eqs. (2-158) vanish. In the second equation we eliminate $\partial L/\partial q_j$ through Eqs. (2-103) and (2-104). This gives the Hamilton equations of motion.

$$\begin{aligned} \frac{\partial H}{\partial \mathfrak{p}_j} &= \dot{q}_j \\ \frac{\partial H}{\partial q_j} &= -\dot{\mathfrak{p}}_j - Q'_j \end{aligned} \tag{2-159}$$

To establish the physical significance of the hamiltonian we relate the quantities on the right-hand side of Eq. (2-157) to the kinetic and potential energies of the system. In cartesian coordinates the kinetic energy is

$$T = \sum_k \tfrac{1}{2} m_k \dot{x}_k^2 = \tfrac{1}{2} p_k \dot{x}_k$$

Using Eqs. (2-91) and (2-94), the kinetic energy can be rewritten in the form

$$T = \tfrac{1}{2} p_k \left(\frac{\partial x_k}{\partial q_j}\dot{q}_j + \frac{\partial x_k}{\partial t}\right) = \tfrac{1}{2}\mathfrak{p}_j \dot{q}_j + \tfrac{1}{2} p_k \frac{\partial x_k}{\partial t} \tag{2-160}$$

Solving for $\mathfrak{p}_j \dot{q}_j$ from Eq. (2-160),

$$\mathfrak{p} \dot{q}_j = 2T - p_k \frac{\partial x_k}{\partial t} \tag{2-161}$$

and substituting the result into Eq. (2-157), we obtain

$$H = 2T - p_k \frac{\partial x_k}{\partial t} - L$$

$$= T + V - p_k \frac{\partial x_k}{\partial t} \tag{2-162}$$

In the event that the transformation between the cartesian and generalized coordinates in Eqs. (2-89) has no explicit time dependence,

$$\frac{\partial x_k}{\partial t}(q_1, q_2, \ldots, q_n) = 0 \tag{2-163}$$

the hamiltonian is the total energy of the system,

$$H = T + V \tag{2-164}$$

Most physical applications of hamiltonian methods concern systems for which Eq. (2-164) holds.

To find the conditions under which the hamiltonian is a conserved quantity, we compute the total time derivative.

$$\frac{dH}{dt} = \frac{\partial H}{\partial q_k}\dot{q}_k + \frac{\partial H}{\partial \mathfrak{p}_k}\dot{\mathfrak{p}}_k + \frac{\partial H}{\partial t} \tag{2-165}$$

Upon use of Eqs. (2-159), this reduces to

$$\frac{dH}{dt} = -Q'_k \dot{q}_k + \frac{\partial H}{dt} \tag{2-166}$$

Thus, if the forces are derivable from a potential ($Q'_j = 0$) and H has no explicit time dependence ($\partial H/\partial t = 0$), the hamiltonian is a constant of the motion. This constant is the total energy of the system if Eq. (2-163) holds.

As an elementary example of the hamiltonian method, we consider the one-dimensional harmonic oscillator for which

$$\begin{aligned} T &= \tfrac{1}{2}m\dot{x}^2 \\ V &= \tfrac{1}{2}kx^2 \\ L &= T - V = \tfrac{1}{2}m\dot{x}^2 - \tfrac{1}{2}kx^2 \end{aligned} \tag{2-167}$$

The momentum is found by differentiation according to Eq. (2-155).

$$p_x = \frac{\partial T}{\partial \dot{x}} = m\dot{x} \tag{2-168}$$

The hamiltonian from Eq. (2-157) is

$$H = p_x\dot{x} - (\tfrac{1}{2}m\dot{x}^2 - \tfrac{1}{2}kx^2)$$

$$= \frac{p_x{}^2}{2m} + \tfrac{1}{2}kx^2 \tag{2-169}$$

where Eq. (2-168) has been used to eliminate $\dot{x}$ in favor of p_x. The hamiltonian in this case is immediately recognizable as the total energy of the oscillator.

Hamilton's equations of motion from Eqs. (2-159),

$$\frac{\partial H}{\partial p_x} = \dot{x} \qquad \frac{\partial H}{\partial x} = -\dot{p}_x$$

yield

$$\frac{p_x}{m} = \dot{x} \qquad kx = -\dot{p}_x \tag{2-170}$$

When these two first-order equations are combined, we obtain the usual second-order differential form of Newton's second law.

$$m\frac{d^2x}{dt^2} = -kx$$

PROBLEMS

1. A spring of mass M (assume constant mass density along its length) has one end attached to a fixed support. The other end is given a small displacement and released. Show that the frequency of oscillation is the same as that of a massless spring (with the same spring constant) with a mass m at its free end, and find that mass. (*Hint:* Compute the kinetic energy of the spring.)

2. A particle of mass m moves under the action of a force

$$F = -F_0 \sinh ax = -\frac{F_0}{2}(e^{ax} - e^{-ax})$$

where $a > 0$. Sketch the potential, discuss the motion, and solve for the frequency of small oscillation if there exists a point of stability.

3. A particle moves in the region $x > 0$ subject to the potential energy

$$V(x) = V_0\left(\frac{a}{x} + \frac{x}{a}\right)$$

where $V_0 a > 0$. Locate any stable-equilibrium points and obtain the frequency of small oscillations about those points. Estimate the displacement from equilibrium for which the motion ceases to be harmonic.

4. The potential energy of a mass element dm at a height z above the earth's surface is $dV = dmgz$. Compute the potential energy in a pyramid of height h, square base $b \times b$, and mass density ρ. The Great Pyramid of Khufu is 147 m high and has a base 234 × 234 m. Estimate its potential energy using $\rho = 2.5\ \text{g/cm}^3$ for the density of building materials. If an average man lifts 50 kg through a distance of 2 m each minute of a 10-hr workday, estimate the man-years of labor expended in the construction of the Great Pyramid.

5. Find the escape velocity for an object of mass 4 g on the surface of the moon.

Gravitational acceleration:

$$g(\text{moon}) \approx \frac{1}{6}\, g(\text{earth})$$

Radius:

$$R(\text{moon}) \approx \frac{1}{3.6}\, R(\text{earth})$$

Escape velocity on earth:

$$v_{\text{esc}}(\text{earth}) \approx 11.2\ \text{km/s}$$

6. A projectile is fired from the surface of the earth to the moon. Neglecting the orbital motion of the moon, what is the minimum velocity of impact on the surface of the moon?

7. Prove the vector-product identities of Eqs. (2-44*a*) and (2-44*b*).

8. A force field is given by

$$F_x = kyz \sin kxy$$
$$F_y = kxz \sin kxy$$
$$F_z = -\cos kxy$$

a. Evaluate $\nabla \times \mathbf{F}$ to determine if $\mathbf{F}$ is conservative.

b. If the reference potential energy at $(x = 0,\ y = 0,\ z = 0)$ is zero, compute the potential energy at the point $(x = 1.0,\ y = 1.0,\ z = 1.0)$. Use any convenient path.

c. Compute the potential energy at the same point, but use a different path to verify path independence.

9. Consider the following force:

$$\mathbf{F} = -K(x - z)^2(\hat{\mathbf{i}} - \hat{\mathbf{k}})$$

a. Show that it is conservative.

b. Find the potential energy $V(\mathbf{r})$.

c. Calculate $\nabla V(\mathbf{r})$ to verify that it gives $\mathbf{F}$ correctly.

10. Determine whether or not the force $\mathbf{F} = \mathbf{r} \times \mathbf{a}$ (where $\mathbf{a}$ is a constant vector) leads to a conservative potential. Compute $\int \mathbf{F} \cdot d\mathbf{r}$ around a circle of radius R in the xy plane.

11. Show that the ∇ operator can be expressed in spherical coordinates as

$$\nabla = \hat{r}\frac{\partial}{\partial r} + \hat{l}\frac{1}{r}\frac{\partial}{\partial \theta} + \frac{\hat{m}}{r \sin\theta}\frac{\partial}{\partial \phi}$$

where ($\hat{\mathbf{r}}$, $\hat{\mathbf{l}}$, $\hat{\mathbf{m}}$) are perpendicular unit vectors in the direction of increasing (r, θ, ϕ). (*Hint:* Use $df = d\mathbf{r} \cdot \nabla\mathbf{f}$, where $d\mathbf{r}$ is given by $d\mathbf{r} = \hat{\mathbf{r}}\, dr + \hat{\mathbf{l}}r\, d\theta + \hat{\mathbf{m}}r \sin\theta\, d\phi$ and f is an arbitrary scalar function. Express df in terms of partial derivatives.) Show that the ∇ operator in cylindrical coordinates (ρ, ϕ, z) is

$$\nabla = \hat{\boldsymbol{\rho}}\frac{\partial}{\partial \rho} + \frac{\hat{\mathbf{m}}}{\rho}\frac{\partial}{\partial \phi} + \hat{\mathbf{k}}\frac{\partial}{\partial z}$$

12. A spring pendulum consists of a mass m attached to one end of a massless spring with spring constant k. The other end of the spring is tied to a fixed support. When no weight is on the spring, its length is l. Assume that the motion of the system is confined to a vertical plane. Derive the equations of motion first by lagrangian methods and then by direct application of Newton's law. Solve the equations of motion in the approximation of small angular and radial displacements from equilibrium.

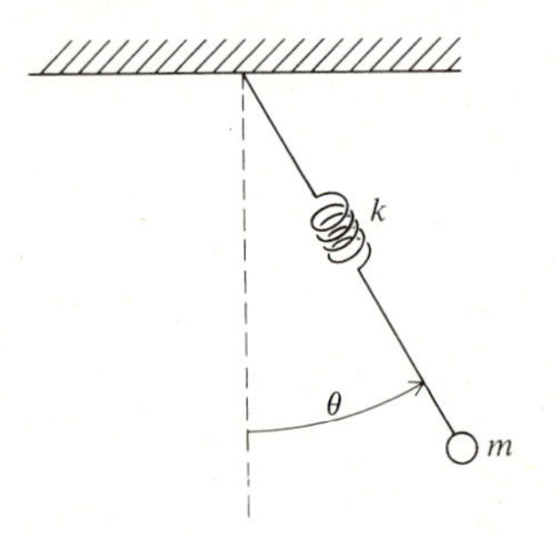

13. A hemispherical thin glass goblet of radius $R = 5$ cm will withstand a perpendicular force up to 2 N. If a 100-g steel ball is released from rest at the lip of the goblet and allowed to slide down the inside, at what point on the goblet will the ball break through? Neglect the radius of the ball.

14. A mass m is attached at one end of a massless rigid rod of length l, and the rod is suspended at its other end by a frictionless pivot, as illustrated. The rod is released from rest at an angle $\alpha_0 < \pi/2$ with the vertical. At what angle α does the force in the rod change from compression to tension?

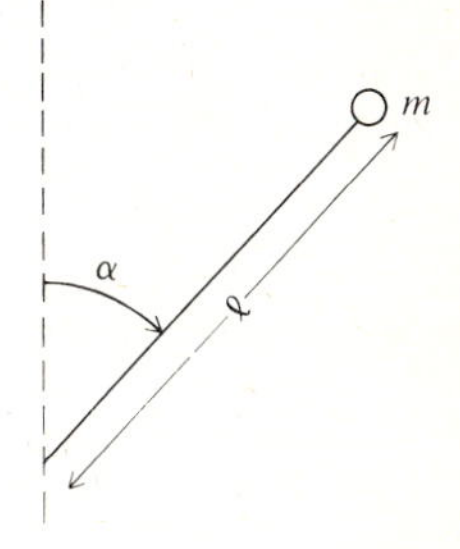

15. A ball of mass m is suspended from a horizontal rod by a string of length l. For what ranges of the total energy will the string remain taut when the ball swings in an arc in a vertical plane perpendicular to the rod? Choose the lowest point on the arc as the reference point for the potential energy.

16. The bob of a simple pendulum moves in a horizontal circle as illustrated. Find the angular frequency of the circular motion in terms of the angle θ and the length l of the rod.

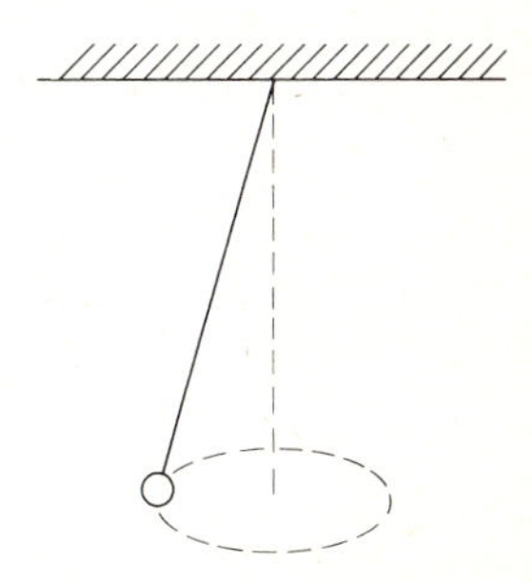

17. Two equal masses are constrained by the spring-and-pulley system shown in the accompanying sketch. Assume a massless pulley

and a frictionless surface. Let x be the extension of the spring from its relaxed length. Derive the equations of motion by lagrangian methods. Solve for x as a function of time with the boundary conditions $x = 0$, $\dot{x} = 0$ at $t = 0$.

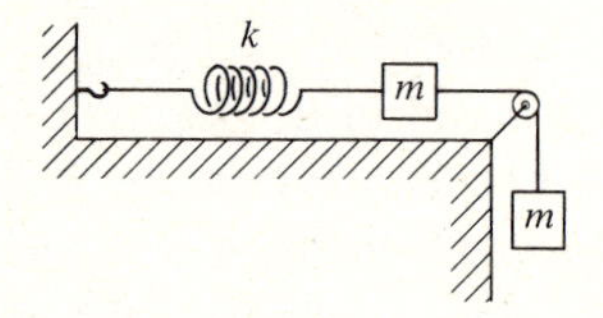

18. A body of mass M is suspended from the ceiling by a spring with constant k. There is also a spring with constant κ connecting the body to the floor, as illustrated. Derive the equation for vertical motion of M and find the solution.

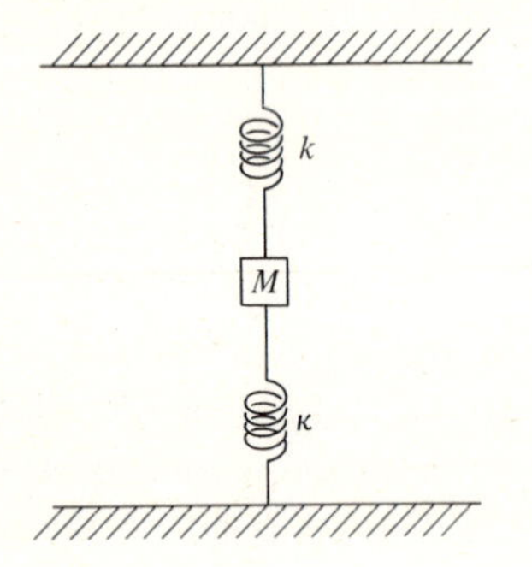

19. A mass $2m$ is suspended from a fixed support by a spring with spring constant $2k$. A second mass m is suspended from the first mass by a spring of constant k. Find the equation of motion for this coupled system and determine the frequencies of oscillation of normal modes. Neglect the masses of the springs. (*Hint:* First form an arbitrary linear combination of the two equations of motion.)

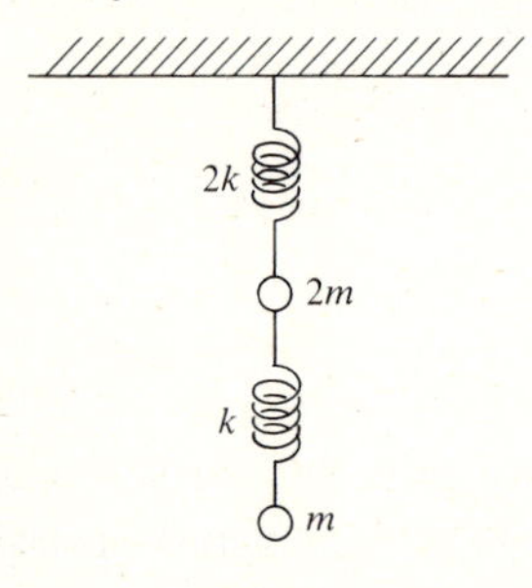

20. A double pendulum consists of two weightless rods connected to each other and a point of support, as illustrated. The masses m_1 and m_2 are not equal, but the lengths of the rods are. The pendulum shafts are free to swing only in one vertical plane.

a. Set up the lagrangian of the system for arbitrary displacements and derive the equations of motion from it.

b. Find the normal-mode frequencies of the system when the oscillations are small.

c. Show that the frequencies become approximately equal if $m_1 \gg m_2$ and interpret this. For $m_2 \gg m_1$ interpret the normal-mode frequencies and describe the motion of each mass.

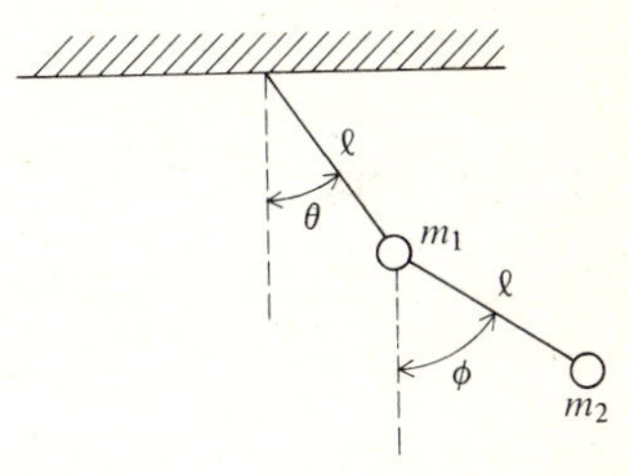

21. A mass m is suspended from a support by a spring with spring constant $m\omega_1^2$. A second mass m is suspended from the first by a spring with spring constant $m\omega_2^2$. A vertical harmonic force $F_0 \cos \omega t$ is applied to the upper mass. Find the steady-state motion for each mass. Examine what happens when $\omega = \omega_2$.

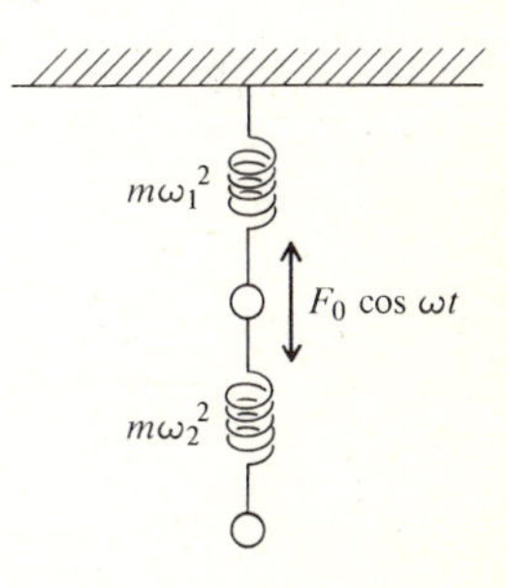

22. For a particle moving in a plane under the influence of a central potential energy $V(r)$, find the hamiltonian as a function of r, θ, $\mathfrak{p}_r$, and $\mathfrak{p}_\theta$. Solve for the four Hamilton equations of motion. Show that the results are equivalent to Eqs. (2-108).

CHAPTER 3

Momentum Conservation

The conservation of linear momentum is a universal law for all physics. In classical mechanics this conservation law is a direct consequence of Newton's laws. In the absence of external forces, the equation of motion

$$\frac{d\mathbf{p}}{dt} = \mathbf{F} = 0 \tag{3-1}$$

implies that $\mathbf{p}$ is independent of time. In other words, a particle with definite mass moves with constant velocity $\mathbf{v}$ in a force-free region. The most interesting ramifications of momentum conservation concern systems of more than one particle.

3-1 ROCKET MOTION

For a two-particle system, both internal forces $\mathbf{F}^{\text{int}}$ between the particles and external forces $\mathbf{F}^{\text{ext}}$ can be present. The laws of motion for particles 1 and 2 are

$$\frac{d\mathbf{p}_1}{dt} = \mathbf{F}_1^{\text{int}} + \mathbf{F}_1^{\text{ext}}$$
$$\frac{d\mathbf{p}_2}{dt} = \mathbf{F}_2^{\text{int}} + \mathbf{F}_2^{\text{ext}} \tag{3-2}$$

The total momentum of the system

$$\mathbf{P} = \mathbf{p}_1 + \mathbf{p}_2 \tag{3-3}$$

obeys an equation given by the sum of Eqs. (3-2).

$$\frac{d\mathbf{P}}{dt} = (\mathbf{F}_1^{\text{int}} + \mathbf{F}_2^{\text{int}}) + (\mathbf{F}_1^{\text{ext}} + \mathbf{F}_2^{\text{ext}}) \tag{3-4}$$

If the total external force is zero,

$$\mathbf{F}^{\text{ext}} = \mathbf{F}_1^{\text{ext}} + \mathbf{F}_2^{\text{ext}} = 0$$

and the internal forces cancel,

$$\mathbf{F}_1^{\text{int}} = -\mathbf{F}_2^{\text{int}} \tag{3-5}$$

as implied by the third law, then the total momentum is conserved.

$$\frac{d\mathbf{P}}{dt} = 0 \tag{3-6}$$

The preceding argument can be generalized to demonstrate the conservation of total momentum for a system consisting of an arbitrary number of particles.

As an illustration of momentum-conservation methods, we apply Eq. (3-6) to the following problem with a time-varying mass. An open gondola freight car of mass m_0 coasts along a level straight frictionless track with initial velocity v_0. Rain starts falling straight down at time $t_0 = 0$, and water accumulates in the gondola at the rate (mass per unit time) σ. The problem is to find the velocity of the gondola.

Since the falling rain has no horizontal component of velocity, it has no component of momentum along the track. Hence the momentum along the track of the gondola is unchanged by the accumulating rain.

$$P = mv = \text{constant} = m_0 v_0 \tag{3-7}$$

Here m is the mass of the total system, gondola, and accumulated rain. Since the time rate of change of m is given by

$$\frac{dm}{dt} = \sigma \tag{3-8}$$

we have

$$m = m_0 + \sigma t \tag{3-9}$$

From Eqs. (3-7) and (3-9), we find the velocity of the gondola at time t to be

$$v = v_0 \frac{m_0}{m_0 + \sigma t} \tag{3-10}$$

More frequently, situations involving a time-varying mass are encountered in conjunction with an externally applied force. Rocket motion is such an example. The time variation of the mass in rocket motion is due to the expulsion of the exhaust. The external forces are primarily due to gravity and air resistance. To derive the fundamental equation for linear rocket motion, we treat the rocket as a system of a large number of particles. The generalized form of Eqs. (3-4) and (3-5) is

$$\frac{d\mathbf{P}}{dt} = \mathbf{F}^{\text{ext}} \tag{3-11}$$

where $\mathbf{P}$ is the total momentum and $\mathbf{F}^{\text{ext}}$ is the net external force. At time t, the rocket of mass m is moving with a velocity v relative to a fixed coordinate system, as illustrated in Fig. 3-1. The exhaust is ejected with a constant velocity $-u$ relative to the rocket. The velocity of the exhaust relative to the fixed coordinate system is $(v - u)$. The momentum of the rocket at time t is

$$P = mv \tag{3-12}$$

An infinitesimal time dt later the rocket's mass has decreased to $m + dm$, where $-dm$ is the mass of exhaust ejected between times t and $t + dt$. At time $t + dt$, the velocity of the rocket is $(v + dv)$, and the velocity of the exhaust relative to the fixed coordinate system is $(v + \varepsilon\, dv - u)$, where $0 \leq \varepsilon \leq 1$. Thus the momentum at time $t + dt$ of the mass which was in the rocket at time t is

$$P + dP = (m + dm)(v + dv) + (-dm)(v + \varepsilon\, dv - u) \tag{3-13}$$

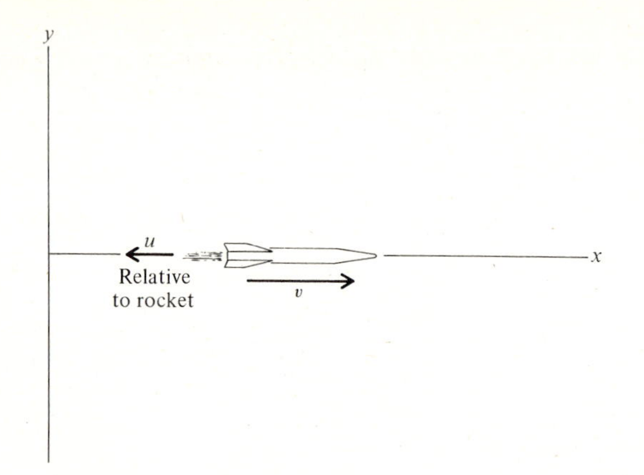

FIGURE 3-1 **Motion of a rocket with velocity v relative to a fixed coordinate system. The velocity of the exhaust relative to the rocket is $-u$.**

The change of momentum dP in the time interval dt is just the difference of Eqs. (3-13) and (3-12). To first order in the differentials, the momentum change is

$$dP = m\,dv + u\,dm \tag{3-14}$$

The time rate of change of momentum is

$$\frac{dP}{dt} = m\frac{dv}{dt} + u\frac{dm}{dt} = F^{\text{ext}} \tag{3-15}$$

where we have used Eqs. (3-11) and (3-14). This fundamental equation of rocket motion is usually written

$$m\frac{dv}{dt} = F^{\text{ext}} - u\frac{dm}{dt} \tag{3-16}$$

Since dm/dt is negative, the term $u(dm/dt)$ increases the velocity v of the rocket. This term is called the *thrust* of the rocket. In the absence of external forces, Eq. (3-16) can be readily solved. Multiplying through by dt and rearranging factors, we have

$$dv = -u\frac{dm}{m} \tag{3-17}$$

The result of integration is

$$v_f - v_i = u \ln\left(\frac{m_i}{m_f}\right) \tag{3-18}$$

where the i and f subscripts label initial and final values. The exhaust velocity u depends on the type of rocket fuel that is burned. Fuels with low molecular weights generally have higher exhaust velocities, and thus yield high rocket velocities. Present rocket technology gives exhaust velocities close to the thermodynamic limit for chemical fuels. The result in Eq. (3-18) places a limit on the velocities which can be reached with a single-stage rocket. Velocities several orders of magnitude greater than u cannot be achieved. The mass of the payload plus empty rocket m_f is an important factor in determining the final velocity that the rocket reaches.

For a rocket fired vertically upward from the surface of the earth, the rocket equation (3-16) is

$$m\frac{dv}{dt} = -mg - u\frac{dm}{dt} \tag{3-19}$$

provided that air resistance is neglected and the earth is regarded as an inertial frame. To achieve a lift-off, it is necessary that $u|dm/dt| > m_i g$. An analytic solution to this equation is possible for constant rate of burn,

$$\frac{dm}{dt} = -\alpha m_i$$

for which

$$m = m_i(1 - \alpha t) \tag{3-20}$$

where α is a constant factor. After substitution of Eq. (3-20) into Eq. (3-19), we rearrange terms and integrate from ignition at $t = 0$ to burnout at $t = t_f$.

$$\int_0^{v_f} dv = \int_0^{t_f} \left(-g + \frac{u\alpha}{1 - \alpha t}\right) dt \tag{3-21}$$

The velocity at the instant the fuel is exhausted is

$$v_f = -gt_f - u \ln(1 - \alpha t_f) \tag{3-22}$$

The burnout time can be expressed in terms of the final mass of the rocket from Eq. (3-20) as

$$t_f = \frac{1}{\alpha}\left(1 - \frac{m_f}{m_i}\right) \tag{3-23}$$

Thus the final velocity can be written

$$v_f = -\frac{g}{\alpha}\left(1 - \frac{m_f}{m_i}\right) + u \ln\left(\frac{m_i}{m_f}\right) \tag{3-24}$$

where $u\alpha \geq g$. Upon comparison with Eq. (3-18), we observe that the second term on the right-hand side of Eq. (3-24) is the final velocity for the rocket in a gravity-free region. The effect of gravity is naturally to reduce the final velocity. The greater the burn-rate factor α, the smaller is the effect of the gravity on the final velocity. The ratio of final rocket velocity to the exhaust velocity can be written in terms of the thrust,

$$T \equiv u\left|\frac{dm}{dt}\right| = u\alpha m_i \tag{3-25}$$

as

$$\frac{v_f}{u} = -\frac{1}{T}(m_i - m_f)g + \ln\left(\frac{m_i}{m_f}\right) \tag{3-26}$$

For the Apollo moon rocket, the initial mass of the vehicle and fuel is

$$m_i = 2.94 \times 10^6 \text{ kg} \tag{3-27}$$

The fuel carried by the first stage of this three-stage Saturn V rocket is 2.15×10^6 kg. Thus the final mass at the instant first-stage burnout is

$$m_f = 0.79 \times 10^6 \text{ kg} \tag{3-28}$$

The thrust of the Saturn V first stage builds from 34×10^6 N at lift-off to 40.4×10^6 N at burnout. For calculations with Eqs. (3-25) and (3-26), we approximate the thrust by its mean value,

$$T = 37.2 \times 10^6 \text{ N} \tag{3-29}$$

For vertical ascent, we can calculate the ratio of rocket to exhaust velocities for the Apollo spaceship from Eq. (3-26) using the mass and thrust values of Eqs. (3-27) to (3-29). We obtain

$$\frac{v_f}{u} = -\frac{2.15(9.8)}{37.2} + \ln\left(\frac{2.94}{0.79}\right) = -0.57 + 1.32 = 0.75 \tag{3-30}$$

The average exhaust velocity u of the Apollo booster can be calculated from Eq. (3-25) using the propellant flow rate

$$\frac{dm}{dt} = 1.33 \times 10^4 \text{ kg/s} \tag{3-31}$$

to the first-stage engines. We find

$$u = \frac{T}{|dm/dt|} = 37.2 \times 10^6 / 1.33 \times 10^4$$

or

$$u = 2.8 \text{ km/s} \tag{3-32}$$

From Eqs. (3-30) and (3-32), we get

$$v_f = 0.75(2.8) = 2.1 \text{ km/s} \tag{3-33}$$

for the final velocity that would be reached by the Apollo moon rocket in vertical flight at the time of first-stage burnout. This value neglects the effects of air resistance and the time variation of the thrust.

The actual trajectory of the Apollo moon rocket has considerable curvature. At first-stage burnout, the space vehicle is 93 km downrange at an altitude of 67 km. The burnout of the booster rocket occurs at t_f = 161 s, and the velocity at that instant is

$$v_f = 2.75 \text{ km/s}$$

which is somewhat greater than the calculated value for a vertical trajectory in Eq. (3-33), as we should expect.

3-2 FRAMES OF REFERENCE

Newton's law stipulates that the equation of motion applies only in an inertial frame. Since a frame moving with a constant velocity relative to an inertial frame is also inertial, considerable latitude exists in choosing a coordinate frame reference. Frequently, the solution to a problem can be simplified by a suitable choice of coordinate frame.

The galilean transformation of classical mechanics relates positions of a point as measured from two coordinate systems in relative translational motion. We take S' to denote a coordinate frame whose origin moves with constant velocity $\mathbf{V}_0$ relative to the origin of a fixed coordinate frame S, as illustrated in Fig. 3-2. If $\mathbf{r}'$ is the location of a point in space as measured in S', and $\mathbf{r}$ is the location of the same point as measured in S, the galilean transformation is

$$\mathbf{r} = \mathbf{r}' + \mathbf{V}_0 t \tag{3-34}$$

When we differentiate Eq. (3-34) with respect to t, we get the velocity-addition law

$$\mathbf{v} = \mathbf{v}' + \mathbf{V}_0 \tag{3-35}$$

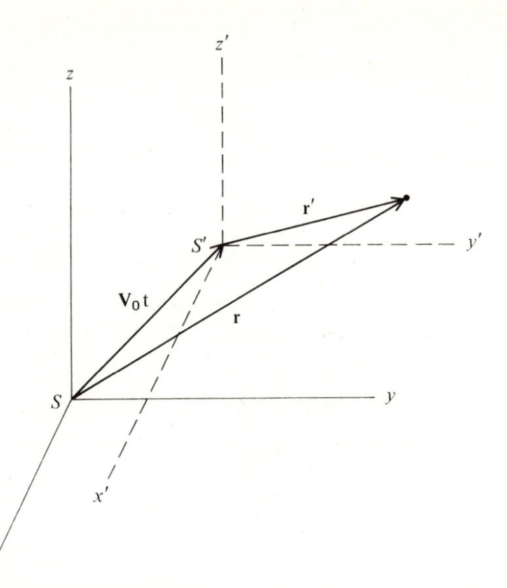

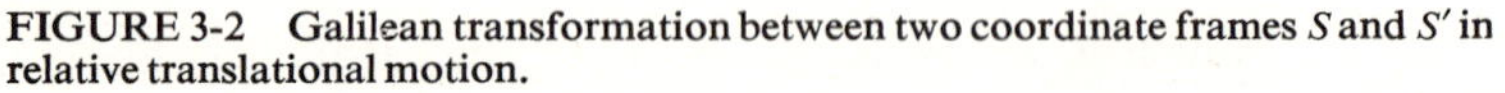
FIGURE 3-2 Galilean transformation between two coordinate frames S and S' in relative translational motion.

A second differentiation gives

$$\frac{d^2\mathbf{r}}{dt^2} = \frac{d^2\mathbf{r}'}{dt^2} \tag{3-36}$$

due to the constancy of $\mathbf{V}_0$. Since the mass of a particle is a constant (in particular, it is independent of the velocity), m is unchanged under a galilean transformation. Combining the invariance of the mass with the invariance of acceleration [Eq. (3-36)], we see that the ma term of Newton's second law is invariant under galilean transformations. The force $\mathbf{F}'(\mathbf{r}', \mathbf{v}', t)$ on the particle in the S' frame is the same as the force $\mathbf{F}(\mathbf{r}, \mathbf{v}, t)$ in the S frame if the force is velocity-independent:

$$\mathbf{F}'(\mathbf{r}', t) = \mathbf{F}(\mathbf{r}, t) \tag{3-37}$$

This follows from the fact that $\mathbf{r}$ and $\mathbf{r}'$ denote the same physical location in space. Thus, if the force is independent of velocity, the equation of motion is galilean-invariant.

Even though the galilean transformation of Eqs. (3-34) and (3-35) seems self-evident, it must be regarded as a postulate. In fact, for a velocity $\mathbf{v}$ comparable with the speed of light, the equations of motion are not invariant to a galilean transformation. The Lorentz transformation of Einstein's special relativity theory to which the equations of motion are always invariant reduces to the galilean transformation in the limit of small velocity.

To show the advantage of a judicious choice of coordinate frame, we discuss the lift experienced by seagulls rising in a uniform wind. On a windy day along the beach, seagulls on the ground are often observed to extend their wings and to be carried aloft without the need of flapping. Since the seagull does no work in the ascent, energy-conservation methods should apply. However, in a coordinate frame which is at rest with respect to the ground, the gull gains energy, both kinetic and potential! The energy gained by the gull is given up by the wind. The change ΔT_w in the kinetic energy $T_w = \frac{1}{2}M_w v_w^2$ of the wind is

$$\Delta T_w = M_w v_w \, \Delta v_w$$

where Δv_w is the small change in the wind's velocity due to the interaction with the gull. Now, if we make a galilean transformation to a frame which moves with the air, $v_w' = 0$, $T_w' = 0$, and $\Delta T_w' = \frac{1}{2}M_w(\Delta v_w')^2$ is vanishingly small. Thus, in the frame moving with the wind, the energy of the gull is conserved, and energy-conservation methods can be used to find the height to which the gull can ascend. The initial mechanical energy of the seagull in the moving frame is

$$T = \tfrac{1}{2}mv_w^2$$

where m is the mass of the gull. The kinetic energy is converted to potential energy by the gull's rising. At height x above the beach the potential energy is

$$V(x) = mgx$$

At the maximum height h which the gull can attain, the kinetic energy is entirely converted to potential energy. Hence we obtain

$$mgh = \tfrac{1}{2}mv_w^2$$

or

$$h = \frac{v_w^2}{2g} \qquad \text{(3-38)}$$

In a brisk wind of 40 km/h an ideal seagull can glide up to a height of

$$h = \frac{[(40(1{,}000/3{,}600)]^2}{2(9.8)} = 6.3 \text{ m}$$

3-3 ELASTIC COLLISIONS: LABORATORY AND CENTER-OF-MASS SYSTEMS

Collisions provide especially interesting examples of momentum-conservation methods. In collisions where no external forces are involved and the internal forces satisfy the third law, the total momentum of the colliding objects is conserved, as in Eq. (3-6). The following discussion, which is based on momentum conservation, applies to all collisions, no matter what the detailed interactions are, so long as the interactions are sufficiently short-ranged so that the bodies can be treated as free before and after the collision.

In the description of two-particle collisions, the most common choices of coordinate frames of reference are the laboratory (lab) frame and the center-of-mass (CM) frame. Hereafter, we label the lab coordinate system by S and the CM coordinate system by S'. In the lab system the target particle m_2 is initially at rest, $\mathbf{v}_{2i} = 0$, and the incident particle m_1 has velocity $\mathbf{v}_{1i}$. This system is so named because most experiments in the laboratory are performed with these initial conditions. After the collision, when the forces are no longer acting, the final lab velocities are $\mathbf{v}_{1f}$ and $\mathbf{v}_{2f}$. Conservation of total momentum in the lab frame implies

$$\mathbf{P} = m_1\mathbf{v}_{1i} = m_1\mathbf{v}_{1f} + m_2\mathbf{v}_{2f} \tag{3-39}$$

where we assume that the masses are unchanged by the collision.

The center-of-mass frame is the system of coordinates for which the total momentum of the two particles is zero.

$$\mathbf{P}' = \mathbf{P}'_{1i} + \mathbf{P}'_{2i} = 0 = \mathbf{P}'_{1f} + \mathbf{P}'_{2f} \tag{3-40}$$

In this frame the CM is at rest; the CM frame is also sometimes called the *center-of-momentum frame*. To determine the galilean transformation velocity $\mathbf{V}_0$ between the lab and CM frames, we use Eq. (3-35) to relate the initial velocities.

$$\begin{aligned} \mathbf{v}'_{1i} &= \mathbf{v}_{1i} - \mathbf{V}_0 \\ \mathbf{v}'_{2i} &= \mathbf{v}_{2i} - \mathbf{V}_0 = -\mathbf{V}_0 \end{aligned} \tag{3-41}$$

When we impose the $\mathbf{P}' = 0$ requirement of Eq. (3-40) for the initial velocities in Eq. (3-41),

$$m_1(\mathbf{v}_{1i} - \mathbf{V}_0) + m_2(-\mathbf{V}_0) = 0$$

the transformation velocity is obtained as

$$\mathbf{V}_0 = \frac{m_1}{m_1 + m_2} \mathbf{v}_{1i} \tag{3-42}$$

Of course, $\mathbf{V}_0$ is the velocity of the CM in the lab frame. From Eqs. (3-41) and (3-42), we find

$$\begin{aligned} \mathbf{v}'_{1i} &= \frac{m_2}{m_1 + m_2} \mathbf{v}_{1i} \\ \mathbf{v}'_{2i} &= -\frac{m_1}{m_1 + m_2} \mathbf{v}_{1i} \end{aligned} \tag{3-43}$$

for the initial velocities of the particles in the CM frame. A kinematical diagram of the initial and final velocities in the lab and CM frames is given in Fig. 3-3.

Collisions in which kinetic energy as well as momentum is conserved are called *elastic* collisions. In an elastic collision of two particles, the momentum-conservation condition in Eq. (3-39) or (3-40) is supplemented by the energy-conservation condition,

$$\tfrac{1}{2}m_1(\mathbf{v}_{1i})^2 = \tfrac{1}{2}m_1(\mathbf{v}_{1f})^2 + \tfrac{1}{2}m_2(\mathbf{v}_{2f})^2 \tag{3-44}$$

in the lab, or equivalently,

$$\tfrac{1}{2}m_1(\mathbf{v}'_{1i})^2 + \tfrac{1}{2}m_2(\mathbf{v}'_{2i})^2 = \tfrac{1}{2}m_1(\mathbf{v}'_{1f})^2 + \tfrac{1}{2}m_2(\mathbf{v}'_{2f})^2 \tag{3-45}$$

in the CM. The momentum- and energy-conservation conditions lead to simple relations between the velocities in the CM system. From the momentum-conservation condition of Eq. (3-40), we have

$$\begin{aligned} \mathbf{v}'_{2i} &= -\frac{m_1}{m_2} \mathbf{v}'_{1i} \\ \mathbf{v}'_{2f} &= -\frac{m_1}{m_2} \mathbf{v}'_{1f} \end{aligned} \tag{3-46}$$

By substitution of this result into the energy-conservation relation of Eq. (3-45), we obtain

$$\begin{aligned} v'_{1i} &= v'_{1f} \\ v'_{2i} &= v'_{2f} \end{aligned} \tag{3-47}$$

for the magnitudes of the velocity vectors. Viewed in the CM frame, the energy of each particle is unchanged by an elastic collision.

The significance of the CM frame for theoretical analyses of collision problems can be summarized as follows:

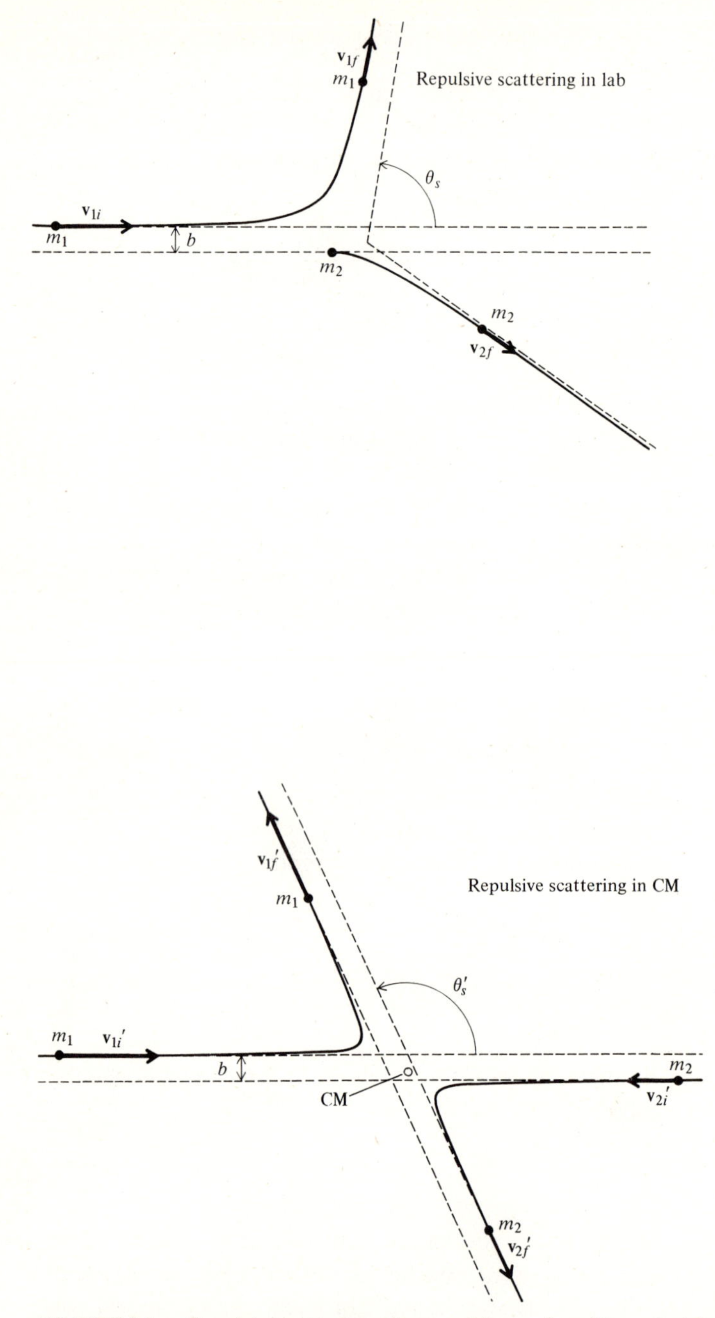

FIGURE 3-3*a* **Repulsive scattering of two particles as viewed from the laboratory and center-of-mass coordinate frames.**

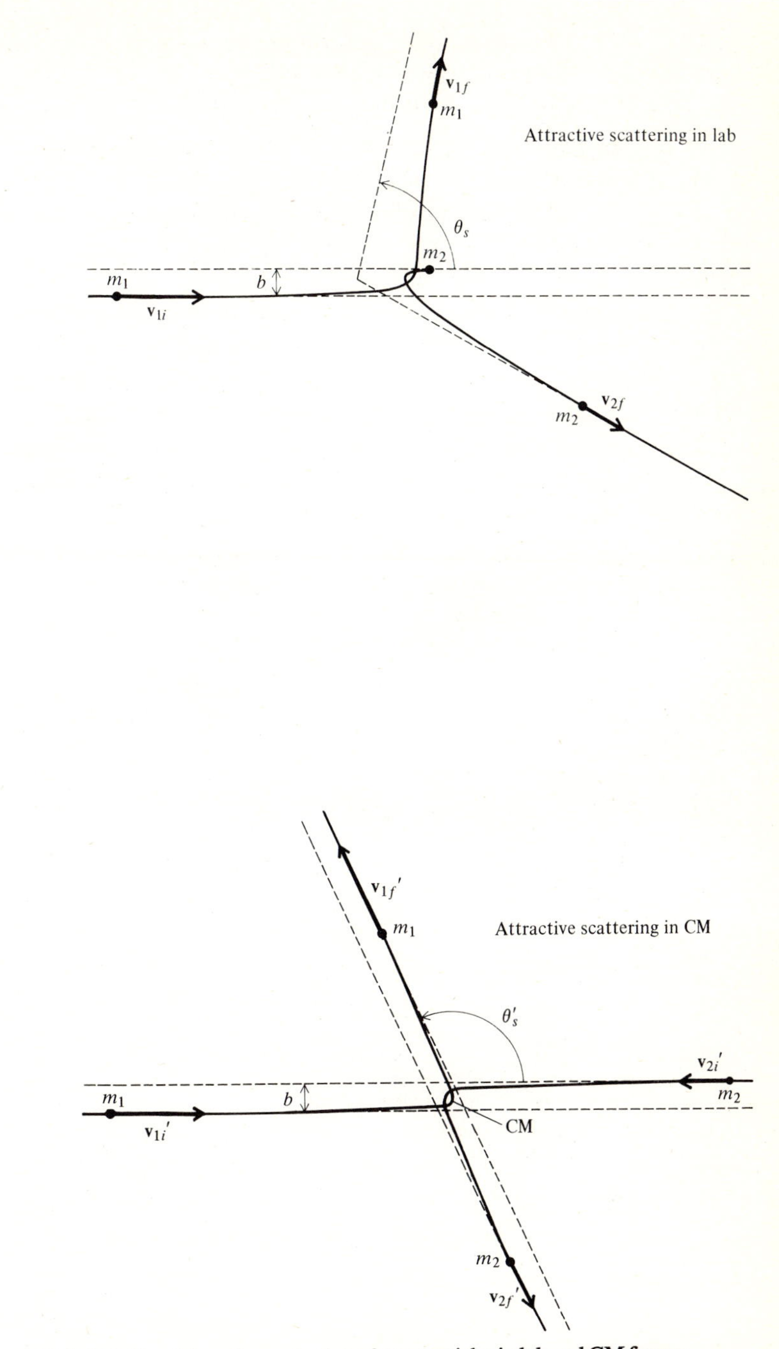

FIGURE 3-3*b* **Attractive scattering of two particles in lab and CM frames.**

1. The motion both before and after the collision is *collinear* in the CM frame, and thus simpler to discuss.
2. For elastic collisions, the magnitude of the velocity and the energy of each particle are the same before and after the collision in the CM frame.
3. For inelastic collisions, *all* the initial energy in the CM frame is available for inelastic processes, as discussed in Sec. 3-5.

Together, the momentum- and energy-conservation conditions provide four relations between the initial and final velocities. Thus two out of the six final velocity components are unspecified by the initial velocities. This indeterminacy is easily visualized in the CM system. The two final velocities must be opposite in direction, with magnitudes in ratio $v'_{2f}/v'_{1f} = m_1/m_2$ by Eqs. (3-46). Thus momentum conservation leaves only one independent final-velocity vector, say $\mathbf{v}'_{1f}$. The magnitude of this velocity vector is specified by the energy-conservation condition of Eqs. (3-47) as $v'_{1f} = v'_{1i}$, leaving the two orientation angles (θ', ϕ') of this velocity as undetermined by the initial conditions.

The angle between the final and initial velocities of particle 1 is defined to be the *scattering angle*, as illustrated in Fig. 3-3. The scattering angle θ' in the CM system can be related to the lab scattering angle θ through the velocity-transformation equation

$$\mathbf{v}_{1f} = \mathbf{v}'_{1f} + \mathbf{V}_0$$

of Eq. (3-35). When we take components of this equation along directions parallel and perpendicular to the initial velocities as shown in Fig. 3-4, we obtain

$$v_{1f} \sin\theta = v'_{1f} \sin\theta'$$

$$v_{1f} \cos\theta = v'_{1f} \cos\theta' + V_0$$

The ratio of these equations is

$$\tan\theta = \frac{\sin\theta'}{\cos\theta' + V_0/v'_{1f}} \tag{3-48}$$

From Eqs. (3-41), (3-46), and (3-47) the velocity ratio V_0/v'_{1f} is given by

$$\frac{V_0}{v'_{1f}} = \frac{v'_{2i}}{v'_{2f}} = \frac{m_1}{m_2}\frac{v'_{1i}}{v'_{1f}} = \frac{m_1}{m_2}$$

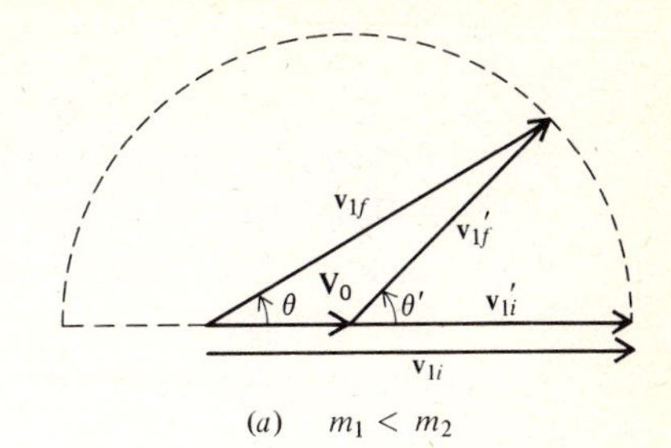

(a) $m_1 < m_2$

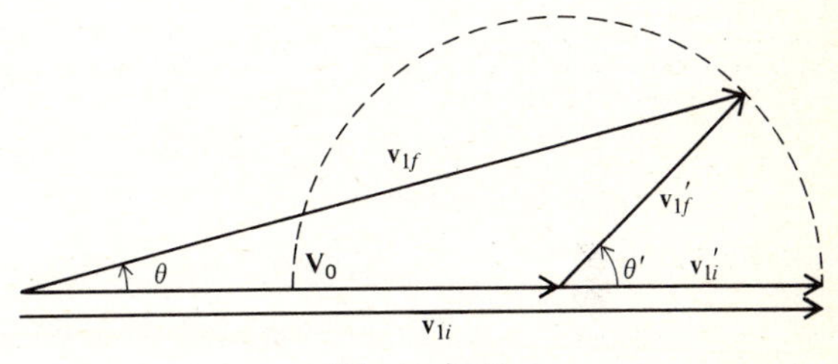

(b) $m_1 > m_2$

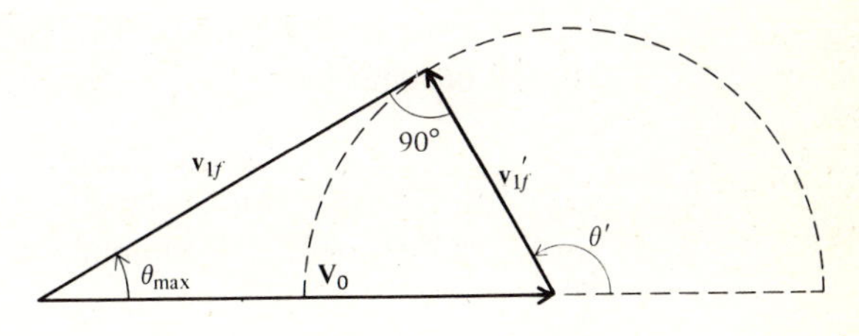

(c) $m_1 > m_2$ (maximum θ)

FIGURE 3-4 Velocity diagram illustrating the lab and CM quantities for elastic scattering.

With this substitution in Eq. (3-48) we arrive at the desired relation between the lab and CM scattering angles.

$$\tan\theta = \frac{\sin\theta'}{\cos\theta' + m_1/m_2} \tag{3-49}$$

For a fixed-target particle (that is, $m_2 = \infty$), the lab and CM scattering angles are equal. For equal masses the lab angle is half the CM angle, $\theta = \theta'/2$.

As the CM scattering angle θ' varies from 0 to π, the vector $\mathbf{v}'_{1f}$ traces out the dashed circles in the velocity diagrams of Fig. 3-4.

If $m_1 < m_2$, the lab angle also varies from 0 to π, as can be deduced from Fig. 3-4*a* or from Eq. (3-49). For $m_1 > m_2$, illustrated in Fig. 3-4*b*, θ increases from 0 to a maximum value θ_{max} and then decreases back to 0 as θ' goes from 0 to π. The maximum lab angle occurs when $\mathbf{v}_{1f}$ and $\mathbf{v}'_{1f}$ are perpendicular ($\mathbf{v}_{1f} \cdot \mathbf{v}'_{1f} = 0$), as indicated in Fig. 3-4*c*. This orthogonality condition leads to the following expression for θ_{max}:

$$\sin \theta_{max} = \frac{m_2}{m_1} \qquad m_1 > m_2 \tag{3-50}$$

For example, the maximum laboratory scattering angle for a proton incident on an electron is 0.03°.

In the lab system the kinetic energy transferred to the target particle is given by

$$\begin{aligned} T_{2f} &= \tfrac{1}{2} m_2 v_{2f}^2 = \tfrac{1}{2} m_2 (\mathbf{v}'_{2f} - \mathbf{v}'_{2i})^2 \\ &= m_2 v_{2i}'^2 (1 - \cos \theta') = 2 m_2 v_{2i}'^2 \sin^2 \frac{\theta'}{2} \end{aligned}$$

Here we have made use of Eqs. (3-35) and (3-47). This energy transfer is a maximum for $\theta' = \pi$, which corresponds to backward scattering in the CM system. The ratio of the final energy of the target particle T_{2f} to the incident energy T_{1i} of the projectile is

$$\frac{T_{2f}}{T_{1i}} = \frac{2 m_2 v_{2i}'^2 \sin^2 (\theta'/2)}{\tfrac{1}{2} m_1 v_{1i}^2}$$

Using Eqs. (3-43), this ratio simplifies to

$$\frac{T_{2f}}{T_{1i}} = \frac{4 m_1 m_2}{(m_1 + m_2)^2} \sin^2 \frac{\theta'}{2} \tag{3-51}$$

Thus the greatest energy transfer in the lab frame occurs when target and projectile have equal mass.

3-4 COLLISIONS OF BILLIARD BALLS

As an illustration of scattering, we consider the collisions of billiard balls on a smooth table. The initial conditions simulate the lab frame since the target ball is at rest and the cue ball in motion with velocity $\mathbf{v}_{1i}$. For a head-on collision the velocity of the cue ball is directed at the center of the target ball. To analyze the subsequent collision we transform to the CM frame. In a frame moving with velocity $\mathbf{V}_0$ relative to the lab system, the velocities of the balls are

$$\begin{aligned} \mathbf{v}'_{1i} &= \mathbf{v}_{1i} - \mathbf{V}_0 \\ \mathbf{v}'_{2i} &= -\mathbf{V}_0 \end{aligned} \tag{3-52}$$

For the CM frame the vanishing of the total momentum requires

$$\mathbf{v}'_{1i} = -\mathbf{v}'_{2i} \tag{3-53}$$

because of the equality of the masses. By comparison of Eqs. (3-52) and (3-53) we obtain

$$\mathbf{V}_0 = \frac{\mathbf{v}_{1i}}{2} = \mathbf{v}'_{1i} = -\mathbf{v}'_{2i} \tag{3-54}$$

After the collision, the balls are moving away from each other with equal velocities in the CM frame.

$$\mathbf{v}'_{1f} = -\mathbf{v}'_{2f} \tag{3-55}$$

Billiard balls have both linear and rotational kinetic energy.

$$T = \tfrac{1}{2}mv^2 + \tfrac{1}{2}I\omega^2 \tag{3-56}$$

where I is the moment of inertia and ω is the angular velocity of rotation (i.e., the spin). Since the balls are assumed to be smooth, no spin transfer occurs in the collision, and the rotational energies of the individual balls remain unchanged by the collision. Assuming that the collision is elastic, the translational kinetic energies must satisfy the conservation condition

$$\tfrac{1}{2}m[(v'_{1i})^2 + (v'_{2i})^2] = \tfrac{1}{2}m[(v'_{1f})^2 + (v'_{2f})^2] \tag{3-57}$$

From Eqs. (3-53), (3-55), and (3-57), the magnitudes of the initial and final velocities must be equal.

$$v'_{1f} = v'_{1i} = v'_{2f} = v'_{2i} \tag{3-58}$$

In vector form the velocities after the collision are related as

$$\begin{aligned} \mathbf{v}'_{1f} &= -\mathbf{v}'_{1i} = -\frac{\mathbf{v}_{1i}}{2} \\ \mathbf{v}'_{2f} &= -\mathbf{v}'_{2i} = \frac{\mathbf{v}_{1i}}{2} \end{aligned} \tag{3-59}$$

To express this result in terms of what is observed on the billiard table, we transform back to the lab frame, using Eqs. (3-35) and (3-59).

$$\begin{aligned} \mathbf{v}_{1f} &= \mathbf{v}'_{1f} + \mathbf{V}_0 = 0 \\ \mathbf{v}_{2f} &= \mathbf{v}'_{2f} + \mathbf{V}_0 = \mathbf{v}_{1i} \end{aligned} \tag{3-60}$$

The cue ball stops, and the target ball moves forward with the initial velocity of the cue ball as a result of this head-on collision. Of course,

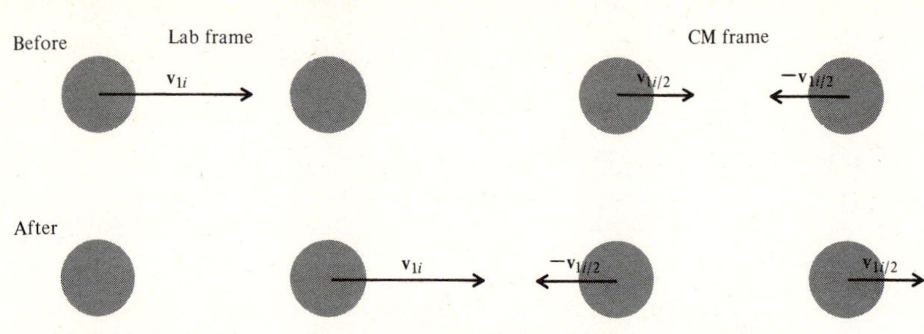

FIGURE 3-5 Head-on collisions of billiard balls in the lab and CM frames.

we could have solved this collision problem directly in the lab frame, but the physical interpretation is more transparent in the CM frame.

In noncentral collisions, as any good billiard player knows, the balls leave the point of impact at right angles. We can prove this fact directly by momentum and energy conservation in the lab system. From momentum conservation,

$$(\mathbf{p}_{1i})^2 = (\mathbf{p}_{1f} + \mathbf{p}_{2f})^2 = (\mathbf{p}_{1f})^2 + (\mathbf{p}_{2f})^2 + 2\mathbf{p}_{1f} \cdot \mathbf{p}_{2f} \tag{3-61}$$

and from energy conservation,

$$(\mathbf{p}_{1i})^2 = (\mathbf{p}_{1f})^2 + (\mathbf{p}_{2f})^2 \tag{3-62}$$

we obtain

$$\mathbf{p}_{1f} \cdot \mathbf{p}_{2f} = 0 \tag{3-63}$$

from the difference of Eqs. (3-61) and (3-62). This establishes that the final velocities are at right angles if neither velocity is zero.

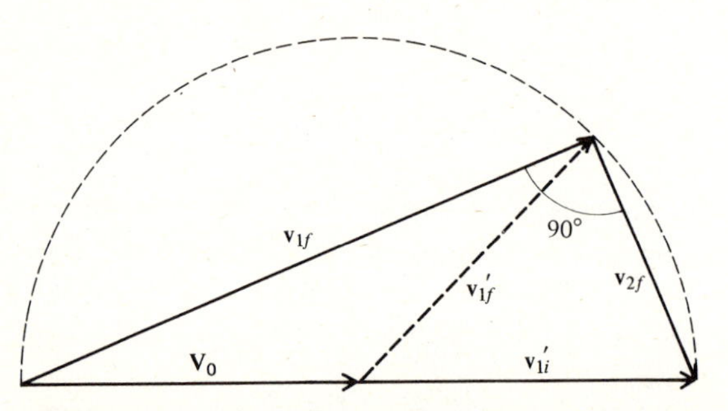

FIGURE 3-6 Velocity diagram for elastic scattering of two equal-mass particles. Since $\mathbf{v}_{1f}$ and $\mathbf{v}_{2f}$ form the sides of a triangle inscribed in a semicircle, these two vectors are perpendicular by geometry.

3-5 INELASTIC COLLISIONS

In many instances energy is not conserved in the collision and the individual masses of the final particles may be different from the masses of the initial particles. In this more complex situation we can write an energy-balance equation for the colliding system and surroundings of the form

$$T_1 = T_3 + T_4 + Q \tag{3-64}$$

in the lab system, or

$$T_1' + T_2' = T_3' + T_4' + Q \tag{3-65}$$

in the CM system. In these equations T designates particle kinetic energy and Q is the energy released ($Q > 0$) or absorbed ($Q < 0$) by the collision in the form of heat. The value of Q is independent of reference frame. Collisions in which energy is transferred ($Q \neq 0$), or the final masses are different from the initial masses, are called *inelastic*. In elastic processes $Q = 0$, and the masses are not changed by the collision.

As an example we consider an inelastic collision between two equal-mass putty balls. In the lab system the projectile putty ball has velocity $\mathbf{v}_{1i}$ and the target putty ball is at rest. In the center-of-mass system the initial velocities are

$$\begin{aligned} \mathbf{v}_{1i}' &= \frac{\mathbf{v}_{1i}}{2} \\ \mathbf{v}_{2i}' &= -\frac{\mathbf{v}_{1i}}{2} \end{aligned} \tag{3-66}$$

Upon impact the two putty balls stick together to form a mass $2m$. The final velocity of the aggregate putty ball is zero in the CM frame. The energy released by the collision is deduced to be

$$\begin{aligned} Q &= T_1' + T_2' - T_3' \\ &= \tfrac{1}{2}m[(\mathbf{v}_{1i}')^2 + (\mathbf{v}_{2i}')^2] - 0 \\ &= \tfrac{1}{4}m(\mathbf{v}_{1i})^2 \end{aligned} \tag{3-67}$$

Since there is no final kinetic energy in the CM system, all the initial kinetic energy has been converted to heat, and the collision is called *completely inelastic*. The final velocity of the aggregate putty ball in the lab system is

$$\mathbf{v}_{3f} = 0 + \frac{\mathbf{v}_{1i}}{2} \tag{3-68}$$

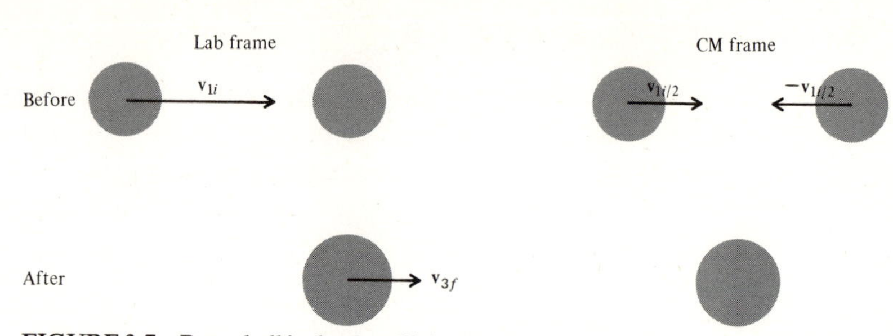

FIGURE 3-7 Putty-ball inelastic collision in lab and CM frames.

3-6 SCATTERING CROSS SECTIONS

In classical mechanics the trajectory of a single particle can be followed from its initial to its final position. However, for experiments involving a large number of incident particles, it is not feasible to follow the trajectory of each individual projectile. Instead, the experimental results and the theoretical analysis are more conveniently expressed in terms of probabilities. As an example, we discuss the results of shooting a **BB** gun at a fixed smooth cylindrical steel pipe. For N shots fired toward the pipe we would like to know the expected number of **BB**s that ricochet in a given direction. We assume that in a large number of shots N, the incident **BB**s are uniformly distributed over an area A which includes the pipe; i.e., the density of shots per unit area incident on the pipe is N/A.

Typical trajectories for **BB**s incident on the can are illustrated in Fig. 3-8. The incident velocities along the x direction are perpendicular to the vertical axis (z axis) of the pipe. Since the force of impact is normal to the surface of the pipe, the final velocities are also perpendicular to the z axis. Thus it is sufficient to consider a circular ring in the xy plane, as in the lower part of Fig. 3-8.

A **BB** incident at a distance $y = b$ from the x axis is scattered by an angle θ from the incident direction. To relate the impact parameter b to the scattering angle θ, we use momentum and energy conservation. The condition for conservation of energy is

$$\frac{p_i^2}{2m} = \frac{p_f^2}{2m} + \frac{p^2}{2M} \tag{3-69}$$

where p is the momentum absorbed by the pipe and M is the mass of the pipe. In the approximation of a very massive pipe, the recoil

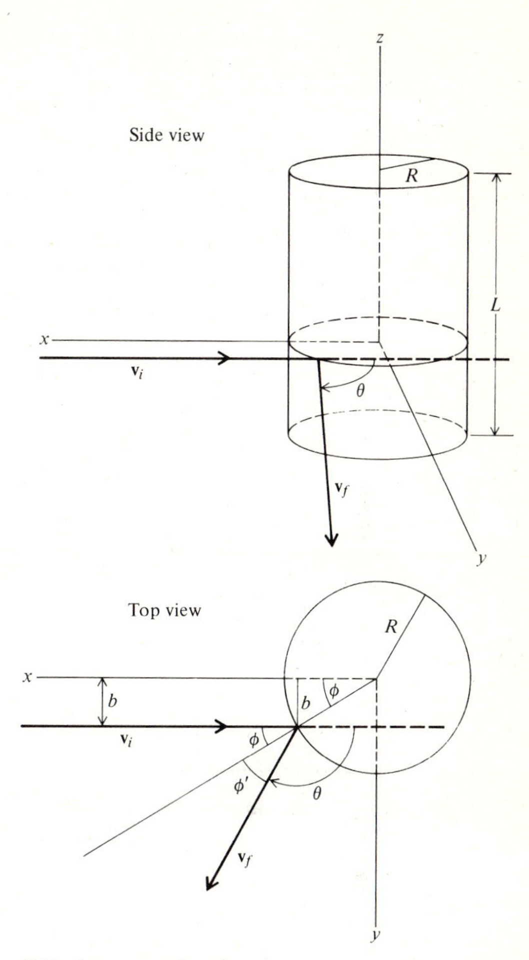

FIGURE 3-8 Scattering of BBs from a massive pipe.

energy of the pipe can be neglected (ideally, $M \to \infty$). Thus the magnitude of the BB's momentum is unchanged by the collision.

$$p_i = p_f \tag{3-70}$$

The tangential component of momentum along the ring is conserved since the force of impact is orthogonal to this direction. This leads to the condition

$$p_i \sin \phi = p_f \sin \phi' \tag{3-71}$$

where the angles are labeled in Fig. 3-8. The combination of Eqs. (3-70) and (3-71) yields

$$\phi = \phi'$$

From the geometry of Fig. 3-8 we conclude that

$$2\phi + \theta = \pi \tag{3-72}$$

The impact parameter b and scattering angle θ are related by

$$b = R \sin \phi = R \sin \left(\frac{\pi - \theta}{2}\right) = R \cos \frac{\theta}{2} \tag{3-73}$$

This equation completely specifies the trajectory of a single BB.

The number of BBs incident in a range $|db|$ of impact parameters (i.e., in a differential area $d\sigma = |db| L$) is

$$dN = d\sigma \frac{N}{A} \tag{3-74}$$

These BBs scatter into a range $d\theta$ of angles determined by the relation (3-73). Thus the number of scattered BBs per unit angle is

$$\frac{dN}{d\theta} = \frac{d\sigma}{d\theta} \frac{N}{A} \tag{3-75}$$

The quantity $d\sigma/d\theta$ is called the *differential-scattering cross section.* Carrying out the calculation of $d\sigma/d\theta$, we differentiate Eq. (3-73) to get

$$|db| = \tfrac{1}{2} R \sin \theta \, d\theta \tag{3-76}$$

and so

$$\frac{d\sigma}{d\theta} = \frac{|db| L}{d\theta} = \tfrac{1}{2} RL \sin \frac{\theta}{2} \tag{3-77}$$

Integration of Eq. (3-74) over all angles gives the expected number of scattered BBs,

$$N_{\mathrm{scatt}} = \sigma \frac{N}{A} \tag{3-78}$$

where the cross section σ is

$$\sigma = \int_0^{2\pi} \frac{d\sigma}{d\theta} \, d\theta = \frac{RL}{2} \int_0^{2\pi} \sin \frac{\theta}{2} \, d\theta = 2RL \tag{3-79}$$

Thus σ is indeed the projected area of the pipe perpendicular to the beam. Even in nonclassical situations where a target area cannot be defined, the proper expression of the scattering probability must have dimensions of area (hence the term *cross section*).

The experimental quantity dN, which is the number of particles scattered into $d\theta$ at θ, is given in terms of the differential cross section by Eqs. (3-75) and (3-77) as

$$dN = \frac{N}{A}\left(\frac{d\sigma}{d\theta}\right) d\theta = \frac{RL}{2}\frac{N}{A}\sin\frac{\theta}{2}\,d\theta \tag{3-80}$$

Out of 100 shots which hit the pipe ($A = 2RL$), the number ricocheting into the backward angular sector, $90° < \theta < 270°$, is

$$n = \frac{100}{4}\int_{\pi/2}^{3\pi/2} \sin\frac{\theta}{2}\,d\theta \approx 71 \tag{3-81}$$

The concept of a scattering cross section is basic to the description of phenomena in atomic, nuclear, and elementary physics, where the location of the scattering center is not well-determined and individual particle trajectories cannot be followed.

PROBLEMS

1. Material drops from a hopper at a constant rate $dm/dt = \sigma$ onto a conveyor belt moving with constant velocity v parallel to the ground. What power motor would be needed to drive the belt? Why is this required power double the time rate of change of mechanical energy? (*Note:* Power is dW/dt, where $W = \int \mathbf{F} \cdot d\mathbf{r}$.)

2. A vertical drain in the floor of the gondola car discussed in Sec. 3-1 is opened to keep rainwater from accumulating in the car. Find the velocity of the gondola at time t in terms of the initial velocity v_0, the mass of the car m_0, and the rate σ at which rain enters and leaves the gondola.

3. During a rocket burn in free space, at what residual mass is the momentum a maximum if the rocket starts from rest?

4. At cruising velocity of 1,000 km/h, each of the four fan jet engines on a Boeing 747 plane burns fuel at a rate of 0.3 kg/s. The airflow through each engine turbine of 50 kg/s has an exhaust velocity of 2,100 km/h. The airflow of 250 kg/s through the fans around each turbine is exhausted with a velocity of 1,300 km/h. Calculate the forward thrust on the plane.

5. It is difficult to construct a rocket which, even if it carries no payload, has a mass ratio $r = m_{\text{initial}}/m_{\text{final}}$ as large as 10. The final

velocity in free space for a single-stage rocket is then $v_f \leq u \ln 10 = 2.3u$. Since the exhaust velocity u for chemical rockets is less than about 5 km/s, it is impossible to send a payload to the moon using a single rocket. Fortunately, the device of "staging" circumvents this difficulty. If a large rocket carries as its payload a smaller rocket which fires after the first burns out, a considerable increase in final velocity can be achieved; explain. Show that if n rockets are staged so that each has the same mass ratio r (where the masses include all the upper stages), the final velocity of the last stage is $nu \ln r$. (Each stage has the same exhaust velocity u.)

6. Show that the maximum altitude which can be attained by a rocket shot vertically from the surface of the earth is

$$h = \frac{u^2}{2g} \ln^2 \left(\frac{m_i}{m_f}\right) - \frac{u}{\alpha}\left[\ln\left(\frac{m_i}{m_f}\right) - \left(1 - \frac{m_f}{m_i}\right)\right]$$

Calculate the numerical value of h for the first stage of the Saturn V rocket. Compare the result with the altitude of the Apollo rocket which could be reached if the entire fuel burn occurred instantaneously (that is, $\alpha = \infty$).

7. A mass m moving horizontally with velocity v_0 strikes a pendulum of mass m as shown.

a. If the two masses stick together, find the maximum height reached by the pendulum.

b. If the masses scatter elastically along the line of the initial motion, find the resulting maximum height.

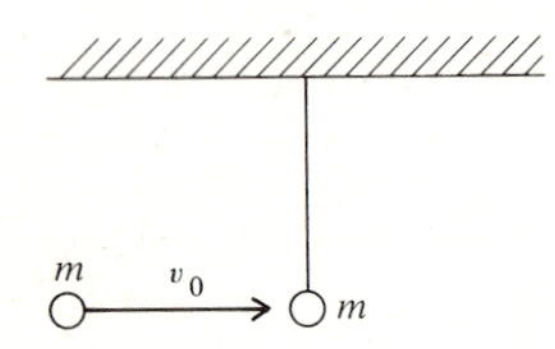

8. A beam of hydrogen molecules moves along the $\hat{\mathbf{z}}$ direction with a kinetic energy of 1 electron volt (eV) per molecule. The molecules are in an excited state, from which they can decay and dissociate into two hydrogen atoms. When the velocity of a dissociation atom is perpendicular to the $\hat{\mathbf{z}}$ direction, its energy is always 0.8 eV. Calculate the energy per molecule released in the dissociative reaction.

9. A proton of energy 4 MeV scatters off a second proton at rest. One proton comes off at an angle of 30° in the lab system. What is its

energy? What is the energy and scattering angle of the second proton?

10. In a collision with a nucleus of unknown mass, an α particle scatters directly backward and loses 75 percent of its energy. What is the mass of the nucleus, assuming that the scattering is elastic?

11. Two balls of unequal mass moving with equal velocities in opposite directions collide. One ball is stationary after the collision. If the collision is elastic, what is the ratio of the masses?

12. For elastic scattering of two particles with masses m_1 and m_2, develop an expression relating the lab and CM scattering angles for the target particle. (Define the angles with respect to the direction of incident projectile.)

13. A mass m moving with velocity v strikes and sticks to a spring system of length l and spring constant k with masses m at each end, as shown. During the subsequent motion, what is the maximum compression of the spring?

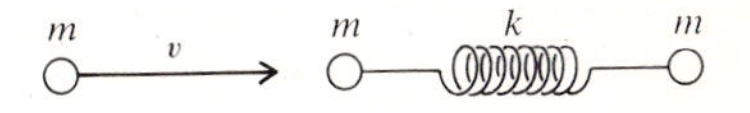

14. A group of n_1 identical smooth billiard balls moving along a line in contact collide elastically with a group of n_2 stationary balls also lined up in contact. How many masses will come out? (*Hint:* Assume that the collision forces propagate with finite speed and treat the collisions successively.)

15. Show that the drag force on a satellite moving with velocity v in the earth's upper atmosphere is approximately

$$f_D = \rho A v^2$$

where ρ is the atmospheric density and A is the cross-sectional area perpendicular to the direction of motion. Assume that the air molecules are moving slowly compared with v and that their collisions with the satellite are completely inelastic.

16. A neutron with kinetic energy T_0 collides with a proton at rest. If the differential cross section for elastic scattering in the laboratory system is $d\sigma/d\theta = A \sin 2\theta$, what is the distribution $d\sigma/dT$, where T is the recoil kinetic energy of the proton? Assume equal mass for the proton and neutron.

17. Calculate the differential cross section $d\sigma/d\theta$ for the scattering of BBs from a steel sphere of radius R. Ignore the effects of gravity.

CHAPTER 4

Angular-momentum Conservation

A conservation law pertaining to angular motion can also be derived from Newton's law. As is the case for the momentum- and energy-conservation laws, the validity of angular-momentum conservation extends far beyond the domain of classical mechanics as a universal truth in physics.

4-1 CENTRAL FORCES

For a single particle of mass m at a distance $\mathbf{r}$ from the origin 0 of a fixed coordinate system, the law of angular-momentum conservation can be derived by taking the cross product of $\mathbf{r}$ with the equation of motion

$$\mathbf{r} \times \mathbf{F} = \mathbf{r} \times \frac{d\mathbf{p}}{dt} \tag{4-1}$$

The left-hand side of this equation is known as the torque,

$$\mathbf{N} = \mathbf{r} \times \mathbf{F} \tag{4-2}$$

The right-hand side of Eq. (4-1) can be written

$$\begin{aligned} \mathbf{r} \times \frac{d\mathbf{p}}{dt} &= \frac{d}{dt}(\mathbf{r} \times \mathbf{p}) - \frac{d\mathbf{r}}{dt} \times \mathbf{p} \\ &= \frac{d}{dt}(\mathbf{r} \times \mathbf{p}) - m\mathbf{v} \times \mathbf{v} \\ &= \frac{d}{dt}(\mathbf{r} \times \mathbf{p}) \end{aligned} \tag{4-3}$$

The quantity

$$\mathbf{L} = \mathbf{r} \times \mathbf{p} \tag{4-4}$$

is called the *angular momentum* about the origin 0. Equations (4-1) to (4-4) relate the angular momentum and the torque.

$$\mathbf{N} = \frac{d\mathbf{L}}{dt} \tag{4-5}$$

If $\mathbf{N} = 0$, then $\mathbf{L}$ is constant in time. From Eq. (4-2), the angular momentum is conserved if either $\mathbf{F} = 0$ or $\mathbf{F}$ is proportional to $\mathbf{r}$. The latter case is of particular interest since it corresponds to a central force, and the fundamental forces of classical mechanics are central.

The angular momentum for motion in a plane is

$$\mathbf{L} = mrv_\theta \hat{\mathbf{z}} = mr^2\dot{\theta}\hat{\mathbf{z}} \tag{4-6}$$

where $\hat{\mathbf{z}}$ is perpendicular to the (r, θ) directions in a right-hand sense. The time rate of change of L_z is determined by the torque N_z about the origin.

$$\frac{d}{dt}(mr^2\dot{\theta}) = N_z = rF_\theta \tag{4-7}$$

This equation is identical with Eqs. (2-72) for angular motion in a plane. For central forces directed toward the origin of the coordinate system, $F_\theta = 0$ and

$$\frac{d}{dt}(mr^2\dot{\theta}) = 0$$

as in Eq. (2-108).

From the conservation of angular momentum, we can prove that the motion of a particle under a central force occurs in a plane. From the properties of the mixed vector product in Eq. (2-43), the dot product of $\mathbf{L} = m\mathbf{r} \times \mathbf{v}$ with either $\mathbf{r}$ or $\mathbf{v}$ vanishes.

$$\mathbf{r} \cdot \mathbf{L} = 0$$
$$\mathbf{v} \cdot \mathbf{L} = 0$$

Inasmuch as $\mathbf{L}$ is constant, its direction is fixed. The position and velocity of the particle thus remain perpendicular to the direction of $\mathbf{L}$ in the course of the motion. The trajectory is therefore confined to a plane passing through the origin and perpendicular to $\mathbf{L}$.

When the angular momentum is conserved, the dependence of the motion on the r and θ variables in Eqs. (2-72) can be separated, using Eq. (4-6) as

$$m\ddot{r} - mr\left(\frac{|\mathbf{L}|}{mr^2}\right)^2 = F_r \tag{4-8}$$

and

$$|\mathbf{L}| = mr^2\dot{\theta} = \text{constant} \tag{4-9}$$

In this form the radial equation is to be solved first for $r(t)$. Then the angular dependence can be obtained from integration of Eq. (4-9).

$$\int d\theta = \int \frac{|\mathbf{L}|}{mr^2}\,dt \tag{4-10}$$

The angular momentum $|\mathbf{L}|$ is specified by the boundary conditions on $\mathbf{r}$ and $\mathbf{v}$, and can be used in place of one of the boundary conditions.

The radial equation is sometimes written in analogy to a one-dimensional equation of motion as

$$m\ddot{r} = F_r + \frac{|\mathbf{L}|^2}{mr^3} \tag{4-11}$$

In a one-dimensional interpretation of the radial motion, the quantity

$$F_{cf} = \frac{|\mathbf{L}|^2}{mr^3} = mr\dot{\theta}^2 \tag{4-12}$$

is a "fictitious" centrifugal force that must be added to F_r. The fictitious force can be derived from the fictitious potential energy.

$$V_{cf}(r) = \frac{|\mathbf{L}|^2}{2mr^2}$$

Since this potential energy is repulsive and large at small radial distances, the particle is repelled from the vicinity of $r = 0$. Consequently the term *centrifugal barrier* is often used in reference to this potential energy. The equivalent one-dimensional representation of the radial motion can be expressed in terms of an effective potential energy

$$E = \tfrac{1}{2}m\dot{r}^2 + V_{\text{eff}}(\mathbf{r})$$

where

$$V_{\text{eff}}(\mathbf{r}) = V(\mathbf{r}) + \frac{|\mathbf{L}|^2}{2mr^2} \tag{4-13}$$

Since Eq. (4-11) can be written in the form of Newton's law in one dimension, the techniques introduced in Chaps. 1 and 2 can be employed to find the radial solution.

As an example of a system in which angular-momentum conservation plays a primary role, we consider two equal point masses connected by a string which passes through a small hole on a frictionless table, as illustrated in Fig. 4-2. We shall solve for the trajectory of the mass on the table for vertical up-and-down motion of the suspended mass. The equations are most simply formulated in a cylindrical coordinate system (r, θ, z) with the origin at the hole in the table and the positive z axis upward. Because the length l of the string is fixed, only two of the three coordinates are independent. We represent these two degrees of freedom by r and θ since we are primarily interested in the motion of the mass on the table. The variables r and z are related by

$$r - z = l \tag{4-14}$$

The gravitational potential energy associated with the mass on the table is constant; that of the suspended mass is

$$V = mgz = mg(r - l) \tag{4-15}$$

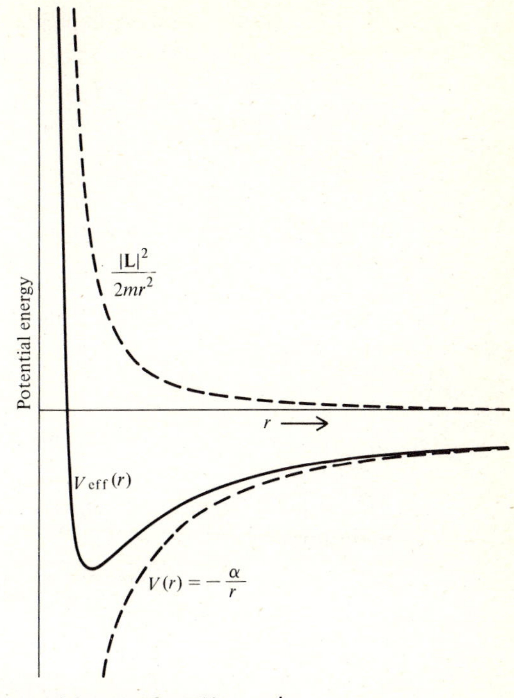

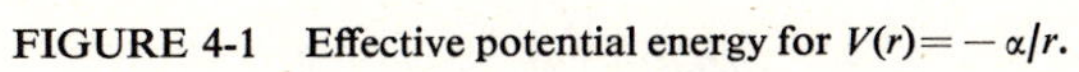
FIGURE 4-1 Effective potential energy for $V(r) = -\alpha/r$.

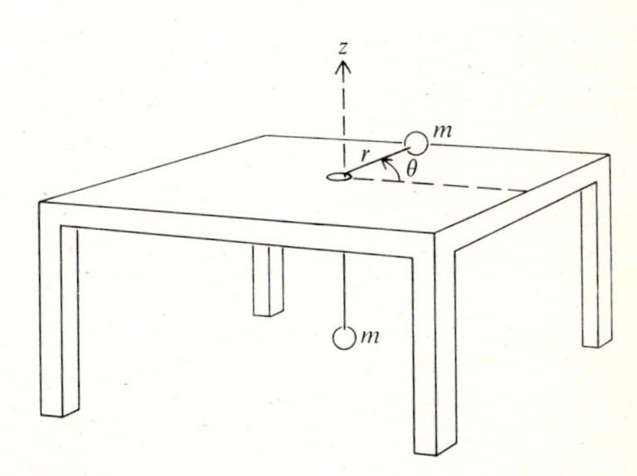

FIGURE 4-2 Central-force example.

Since the potential energy depends only on the variable r, the force is central and the angular momentum is conserved. The kinetic energy of the suspended mass is $T = \frac{1}{2}m\dot{z}^2 = \frac{1}{2}m\dot{r}^2$. From Eqs. (2-105) and (4-15), the lagrangian for the two-mass system is

$$L = \tfrac{1}{2}m\dot{r}^2 + \tfrac{1}{2}m(\dot{r}^2 + r^2\dot{\theta}^2) - mg(r - l) \tag{4-16}$$

With this expression for the lagrangian in Eq. (2-107), we find the radial equation

$$2m\ddot{r} - mr\dot{\theta}^2 + mg = 0 \tag{4-17}$$

and the angular equation

$$\frac{d}{dt}(mr^2\dot{\theta}) = 0$$

The angular equation expresses the conservation of angular momentum:

$$\mathbf{L} = mr^2\dot{\theta}\hat{\mathbf{z}} \tag{4-18}$$

In terms of $\mathbf{L}$ the radial equation is

$$2m\ddot{r} - \frac{|\mathbf{L}|^2}{mr^3} + mg = 0 \tag{4-19}$$

This corresponds to Eq. (4-8) with $F_r = -mg$ and radially accelerated mass $2m$. Equation (4-19) is valid only for $r \leq l$.

The radial equation admits a solution with a circular orbit at $r = r_0$:

$$\ddot{r} = 0 \qquad |\mathbf{L}|^2 = gm^2r_0^3 \tag{4-20}$$

A circular orbit of radius r_0 will be realized if the initial velocities satisfy the conditions

$$\dot{r} = 0 \qquad \dot{\theta} = \frac{|\mathbf{L}|}{mr_0^2} = \sqrt{\frac{g}{r_0}} \equiv \Omega \tag{4-21}$$

as found from Eq. (4-20). In a circular orbit the centrifugal force $mr_0\dot{\theta}^2$ exactly balances the tension mg in the string.

By considering small radial deviations from the circular orbits, we can determine whether or not the motion is stable. We substitute

$$r = r_0 + \varepsilon \tag{4-22}$$

where $\varepsilon \ll r_0$ in Eq. (4-19), and make a power series expansion in ε/r_0. When we retain only constant terms and linear terms in ε, we find

$$2m\ddot{\varepsilon} - \frac{|\mathbf{L}|^2}{mr_0^3}\left(1 - \frac{3\varepsilon}{r_0}\right) + mg = 0 \tag{4-23}$$

After we insert the circular-orbit condition from Eq. (4-20), the equation of motion for small radial deviations reduces to

$$\ddot{\varepsilon} + \frac{3g}{2r_0}\,\varepsilon = 0 \tag{4-24}$$

This describes simple harmonic motion in ε with angular frequency

$$\omega = \sqrt{\frac{3g}{2r_0}} = \sqrt{\frac{3}{2}}\,\Omega \tag{4-25}$$

Thus the particle makes small oscillations about the circular orbit, indicating a stable configuration. The general solution to Eq. (4-24) for the radial-displacement parameter ε is

$$\varepsilon(t) = \varepsilon_0 \cos(\omega t + \alpha_0) \tag{4-26}$$

so that

$$r(t) = r_0 + \varepsilon_0 \cos(\omega t + \alpha_0) \tag{4-27}$$

When we impose the initial conditions

$$r(0) = r_0 + \varepsilon_0$$
$$\dot{r}(0) = 0$$

at $t = 0$, the solution becomes

$$\begin{aligned} \varepsilon(t) &= \varepsilon_0 \cos \omega t \\ r(t) &= r_0 + \varepsilon_0 \cos \omega t \end{aligned} \tag{4-28}$$

The angular motion for circular orbits with small radial oscillations can be calculated from the angular-momentum-conservation relation in Eq. (4-10). Expanding the $1/r^2$ factor in powers of ε_0/r_0, we have

$$\int_{\theta_0}^{\theta} d\theta = \int_0^t \frac{|\mathbf{L}|}{mr(t)^2}\,dt \approx \int_0^t \frac{|\mathbf{L}|}{mr_0^2}\left[1 - 2\,\frac{\varepsilon(t)}{r_0}\right] dt \tag{4-29}$$

When we substitute $\varepsilon(t)$ from Eqs. (4-28), the integration can be carried out. We find

$$\theta(t) = \theta_0 + \frac{|\mathbf{L}|}{mr_0^2}\left(t - \frac{2\varepsilon_0}{\omega r_0}\sin \omega t\right) \tag{4-30}$$

As initial conditions on the angular coordinate, we choose $\theta = 0$ at $t = 0$ and $|\mathbf{L}| = mr_0^2\Omega$. With this choice, $|\mathbf{L}|$ is the angular momentum in a circular orbit with radius r_0 and angular frequency Ω. The solution for the angular position of the particle becomes

$$\theta(t) = \Omega\left(t - \frac{2\varepsilon_0}{\omega r_0}\sin\omega t\right) \tag{4-31}$$

We have thus far considered only a limited class of possible solutions to Eq. (4-19). The complete solution to the radial equation can be discussed qualitatively in terms of the effective one-dimensional potential energy, using the methods of Sec. 2-1. From Eqs. (4-13) and (4-15), the effective one-dimensional potential energy for this system is

$$V_{\text{eff}}(r) = mg(r - l) + \frac{|\mathbf{L}|^2}{2mr^2} \tag{4-32}$$

A sketch of $V_{\text{eff}}(r)$ and its two components in the physical range $0 \leq r \leq l$ is given in Fig. 4-3. A typical energy E is denoted by the dashed line in the figure. Since the system is conservative, the total energy E is given by Eqs. (4-15), (4-16), and (4-18) as

$$E = T + V = m\dot{r}^2 + \tfrac{1}{2}mr^2\dot{\theta}^2 + V(r) = m\dot{r}^2 + \frac{|\mathbf{L}|^2}{2mr^2} + V(r)$$

or

$$E = m\dot{r}^2 + V_{\text{eff}}(r) \tag{4-33}$$

The radial velocity $\dot{r}$ is

$$\dot{r} = \pm\sqrt{\frac{1}{m}[E - V_{\text{eff}}(r)]} \tag{4-34}$$

The allowed physical region for motion is determined by

$$V_{\text{eff}}(r) \leq E \tag{4-35}$$

For the energy given by the dashed line in Fig. 4-3, the radial motion of the particle is bounded by maximum and minimum radii at the turning points, where $E = V_{\text{eff}}$. For an energy $E \geq |\mathbf{L}|^2/2ml^2$, the maximum radius is unbounded and both masses end up on the table when r exceeds l. Of course, for $r \geq l$ the original lagrangian must be modified.

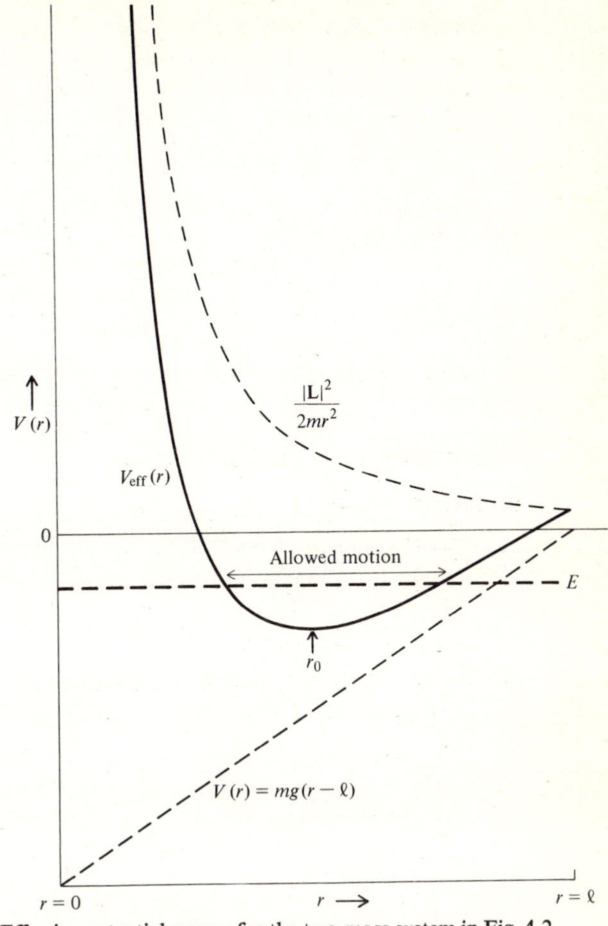

FIGURE 4-3 Effective potential energy for the two-mass system in Fig. 4-2.

The effective potential energy in Fig. 4-3 has a minimum at

$$\left(\frac{dV_{eff}}{dr}\right)_{r=r_0} = 0 = mg - \frac{|\mathbf{L}|^2}{mr_0^3}$$

or

$$|\mathbf{L}|^2 = gm^2 r_0^3 \tag{4-36}$$

For $E = V_{eff}(r_0)$, $\dot{r} = 0$ and the rotating mass moves in a circle, as deduced previously from Eq. (4-19). For E slightly larger than $V_{eff}(r_0)$ the mass on the table undergoes small radial oscillations in the $V_{eff}(r)$

potential well about $r = r_0$. The frequency of these oscillations can be found from a series expansion as in Eq. (2-17). The spring constant k is

$$k = \left(\frac{d^2 V_{\text{eff}}}{dr^2}\right)_{r=r_0} = \frac{3mg}{r_0} \tag{4-37}$$

Since the total mass in the radial-kinetic-energy term is $2m$, the frequency of radial oscillation is

$$\omega = \sqrt{\frac{k}{2m}} = \sqrt{\frac{3g}{2r_0}} \tag{4-38}$$

in agreement with the result given in Eq. (4-25).

4-2 PLANETARY MOTION

The most profound applications of classical mechanics involve the gravitational and Coulomb forces. The potential energy for these conservative forces can be written

$$V(r) = -\frac{\alpha}{r} \tag{4-39}$$

where $\alpha = Gm_1m_2$ for the gravity force and $\alpha = -e_1e_2$ for the static Coulomb force. The angular momentum $\mathbf{L} = \mathbf{r} \times \mathbf{p}$ is conserved, since the force is central.

The orbit equation relating r and θ can be found from the conservation equations for angular momentum and energy. From Eqs. (4-10) and (4-13) we have

$$\begin{aligned} d\theta &= \frac{|\mathbf{L}|}{mr^2}\, dt \\ dr &= \pm\sqrt{\frac{2}{m}\,[E - V_{\text{eff}}(r)]}\, dt \end{aligned} \tag{4-40}$$

where $V_{\text{eff}}(r) = V(r) + |\mathbf{L}|^2/2mr^2$. We eliminate dt from these equations and integrate.

$$\int^{\theta} d\theta = \pm|\mathbf{L}| \int^{r} \frac{dr}{r^2\sqrt{2m[E - V(r) - |\mathbf{L}|^2/2mr^2]}} \tag{4-41}$$

Eq. (4-41) is the general solution for the orbit of a particle moving in a central potential. For the $1/r$ potential of Eq. (4-39), the orbit is determined by

$$\theta = \pm|\mathbf{L}| \int^{r} \frac{dr}{r\sqrt{2mEr^2 + 2m\alpha r - |\mathbf{L}|^2}} \tag{4-42}$$

This indefinite integral is a standard form which can be found in integral tables. The result of integration is

$$\theta - C = \pm \arcsin \left(\frac{m\alpha r - |\mathbf{L}|^2}{\varepsilon m \alpha r} \right) \tag{4-43}$$

where C is the arbitrary constant of integration and

$$\varepsilon \equiv \sqrt{1 + \frac{2E|\mathbf{L}|^2}{m\alpha^2}} \tag{4-44}$$

Taking the sine of both sides, we can directly solve for r as a function of θ.

$$r = \frac{|\mathbf{L}|^2/m\alpha}{1 \pm \varepsilon \sin (\theta - C)} \tag{4-45}$$

For $\alpha > 0$ the minimum value of the radial coordinate on the orbit in Eq. (4-45) is attained at the angle

$$\theta_0 = C \pm \frac{\pi}{2}$$

In terms of θ_0, the final form of the orbit equation becomes

$$r(\theta) = \frac{\lambda(1 + \varepsilon)}{1 + \varepsilon \cos (\theta - \theta_0)} \tag{4-46}$$

where we have introduced the notation

$$\lambda \equiv \frac{|\mathbf{L}|^2}{m\alpha} \cdot \frac{1}{1 + \varepsilon} \tag{4-47}$$

The orbit is symmetric around the angle θ_0. For convenience we may choose our coordinate system in such a way that $\theta_0 = 0$. In terms of cartesian coordinates ($x = r \cos \theta$, $y = r \sin \theta$) the orbit equation (4-46) with $\theta_0 = 0$ can be cast in the forms

$$\frac{\left(x + \dfrac{\varepsilon}{1 - \varepsilon}\lambda\right)^2}{\left(\dfrac{\lambda}{1 - \varepsilon}\right)^2} + \frac{y^2}{\left(\lambda\sqrt{\dfrac{1 + \varepsilon}{1 - \varepsilon}}\right)^2} = 1 \qquad 0 \le \varepsilon < 1 \tag{4-48a}$$

$$y^2 + 4\lambda x = 4\lambda^2 \qquad \varepsilon = 1 \tag{4-48b}$$

$$\frac{\left(x - \dfrac{\varepsilon}{\varepsilon - 1}\lambda\right)^2}{\left(\dfrac{\lambda}{\varepsilon - 1}\right)^2} - \frac{y^2}{\left(\lambda\sqrt{\dfrac{\varepsilon + 1}{\varepsilon - 1}}\right)^2} = 1 \qquad \varepsilon > 1 \tag{4-48c}$$

Equation (4-46) or Eqs. (4-48) represent a conic section with a focus at $r = 0$. The type of conic section depends on the values of the parameters ε and λ as follows:

$$
\begin{array}{llll}
0 \leq \varepsilon < 1 & \lambda > 0 & \text{ellipse } (\varepsilon = 0 \text{ is a circle}) & \\
\varepsilon = 1 & \lambda > 0 & \text{parabola} & \quad (4\text{-}49) \\
\varepsilon > 1 & \lambda < 0 \text{ or } \lambda > 0 & \text{hyperbola} &
\end{array}
$$

Sketches of these orbits with $\theta_0 = 0$ are shown in Fig. 4-4. The $\lambda > 0$ requirement for the elliptical and parabolic orbits follows from the positivity requirement on r in Eq. (4-46). For $\lambda > 0$, the angle $\theta = \theta_0$ corresponds to the turning point of minimum r, with $r_{\min} = \lambda$. For $\lambda < 0$, this turning point occurs at $\theta = \theta_0 + \pi$, with $r_{\min} = \lambda[(1 + \varepsilon)/(1 - \varepsilon)]$, as shown in the hyperbolic orbits of Fig. 4-4.

From Eq. (4-47) we observe that λ is positive for an attractive potential energy ($\alpha > 0$) and negative for a repulsive potential energy ($\alpha < 0$). By reference to Eqs. (4-44) and (4-49), we conclude that the α and energy ranges for the three types of orbits are

$$
\begin{array}{llll}
\text{Ellipse} & \alpha > 0 & E < 0 & \\
\text{Parabola} & \alpha > 0 & E = 0 & \quad (4\text{-}50) \\
\text{Hyperbola} & \alpha > 0 \text{ or } \alpha < 0 & E > 0 &
\end{array}
$$

where the reference point for the potential energy is $r = \infty$.

The motion about the sun of the planets in our solar system is governed by the gravitational potential energy, which has $\alpha > 0$. Of the conic sections only the ellipse is an orbit of finite extent. Thus, from Eqs. (4-50), all planetary orbits have $E < 0$.

For an elliptic orbit, the semimajor axis a and semiminor axis b are commonly specified in place of λ and ε. From Eq. (4-48*a*) the lengths of these axes are related to λ and ε by

$$
\begin{aligned}
a &= \frac{\lambda}{1 - \varepsilon} \\
b &= \lambda \sqrt{\frac{1 + \varepsilon}{1 - \varepsilon}}
\end{aligned}
\qquad (4\text{-}51)
$$

The ratio of b to a is determined by the eccentricity ε

$$\frac{b}{a} = \sqrt{1 - \varepsilon^2}$$

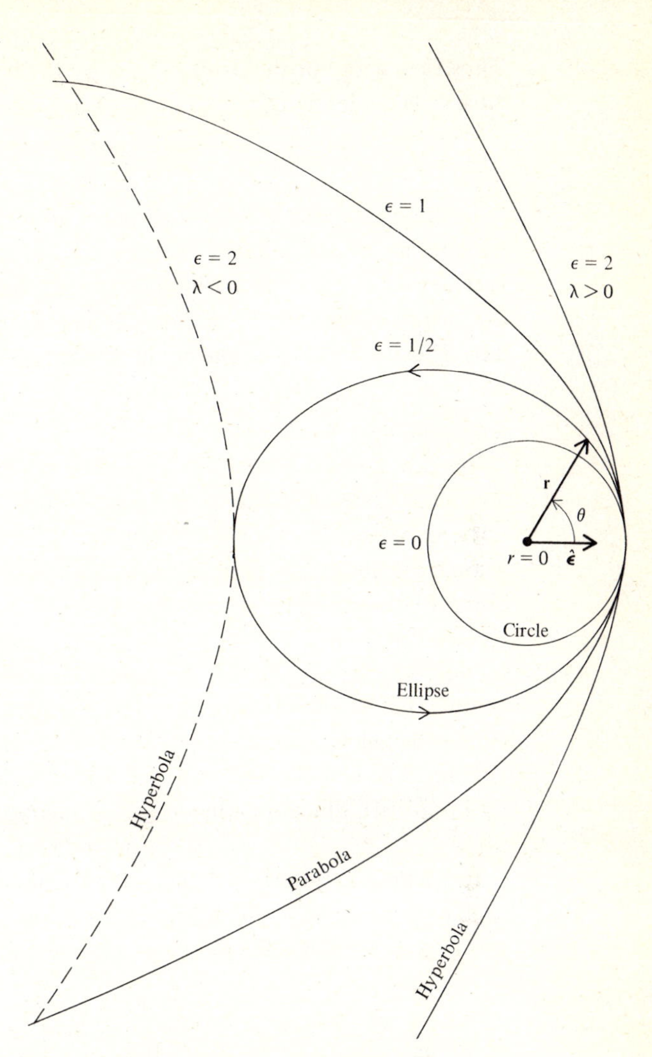

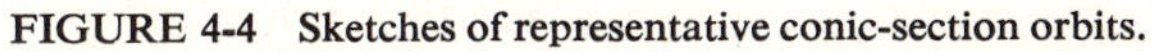

FIGURE 4-4 Sketches of representative conic-section orbits.

In terms of the energy and angular momentum, the semimajor and semiminor axes of the ellipse are given by Eqs. (4-44), (4-47), (4-50), and (4-51) as

$$a = -\frac{\alpha}{2E} = \frac{\alpha}{2|E|}$$
$$b = \frac{|\mathbf{L}|}{\sqrt{2m|E|}} \tag{4-52}$$

Thus the energy of the orbit is fixed by the length of the semimajor axis, independently of the value of $|\mathbf{L}|$

$$E = -\frac{\alpha}{2a} \tag{4-53}$$

In the special case $\varepsilon = 0$, an elliptical orbit reduces to a circle. The eccentricity of the moon's orbit about the earth is $\varepsilon = 0.055$. The eccentricity of the earth's orbit about the sun is $\varepsilon = 0.017$. Both orbits are therefore very nearly circular. In a circular orbit the semimajor axis is just the radius of the circle. From Eq. (4-39) we then find

$$E = \tfrac{1}{2}V(a) = -T \tag{4-54}$$

where we have used $E = T + V$.

We can use the circular-orbit relations of Eq. (4-54) to resolve the so-called "satellite paradox": The effect of the slight atmospheric drag on a satellite in a circular orbit at a height of several hundred kilometers above the earth is to increase the speed of the satellite, contrary to intuition. The atmospheric drag converts mechanical energy into heat. Hence the energy E of the satellite decreases, and so, by Eq. (4-53), the radius a of the orbit decreases. Since the atmosphere is quite thin at this altitude, the satellite makes many orbits before its orbital height is appreciably changed, and the orbit remains nearly circular. The decrease in E must be accompanied by an increase in T by Eq. (4-54). Since the kinetic energy increases, the satellite speeds up.

In Coulomb scattering of charged particles from a fixed scattering center, $E > 0$, and so the particle trajectories are hyperbolic orbits. For hyperbolic orbits the $r \to \infty$ asymptotes of Eq. (4-46) are given by

$$\theta_{\text{asy}}^{\pm} = \pm \arccos\left(-\frac{1}{\varepsilon}\right) \tag{4-55}$$

where again we have taken $\theta_0 = 0$.

4-3 ECCENTRICITY VECTOR

The solution in Eq. (4-46) for the orbit of a particle in an inverse-square force field can be obtained in an elegant alternative manner by vector-analysis techniques. The equation of motion of a particle of mass m in the potential energy of Eq. (4-39) is

$$\dot{\mathbf{p}} = -\alpha\frac{\mathbf{r}}{r^3} \tag{4-56}$$

We take the cross product of both sides of Eq. (4-56) with **L**.

$$\mathbf{L} \times \dot{\mathbf{p}} = -\alpha \frac{1}{r^3} \mathbf{L} \times \mathbf{r} \tag{4-57}$$

Since **L** is constant in time, the left-hand side of Eq. (4-57) can be written

$$\mathbf{L} \times \dot{\mathbf{p}} = \frac{d}{dt}(\mathbf{L} \times \mathbf{p}) \tag{4-58}$$

From the explicit form

$$L = m\mathbf{r} \times \dot{\mathbf{r}}$$

the right-hand side of Eq. (4-57) is

$$-\frac{\alpha}{r^3} \mathbf{L} \times \mathbf{r} = \frac{m\alpha}{r^3} [\mathbf{r}(\dot{\mathbf{r}} \cdot \mathbf{r}) - \dot{\mathbf{r}} r^2] \tag{4-59}$$

where we have used the identity for the triple cross product from Eq. (2-44*b*). The right-hand side of Eq. (4-59) can equivalently be written as a single time derivative.

$$-\frac{\alpha}{r^3} \mathbf{L} \times \mathbf{r} = -m\alpha \frac{d}{dt}\left(\frac{\mathbf{r}}{r}\right) \tag{4-60}$$

Putting Eqs. (4-58) and (4-60) back into Eq. (4-57) leads to

$$\frac{d}{dt}\left(-\mathbf{L} \times \mathbf{p} - m\alpha \frac{\mathbf{r}}{r}\right) = 0 \tag{4-61}$$

The vector quantity

$$\boldsymbol{\varepsilon} = -\frac{\mathbf{L} \times \mathbf{p}}{m\alpha} - \hat{\mathbf{r}} \tag{4-62}$$

is therefore a constant of the motion. The magnitude of $\boldsymbol{\varepsilon}$ is given by

$$\varepsilon^2 = \boldsymbol{\varepsilon} \cdot \boldsymbol{\varepsilon} = \frac{|\mathbf{L}|^2 p^2 - (\mathbf{L} \cdot \mathbf{p})^2}{(m\alpha)^2} + \frac{2\mathbf{L} \times \mathbf{p} \cdot \hat{\mathbf{r}}}{m\alpha} + 1$$

$$= \frac{|\mathbf{L}|^2 p^2}{(m\alpha)^2} - \frac{2|\mathbf{L}|^2}{m\alpha r} + 1$$

$$\varepsilon^2 = \frac{2|\mathbf{L}|^2}{(m\alpha^2)} \left(\frac{p^2}{2m} - \frac{\alpha}{r}\right) + 1 \tag{4-63}$$

The quantity in brackets in Eqs. (4-63) is the energy E of the particle

$$E = T + V = \frac{p^2}{2m} - \frac{\alpha}{r} \tag{4-64}$$

Thus ε is given in terms of $|\mathbf{L}|$ and E by

$$\varepsilon = \sqrt{1 + \frac{2E|\mathbf{L}|^2}{m\alpha^2}} \tag{4-65}$$

which is just the eccentricity of the orbit found previously in Eqs. (4-44) and (4-46). With this association we call $\boldsymbol{\varepsilon}$ the *eccentricity vector*.

We can find the equation for the orbit by taking the dot product of $\boldsymbol{\varepsilon}$ with $\mathbf{r}$.

$$\boldsymbol{\varepsilon} \cdot \mathbf{r} = \frac{-\mathbf{L} \times \mathbf{p} \cdot \mathbf{r}}{m\alpha} - r \tag{4-66}$$

By interchange of dot and cross products according to Eq. (2-43) and use of $\mathbf{L} = \mathbf{r} \times \mathbf{p}$, we obtain

$$\boldsymbol{\varepsilon} \cdot \mathbf{r} = \frac{|\mathbf{L}|^2}{m\alpha} - r \tag{4-67}$$

We define the direction of $\mathbf{r}$ relative to $\boldsymbol{\varepsilon}$ by an angle θ.

$$\boldsymbol{\varepsilon} \cdot \mathbf{r} = \varepsilon r \cos\theta \tag{4-68}$$

From Eqs. (4-67) and (4-68) we get the orbit equation

$$r(\theta) = \frac{|\mathbf{L}|^2/m\alpha}{1 + \varepsilon\cos\theta} \tag{4-69}$$

in agreement with our previous result in Eq. (4-46), with the choice of $\theta_0 = 0$. The orientation of the orbit is specified relative to the eccentricity vector by Eq. (4-68), as indicated in Fig. 4-4. The eccentricity vector lies along the symmetry axis of the conic section.

4-4 KEPLER'S LAWS

The observed data on planetary motion were reduced by Kepler in the early seventeenth century to three empirical laws. These laws played an important role in Newton's discovery of the gravitational-force law. The first law states that the orbit of a planet is an ellipse with the sun at one focus. We have established this law in Sec. 4-2 from the inverse-square nature of the gravitational force. The result neglects perturbations due to the presence of the other planets.

The second law of Kepler states that the time rate of change of area swept out by the radius from the sun to a planet is a constant, as illustrated in Fig. 4-5. In order for this to happen, the planet must

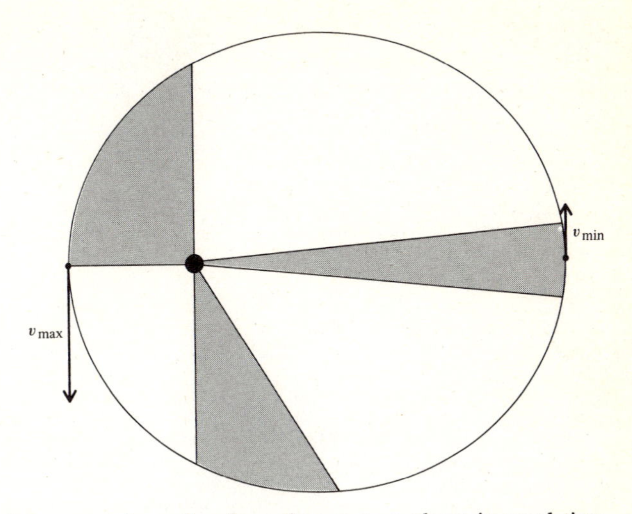

FIGURE 4-5 Areas swept by radius from the sun to a planet in equal time intervals.

have higher tangential velocities at smaller radial distances from the sun. The second law is nothing but angular-momentum conservation. From Fig. 4-6 the element of area swept out in dt is

$$dA = \tfrac{1}{2}r^2\, d\theta = \tfrac{1}{2}r^2\dot{\theta}\, dt$$

so that by Eq. (4-18)

$$\frac{dA}{dt} = \frac{|\mathbf{L}|}{2m} = \text{constant} \tag{4-70}$$

Kepler's third law states that the square of the period of revolution about the sun is proportional to the cube of the semimajor axis of the elliptical orbit. To derive the third law we use the constancy of the angular momentum and integrate Eq. (4-70) over a complete revolution.

$$\begin{aligned} \int_0^\tau dt &= \frac{2m}{|\mathbf{L}|}\int dA \\ \tau &= \frac{2mA}{|\mathbf{L}|} \end{aligned} \tag{4-71}$$

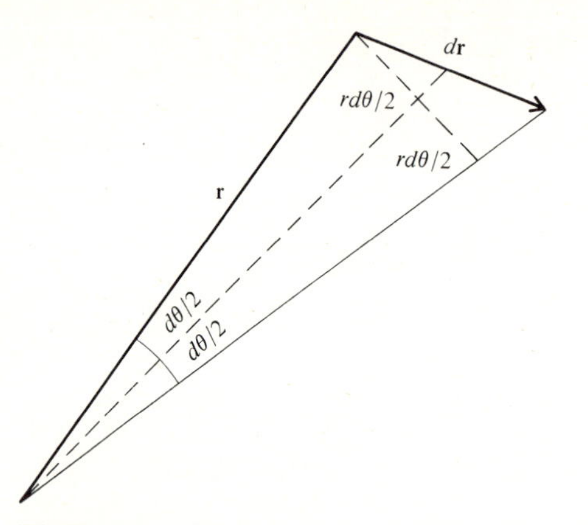

FIGURE 4-6 Element of area swept out by a radius vector in an infinitesimal time interval *dt*.

where A is the area of the elliptical orbit. This area is given in terms of the semimajor and semiminor axes by

$$A = \pi ab$$

Substitution of Eqs. (4-52) in Eq. (4-71) completes the derivation of Kepler's third law.

$$\tau^2 = \left(\frac{4\pi^2 m}{\alpha}\right) a^3 \tag{4-72}$$

Since $\alpha = GmM_\odot$, where $M_\odot$ is the mass of the sun, the ratio

$$\frac{\tau^2}{a^3} = \frac{4\pi^2}{GM_\odot} \tag{4-73}$$

is independent of the mass of the planet.

Kepler's laws neglect the motion of the sun. A correction term to Eq. (4-73) due to the sun's motion will be derived in Sec. 5-1. The correction term is proportional to $m/M_\odot$, and is thus very small.

4-5 SATELLITES AND SPACECRAFT

The orbits of satellites and spacecraft are interesting problems in celestial mechanics. For a satellite in a circular orbit at a distance h above the earth, the period of revolution is found from adaptation of Eq. (4-73) to be

$$\tau = \frac{2\pi}{\sqrt{GM_e}} (R_e + h)^{3/2} \tag{4-74}$$

where M_e is the mass of the earth and R_e is its radius. From Eq. (1-6) we can write the quantity GM_e in terms of the gravitational acceleration at the surface of the earth.

$$GM_e = gR_e^2$$

If $h \ll R_e$, τ is given to a good approximation by

$$\tau \approx 2\pi\sqrt{\frac{R_e}{g}} \approx 2\pi\sqrt{\frac{6{,}371 \times 10^3}{9.8}}$$

$$\approx 5{,}100 \text{ s} \approx 1.4 \text{ h} \tag{4-75}$$

Earth-satellite periods for circular orbits at various heights are given in Table 4-1. The velocity of the satellite in a low-altitude circular orbit about the earth is

$$v_c = (R_e + h)\omega \approx R_e\left(\frac{2\pi}{\tau}\right) \approx (6{,}371)\frac{2\pi}{5{,}100} \approx 7.9 \text{ km/s} \tag{4-76}$$

The velocity is necessarily less than the escape velocity of 11.2 km/s, which was discussed in Sec. 2-2.

To study storm systems, the National Aeronautics and Space Administration (NASA) is now planning a weather satellite which will go out to 7,200 km from the earth's surface and swing in to 200 km. With a circular satellite orbit at an altitude of $h \approx 200$ km as a reference, we can calculate the period of this weather satellite from Kepler's third law in Eq. (4-72).

$$\tau' = \tau\left(\frac{a'}{a}\right)^{3/2} = (1.4)\left(\frac{10{,}100}{6{,}600}\right)^{3/2} \approx 2.65 \text{ h} \approx \frac{24}{9} \tag{4-77}$$

The satellite makes nine orbits in 24 h, and therefore perigee occurs over the same nine points on the earth each day. The velocity at any

TABLE 4-1 EARTH SATELLITE PERIODS

Altitude h, km above surface of earth	Period τ, h
0	1.41
200	1.47
500	1.58
1,680	2.00
35,850	24.00

point on the elliptical orbit of the weather satellite can be calculated from Eqs. (4-53) and (4-64).

$$E = \tfrac{1}{2}mv^2 - \frac{\alpha}{r} = -\frac{\alpha}{2a'} \tag{4-78}$$

Since $\alpha = GM_e m = gR_e^2 m$, the magnitude of the velocity is given by

$$v = \sqrt{\left(\frac{2}{r} - \frac{1}{a'}\right) gR_e^2}$$

Recalling from Eq. (2-14) that the escape velocity from the earth is

$$v =_{\text{esc}} \sqrt{2gR_e} = 11.2 \text{ km/s}$$

we can express the velocity of the satellite at any point of its orbit as

$$v = v_{\text{esc}} \sqrt{\left(\frac{1}{r} - \frac{1}{2a'}\right) R_e} \tag{4-79}$$

At perigee (point of closest approach to the earth's surface), the velocity reaches its maximum value, and at apogee (farthest distance from the earth's surface), the velocity reaches its minimum value. The velocity at perigee ($r = 6{,}600$ km) is $v_p = 9.4$ km/s, and the velocity at apogee ($r = 13{,}600$ km) is $v_a = 4.3$ km/s.

A spacecraft in a circular orbit of radius R_c around the earth can be most economically inserted into an elliptical orbit with distance of closest approach R_c by firing rockets at perigee. A rocket burn at perigee increases the velocity perpendicular to the radius vector, without change in the $\mathbf{v} \cdot \mathbf{r} = 0$ condition as a turning point on the orbit. The increase in velocity is accompanied by an increase in energy and angular momentum. From Eqs. (4-44) and (4-48) the orbit is thereby changed from circular to elliptical. The procedure can be used in reverse to convert an elliptical orbit to a circular one by firing retro-rockets at the distance of closest approach. This technique was followed in the lunar orbit insertion for the Apollo lunar landings. The Apollo spacecraft was first inserted in an elliptical lunar orbit, as illustrated in Fig. 4-7. After two orbits a retrograde burn was used to circularize the orbit prior to the landing on the lunar surface. (See Fig. 4-8 on back endpaper.)

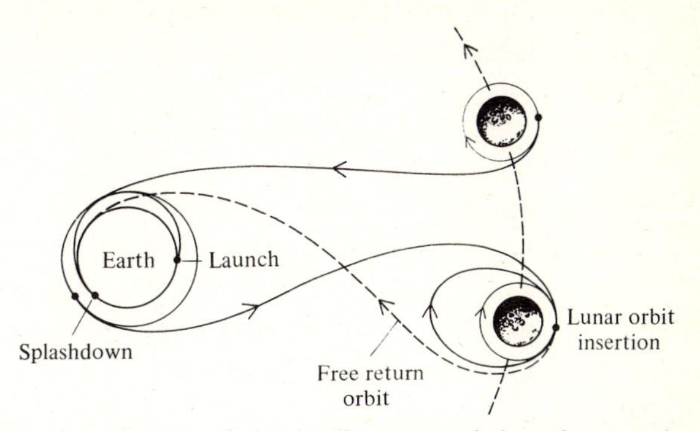

FIGURE 4-7 Schematic trajectory of the Apollo moon missions from earth launch to splashdown.

4-6 GRAND TOURS OF THE OUTER PLANETS

Several spacecraft missions have been made to the two nearest planetary neighbors of Earth—Venus and Mars. Forthcoming flyby missions to the outer planets will also be of great scientific interest. The technical capability for unmanned missions to the outer planets already exists, and these journeys across the solar system will probably be undertaken in this decade. A great difficulty in outer-planet exploration is the long time duration of direct flights. Fortunately, the flight times for outer-planet missions can be considerably shortened by means of gravitational assists as the spacecraft swings by the planets en route. In the 1970s the outer planets will be lined up in a favorable configuration that could permit a single spacecraft to make a "Grand Tour" of the planets Jupiter, Saturn, Uranus, and Neptune.[1] The possibility of this four-planet mission occurs only at 175-year intervals. By utilizing the gravitational energy boost obtained from the Jupiter swingby, the Grand Tour of these four planets can be made in 12 years. In comparison, the flight time for a direct mission to Neptune with equivalent launch energy would take 30 years. The essential aspects of the gravity-assistance trajectory for the Grand Tour can be developed from the planetary-orbit equations derived in preceding sections.

In sending a spacecraft to the outer planets, the launch should be made in the direction of the earth's orbital velocity about the sun, as

[1] W. H. Pickering, *Amer. Sci.*, **58**: 148 (1970).

illustrated in Fig. 4-9. This velocity of the earth, in a nearly circular orbit of radius a_e and period τ_e about the sun, is

$$V_e = \omega_e a_e = \frac{2\pi}{\tau_e} a_e = \frac{(2\pi)(1.5 \times 10^8 \text{ km})}{(365 \times 24 \times 3{,}600 \text{ s})} = 30 \text{ km/s} \tag{4-80}$$

The earth's orbital velocity represents a substantial fraction of the minimum launch velocity needed to send a spacecraft to the outer parts of our solar system. For a spacecraft of mass m at an initial distance a_e from the sun to completely escape the gravitational pull of the sun, the minimum initial velocity is determined by

$$E = 0 = \tfrac{1}{2}m(v_{\text{esc}}^{\odot})^2 - \frac{GmM_{\odot}}{a_e} \tag{4-81}$$

On the other hand, for the circular orbit of the earth about the sun,

$$\frac{m_e(V_e)^2}{a_e} = \frac{Gm_e M_{\odot}}{a_e^2} \tag{4-82}$$

From Eqs. (4-80) to (4-82) we find

$$v_{\text{esc}}^{\odot} = \sqrt{\frac{2GM_{\odot}}{a_e}} = \sqrt{2}V_e = 42 \text{ km/s} \tag{4-83}$$

This relation between the escape velocity at a distance r from a gravitational source and the velocity in a circular orbit at radius r is always true; that is, $v_{\text{esc}}(r) = \sqrt{2}V_c(r)$. By making the launch from the moving earth, the initial velocity required for escape from the gravitational pull of the sun can be reduced to

$$(\sqrt{2} - 1)V_e = 12 \text{ km/s} \tag{4-84}$$

The spacecraft must have additional initial velocity to escape from the gravitational attraction of the earth.

To send a direct mission to the planet Uranus with a minimum amount of propulsion energy, the spacecraft should be launched in the direction of the earth's orbital motion into an elliptical orbit about the sun with perihelion at the earth's orbit and aphelion at the orbit of Uranus, as shown in Fig. 4-9. The launch must be made at the proper time in order that Uranus and the spacecraft arrive together at the aphelion of the spacecraft's orbit. The minimum and maximum values of the distance r from the sun on this spacecraft orbit are

$$r_{\text{min}} = 1 \text{ AU} \qquad \text{(at Earth)}$$
$$r_{\text{max}} = 19.2 \text{ AU} \qquad \text{(at Uranus)}$$

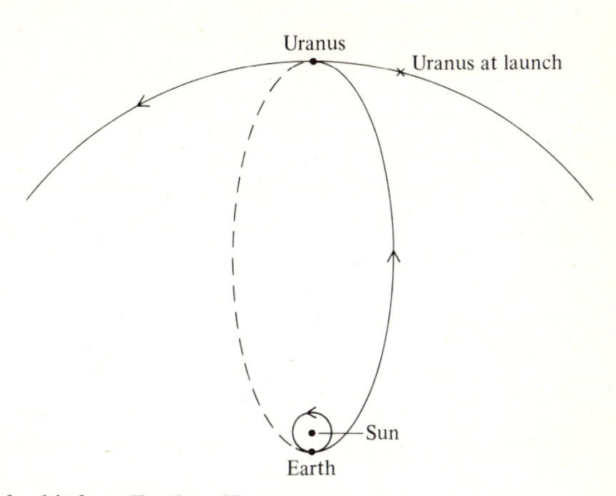

FIGURE 4-9 Elliptical orbit from Earth to Uranus.

where AU stands for the *astronomical unit* of length, namely, the sun-earth distance of 1.5×10^8 km. The parameters λ and ε of the orbit equation (4-46) can be determined from the minimum and maximum values of r.

$$r_{\text{min}} = \lambda$$
$$r_{\text{max}} = \lambda \frac{1+\varepsilon}{1-\varepsilon} \tag{4-85a}$$

giving

$$\lambda = 1 \text{ AU}$$
$$\varepsilon = 0.9 \tag{4-85b}$$

Thus the spacecraft orbit to Uranus is

$$r = \frac{1.9}{1 + 0.9 \cos\theta} \tag{4-86}$$

The semimajor axis of the orbit is

$$a = \tfrac{1}{2}(r_{\text{min}} + r_{\text{max}}) = 10.1 \text{ AU} \tag{4-87}$$

The velocity of the spacecraft at any point on the orbit can be found by use of Eq. (4-53).

$$E = -\frac{GmM_\odot}{2a} = \tfrac{1}{2}mv^2 - \frac{GmM_\odot}{r} \tag{4-88}$$

In terms of Eq. (4-83), the solution for v is

$$v = v_{\text{esc}}^{\odot} \sqrt{\left(\frac{1}{r} - \frac{1}{2a}\right) a_e} \tag{4-89}$$

The perihelion velocity necessary for insertion of the spacecraft at $r = a_e$ into the elliptical orbit to Uranus as calculated from Eqs. (4-83), (4-87), and (4-89) is

$$v_p = v_{\text{esc}}^{\odot} \sqrt{\tfrac{19}{20}} = 41 \text{ km/s}$$

The time after earth launch at which the spacecraft reaches Uranus is just the half period $\tau/2$ of the elliptical orbit. We can use Kepler's third law in Eq. (4-73) to calculate the time duration of this mission from the radius a_e and period τ_e of the earth's orbit about the sun.

$$\frac{\tau}{2} = \frac{\tau_e}{2}\left(\frac{a}{a_e}\right)^{3/2} = \tfrac{1}{2}(10.1)^{3/2} \approx 16 \text{ years}$$

For the same launch energy as needed for the elliptical orbit, the duration of flight to Uranus can be cut from 16 years to about 5 years on a gravity-assistance orbit which swings by Jupiter, as we will now demonstrate. The spacecraft is initially launched from Earth into an elliptical orbit about the sun. To be definite we take the same initial heliocentric orbit as considered in Eq. (4-86). The launch time is chosen such that the spacecraft will make a close encounter with Jupiter, as illustrated in Fig. 4-10. As a result of the encounter the kinetic energy of the spacecraft is changed. In discussing the energetics of the situation, we can neglect the slight change in the direction of Jupiter's velocity during the encounter, since the time duration of the encounter is short compared with Jupiter's period of revolution. If we let $\mathbf{p}_i$ and $\mathbf{p}_f$ denote the spacecraft momenta in the heliocentric (sun-centered) inertial frame just before and just after the Jovian encounter, we can write

$$\begin{aligned} \mathbf{p}_i &= \mathbf{p}_i' + m\mathbf{V}_J \\ \mathbf{p}_f &= \mathbf{p}_f' + m\mathbf{V}_J \end{aligned} \tag{4-90}$$

where $\mathbf{V}_J$ is the velocity of Jupiter about the sun and $\mathbf{p}_i'$, $\mathbf{p}_f'$ are the spacecraft momenta relative to Jupiter (i.e., in the reference frame in which Jupiter is at rest). Equation (4-90) is a galilean transformation. The change in momentum during the encounter,

$$\Delta\mathbf{p} \equiv \mathbf{p}_f - \mathbf{p}_i$$

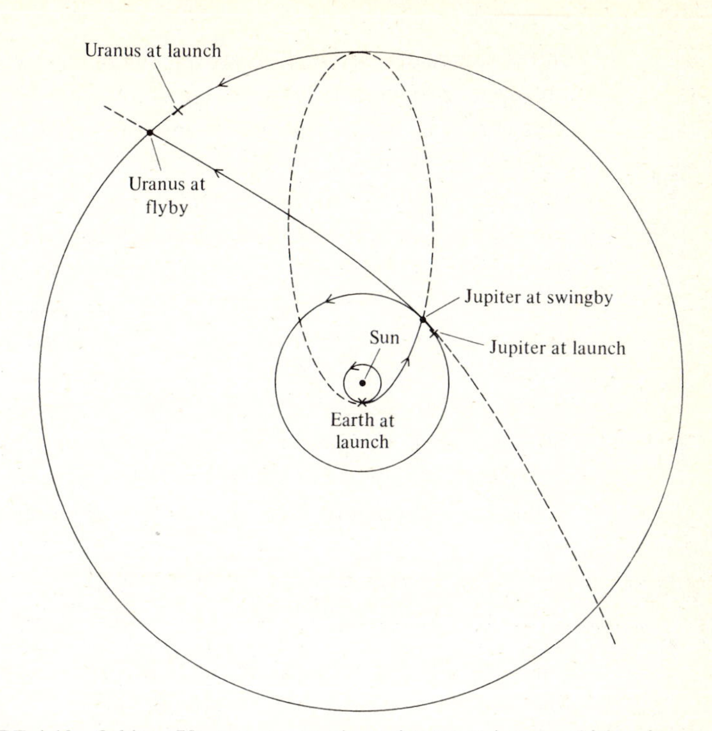

FIGURE 4-10 Orbit to Uranus on a gravity-assistance trajectory which swings past Jupiter.

is the same in both frames.

$$\Delta \mathbf{p} = \Delta \mathbf{p}' \tag{4-91}$$

However, the change in kinetic energy,

$$\Delta T \equiv \frac{p_f^2 - p_i^2}{2m}$$

depends on the frame in which the spacecraft is observed.

$$\Delta T = \Delta T' + \mathbf{V}_J \cdot \Delta \mathbf{p}' \tag{4-92}$$

In the Jupiter-based frame, the encounter is an elastic collision with Jupiter, and so

$$\begin{aligned} \Delta T' &= 0 \\ \Delta T &= \mathbf{V}_J \cdot \Delta \mathbf{p}' = \mathbf{V}_J \cdot \Delta \mathbf{p} \end{aligned} \tag{4-93}$$

In the sun-centered system, however, the scattering of the spacecraft on Jupiter produces a change in the kinetic energy of the spacecraft. This energy is obtained at the expense of Jupiter.

Until the spacecraft reaches the immediate vicinity of Jupiter, the spacecraft orbit is governed by the strong gravitational field of the sun. In the vicinity of Jupiter the gravitational force on the spacecraft due to the sun is relatively constant compared with the gravitational force due to Jupiter. Consequently, as the spacecraft nears Jupiter, its orbit relative to Jupiter is essentially determined by Jupiter's gravitational field. To discuss the motion of the spacecraft relative to Jupiter, we use the reference frame in which Jupiter is initially at rest. We denote the velocities of the spacecraft in the Jupiter rest frame as the spacecraft enters and leaves the region of Jupiter's influence by $\mathbf{u}_i$ and $\mathbf{u}_f$, respectively. The conservation of momentum and energy leads to

$$E' = \tfrac{1}{2}mu_i^2 = \tfrac{1}{2}mu_f^2 + \tfrac{1}{2}M_J\left[\frac{m}{M_J}(\mathbf{u}_i - \mathbf{u}_f)\right]^2 \tag{4-94}$$

Since $m \ll M_J$, the recoil energy of Jupiter is negligible. We therefore find

$$u_i \approx u_f \equiv u \tag{4-95}$$

The energy E' is positive, and so the orbit of the spacecraft about Jupiter is hyperbolic, as illustrated in Fig. 4-11. In the heliocentric inertial frame the spacecraft velocities at the boundary of Jupiter's sphere of influence are

$$\begin{aligned}\mathbf{v}_i &= \mathbf{u}_i + \mathbf{V}_J \\ \mathbf{v}_f &= \mathbf{u}_f + \mathbf{V}_J\end{aligned} \tag{4-96}$$

Here we have again neglected the slight change in the direction of $\mathbf{V}_J$ during the encounter. From Eqs. (4-96) the magnitude of the asymptotic hyperbolic velocity u can be related to the known velocities $\mathbf{v}_i$ and $\mathbf{V}_J$ as

$$\begin{aligned}u &= (v_i^2 + V_J^2 - 2\mathbf{v}_i \cdot \mathbf{V}_J)^{1/2} \\ &= (v_i^2 + V_J^2 - 2v_i V_J \cos\beta_i)^{1/2}\end{aligned} \tag{4-97}$$

where β_i is the angle between $\mathbf{v}_i$ and $\mathbf{V}_J$. The magnitude V_J of Jupiter's circular orbital velocity is

$$V_J = \frac{2\pi}{\tau_J}a_J = \frac{2\pi(5.2 \times 1.5 \times 10^8)}{11.9 \times 365 \times 24 \times 3{,}600} = 13 \text{ km/s} \tag{4-98}$$

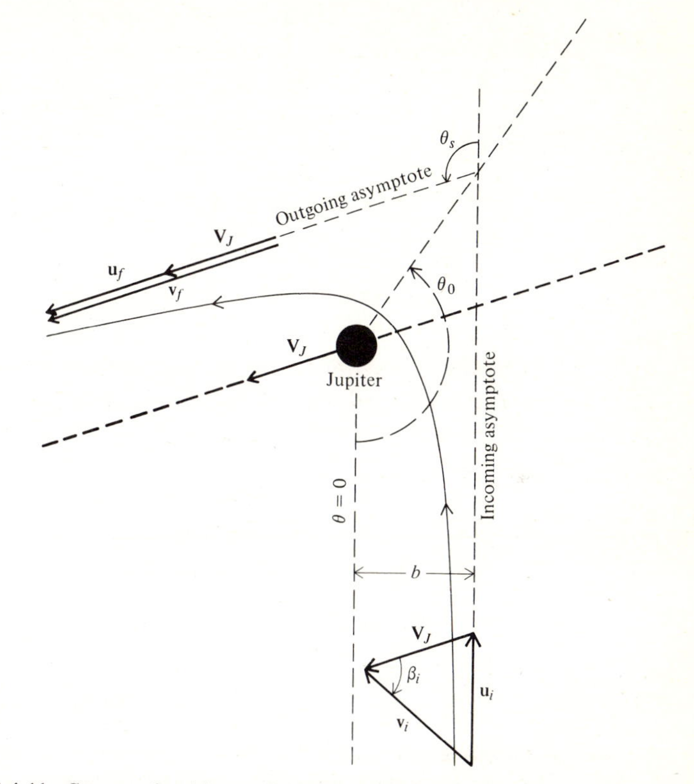

FIGURE 4-11 Spacecraft orbit near Jupiter in a frame moving with Jupiter.

The velocity v_i of the spacecraft as it nears Jupiter can be estimated from Eqs. (4-83) and (4-89), with $r = a_J = 5.2$ AU.

$$v_i = 42\sqrt{\frac{1}{5.2} - \frac{1}{2(10.1)}} = 16 \text{ km/s} \tag{4-99}$$

The intersection angle β_i of the spacecraft's orbit with Jupiter's orbit is determined by

$$\cos\beta_i = \frac{\mathbf{v}_i \cdot \mathbf{V}_J}{v_i V_J} = \frac{(v_i)_\theta}{v_i} \tag{4-100}$$

where $(v_i)_\theta$ is the projection of v_i along the planetary orbit. The component $(v_i)_\theta$ can be found from angular-momentum conservation on the elliptical spacecraft orbit, by equating the angular momentum at the Jupiter encounter with the value at perihelion.

$$L = m(v_i)_\theta a_J = mv_p a_e \tag{4-101}$$

This gives

$$(v_i)_\theta = v_p \left(\frac{a_e}{a_J}\right) = 41 \left(\frac{1}{5.2}\right) = 8 \text{ km/s} \tag{4-102}$$

so

$$\cos \beta_i = \tfrac{8}{16} = \tfrac{1}{2} \tag{4-103}$$

With the numerical values in Eqs. (4-98), (4-99), and (4-103), the asymptotic hyperbolic velocity u in Eq. (4-97) is

$$u = 14.7 \text{ km/s} \tag{4-104}$$

The hyperbolic orbit of the spacecraft relative to Jupiter can be fully specified by the initial energy

$$E' = \tfrac{1}{2}mu^2 \tag{4-105}$$

and the initial angular momentum

$$L' = mub \tag{4-106}$$

The energy is already determined by the asymptotic hyperbolic velocity u in Eq. (4-104). The impact parameter b depends on the location of Jupiter as the spacecraft approaches, which in turn depends on the timing of the launch. We want to arrange the encounter so that the energy boost received by the spacecraft in passing Jupiter is a maximum. By Eqs. (4-93) the energy shift of the spacecraft in the heliocentric system as a result of the Jovian swingby is

$$\Delta E = m\mathbf{V}_J \cdot (\mathbf{u}_f - \mathbf{u}_i) = m\mathbf{V}_J \cdot (\mathbf{v}_f - \mathbf{v}_i) \tag{4-107}$$

Thus the spacecraft receives an energy boost if $\Delta\mathbf{u} = \mathbf{u}_f - \mathbf{u}_i$ has a positive projection on $\mathbf{V}_J$. Since the dot product $\mathbf{V}_J \cdot \mathbf{v}_i$ is fixed by the initial elliptical orbit, we conclude from the geometry of Fig. 4-12 that the greatest energy boost is obtained for $\mathbf{v}_f$ parallel to $\mathbf{V}_J$. By Eqs. (4-96), $\mathbf{u}_f$ is then also parallel to $\mathbf{V}_J$, and we have

$$\mathbf{v}_f = (V_J + u)\hat{\mathbf{V}}_J \tag{4-108}$$

as illustrated in the velocity diagram of Fig. 4-13. From the results in Eqs. (4-98) and (4-104) we can now calculate the final velocity v_f to be

$$v_f = 13 + 14.7 = 27.7 \text{ km/s} \tag{4-109}$$

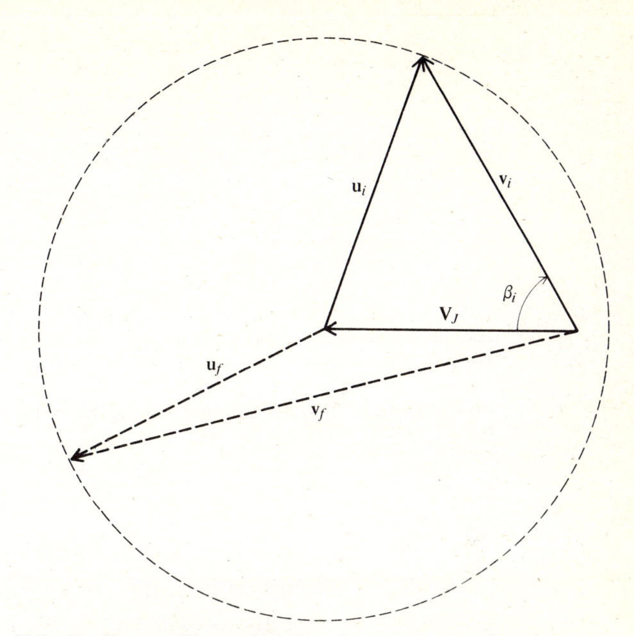

FIGURE 4-12 Velocity diagram illustrating that the spacecraft exit velocity $\mathbf{v}_f$ from the region of Jupiter's influence is maximum for $\mathbf{v}_f$ parallel to $\mathbf{V}_j$.

which is almost a factor of 2 higher than v_i ! The scattering angle θ_s between $\mathbf{u}_i$ and $\mathbf{u}_f$ can be found from the geometry of Fig. 4-13 as follows:

$$V_J = v_i \cos \beta + u_i \cos (\pi - \theta_s) \tag{4-110}$$

$$\cos \theta_s = \frac{v_i \cos \beta_i - V_J}{u}$$

$$= \frac{16(\frac{1}{2}) - 13}{14.7} = -0.34 \tag{4-111}$$

Thus, on this hyperbolic orbit around Jupiter, the spacecraft is deflected through an angle of

$$\theta_s = 109° \tag{4-112}$$

We must find the minimum r of this orbit, since it would be embarrassing if r_{min} were less than the radius of Jupiter. The hyperbolic-orbit equation for the relative motion of the spacecraft about Jupiter is

$$r = \frac{\lambda'(1 + \varepsilon')}{1 + \varepsilon' \cos (\theta - \theta_0)} \tag{4-113}$$

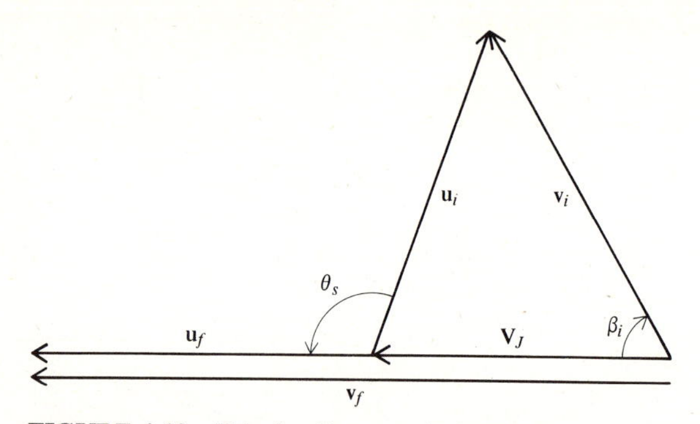

FIGURE 4-13 Velocity diagram relating the scattering parameters for maximum energy boost from Jupiter.

where r is the distance from Jupiter and θ_0 is the symmetry angle of the orbit for the spacecraft entry along $\theta = 0$, as illustrated in Fig. 4-11. To determine the eccentricity ε', we use the initial condition $r = \infty$ at $\theta = 0$ to find

$$\varepsilon' \cos \theta_0 = -1 \tag{4-114}$$

Further, by the geometry of Fig. 4-13, the symmetry angle θ_0 and scattering angle θ_s are related by

$$2\theta_0 - \theta_s = \pi \tag{4-115}$$

Thus ε' can be determined from Eqs. (4-112), (4-114), and (4-115).

$$\varepsilon' = -\frac{1}{\cos(\theta_s/2 + \pi/2)} = \frac{1}{\sin(\theta_s/2)} = 1.23 \tag{4-116}$$

To find λ' we can use Eqs. (4-44) and (4-47) to derive the following relation between λ' and ε':

$$\begin{aligned} (\varepsilon')^2 - 1 &= \frac{2E'L'^2}{m\alpha^2} = \frac{2E'}{\alpha}\lambda'(1+\varepsilon') \\ \lambda' &= \frac{\alpha}{2E'}(\varepsilon' - 1) = \frac{GM_J}{u^2}(\varepsilon' - 1) \end{aligned} \tag{4-117}$$

In terms of the escape velocity from Jupiter,

$$v_{\text{esc}}^J = \sqrt{\frac{2GM_J}{R_J}} = 60 \text{ km/s} \tag{4-118}$$

the result for λ' in Eqs. (4-117) can be written

$$\begin{aligned}\lambda' &= R_J \frac{1}{2}\left(\frac{v_{\text{esc}}^J}{u}\right)^2(\varepsilon' - 1) \\ &= R_J \frac{1}{2}\left(\frac{60}{14.7}\right)^2(1.23 - 1) \\ &= 1.9R_J\end{aligned} \tag{4-119}$$

Since λ' is just the distance of closest approach at $\theta = \theta_0$ on the orbit of Eq. (4-113), the spacecraft passes 0.9 of Jupiter radius above the surface. The orbit by Jupiter is given by

$$r = \frac{4.2R_J}{1 + 1.23 \cos(\theta - 144°)}$$

As the spacecraft leaves the region of Jupiter's influence, the exit velocity $\mathbf{v}_f$ is parallel to Jupiter's velocity. The outgoing orbit of the spacecraft around the sun is another conic section with a turning point at the location of the encounter with Jupiter (that is, $\dot{r} = 0$ at $r = a_J$ since $\mathbf{v}_f$ is parallel to the circular orbit of Jupiter). The type of new heliocentric conic section of the spacecraft orbit depends on the velocity boost as a result of the encounter. Since the escape velocity $V_{\text{esc}}^{\odot}$ from the solar system at Jupiter's orbit is

$$V_{\text{esc}}^{\odot} = \sqrt{2}\, V_J \tag{4-120}$$

the orbit of the spacecraft is related to the velocity v_f as follows:

$$\begin{aligned} v_f &< \sqrt{2}\, V_J \qquad \text{ellipse} \\ v_f &= \sqrt{2}\, V_J \qquad \text{parabola} \\ v_f &> \sqrt{2}\, V_J \qquad \text{hyperbola} \end{aligned}$$

For the encounter considered above,

$$\frac{v_f}{\sqrt{2}\, V_J} = 1.5 \tag{4-121}$$

from Eqs. (4-98) and (4-110), and the new orbit is therefore hyperbolic, as illustrated in Fig. 4-11.

The postencounter hyperbolic orbit about the sun can be cast in the standard form

$$r = \frac{\lambda''(1 + \varepsilon'')}{1 + \varepsilon'' \cos\theta} \tag{4-122}$$

where r is the distance of the spacecraft from the sun. The parameter λ'' is the distance of closest approach to the sun.

$$\lambda'' = a_J = 5.2 \text{ AU} \tag{4-123}$$

To determine the value of ε'' we make use of Eq. (4-47) to write

$$\varepsilon'' = \frac{|\mathbf{L}''|^2}{m\alpha\lambda''} - 1 = \frac{|\mathbf{L}''|^2}{m^2 G M_\odot a_J} - 1 \tag{4-124}$$

The angular momentum $|\mathbf{L}''|$ about the sun is determined by the exit from Jupiter as

$$|\mathbf{L}''| = m v_f a_J \tag{4-125}$$

Then ε'' is given by

$$\varepsilon'' = \frac{v_f^2}{GM_\odot / a_J} - 1 = \frac{v_f^2}{V_J^2} - 1 \tag{4-126}$$

The numerical value of ε'' using Eqs. (4-98), (4-110), and (4-126) is

$$\varepsilon'' = 3.5 \tag{4-127}$$

The orbit from Jupiter to Uranus is thereby completely specified as

$$r = \frac{23}{1 + 3.5 \cos \theta} \tag{4-128}$$

This orbit is plotted in Fig. 4-11. The launch date from earth must be predetermined such that the outer planets are in the proper orientation for the swing by Jupiter and the fly by Uranus. The possibility of a Jupiter-Uranus mission repeats about every 14 years.

We are now in a position to estimate the time duration of the Uranus mission on the gravity-assistance trajectory past Jupiter. The time of travel on a trajectory along a conic section can be determined from the angular-momentum-conservation equation (4-40)

$$\Delta t \equiv \int_{t_1}^{t_2} dt = \frac{m}{|\mathbf{L}|} \int_{\theta_1}^{\theta_2} r^2 \, d\theta \tag{4-129}$$

From Eqs. (4-46) and (4-47) and $\alpha = GmM_\odot$, this integral can be written

$$\Delta t = \frac{[\lambda(1 + \varepsilon)]^{3/2}}{\sqrt{GM_\odot}} \int_{\theta_1}^{\theta_2} \frac{d\theta}{(1 + \varepsilon \cos \theta)^2} \tag{4-130}$$

In order to put the right-hand side in convenient units, we use Eqs. (4-80) and (4-83) to write

$$\sqrt{GM_\odot} = V_e \sqrt{a_e} = \frac{2\pi a_e^{3/2}}{\tau_e}$$

and

$$\Delta t = \left(\frac{\tau_e}{2\pi}\right)\left[\frac{\lambda(1+\varepsilon)}{a_e}\right]^{3/2} \int_{\theta_1}^{\theta_2} \frac{d\theta}{(1+\varepsilon\cos\theta)^2} \qquad (4\text{-}131)$$

For the elliptical orbit, $\theta_1 = 0$ at launch from the earth. The angle θ_2 is determined from the orbit equation (4-86) with $r = a_J$.

$$r = 5.2 = \frac{1.9}{1 + 0.9\cos\theta_2}$$

This yields

$$\theta_2 = 135°$$

Using integral tables, we evaluate the integral in Eq. (4-131) between these limits. With the ε and λ parameters from Eq. (4-85*b*), we find

$$\Delta t = 1.3 \text{ years} \qquad (4\text{-}132)$$

for the flight from earth to Jupiter. For the hyperbolic orbit from Jupiter to Uranus, $\theta_1 = 0$ at Jupiter. θ_2 is determined by setting $r = a_U = 19.2$ AU in the hyperbolic-orbit equation (4-128).

$$r = 19.2 = \frac{23}{1 + 3.5\cos\theta_2}$$

We find in this case

$$\theta_2 = 86.4°$$

When we evaluate Eq. (4-131) with these θ limits and use the λ'' and ε'' parameters from Eqs. (4-123) and (4-127), we obtain

$$\Delta t = 3.7 \text{ years} \qquad (4\text{-}133)$$

for the transit time from Jupiter to Uranus. The total time for the earth-to-Uranus mission on this gravity-assistance trajectory is then about 5 years as compared with the 16 years required for a direct mission. Of course, our numbers are only approximate, since we have treated the gravitational influence of Jupiter and the sun on the spacecraft independently. Numerical methods can be used to make precise calculations of the orbit without such an approximation.

The late 1970s present unusual astronomical opportunities for planetary exploration by sending spacecraft on Grand Tours of the outer planets. Grand Tour missions under consideration by NASA include a Jupiter-Saturn-Pluto flyby with launch in 1977 and a Jupiter-Uranus-Neptune flyby with launch in 1979. The trajectories for the two missions are illustrated in Fig. 4-14. After the last planetary encounter, the spacecraft on these missions will continue to travel away from the sun, escaping the solar gravitational field and entering interstellar space.

4-7 RUTHERFORD SCATTERING

An important physical example in which hyperbolic orbits are realized is the Coulomb scattering of charged particles. For the scattering of a light particle with charge e_1 by a heavy particle of charge e_2, the location of the scattering center can be regarded as essentially fixed at $r = 0$. The potential energy of the interaction is then given by Eq. (4-39), with $\alpha = -e_1 e_2$. For e_1 and e_2 of the same (opposite) sign, the interaction is repulsive (attractive). The boundary conditions can be specified in various ways, as for example by the energy E and the angular momentum L, or by the initial velocity v_0 and impact parameter b.

$$\begin{aligned} E &= \tfrac{1}{2} m v_0^2 \\ L &= m v_0 b \end{aligned} \tag{4-134}$$

Here m is the mass of the light particle.

As in the BB scattering example of Sec. 3-6, we need to find the relation between the impact parameter and the laboratory scattering angle in order to calculate the number of particles scattered into a given angular range. From Eq. (4-55) the asymptotes ($r \to \infty$) of the trajectory occur at angles $\theta_{\text{asy}}^{\pm} = \pm \arccos(-1/\varepsilon)$. From the geometry of Fig. 4-15, the angle of scattering θ_s of the particle from the incident direction is

$$\theta_s = 2 \operatorname{arc\,cos} \left(-\frac{1}{\varepsilon} \right) - \pi \tag{4-135}$$

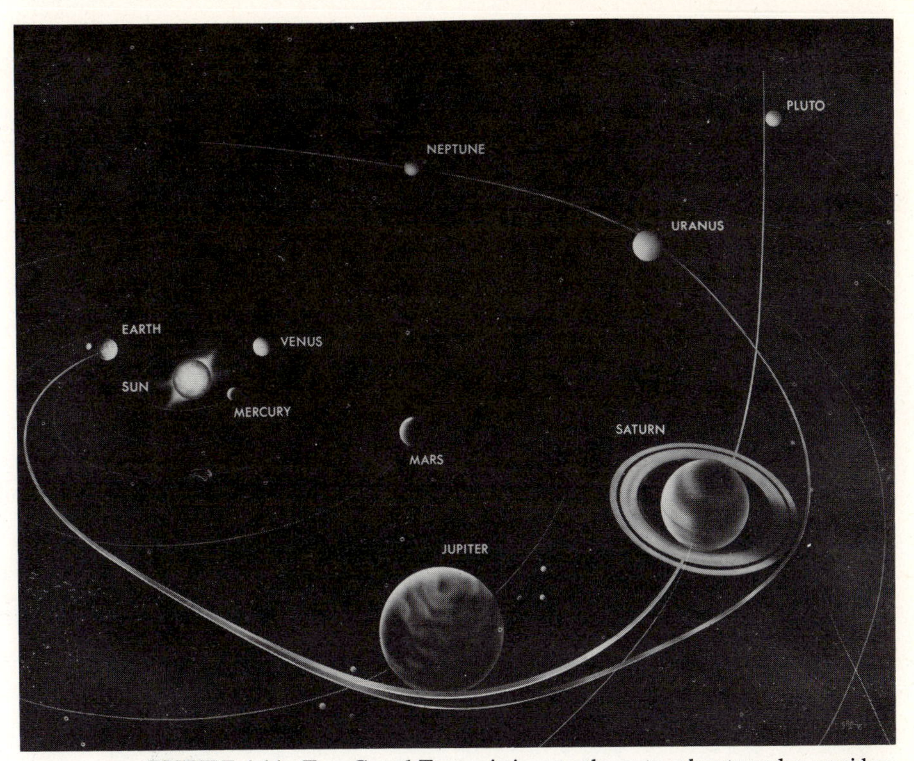

FIGURE 4-14 Two Grand Tour missions to the outer planets under consideration by NASA for the late 1970s. *(Photo provided by courtesy of the Jet Propulsion Laboratory, California Institute of Technology.)*

for both attractive and repulsive potentials. The result can be written

$$\cos\left(\frac{\theta_s}{2} + \frac{\pi}{2}\right) = -\frac{1}{\varepsilon}$$

or

$$\sin\frac{\theta_s}{2} = \frac{1}{\varepsilon} \tag{4-136}$$

From Eqs. (4-44) and (4-134), we can express θ_s in terms of b and v_0.

$$\sin\frac{\theta_s}{2} = \frac{1}{\sqrt{1 + (mv_0{}^2 b/\alpha)^2}} \tag{4-137}$$

When we solve this equation for b, we get

$$b = \frac{|\alpha|}{mv_0{}^2}\cot\frac{\theta_s}{2} \tag{4-138}$$

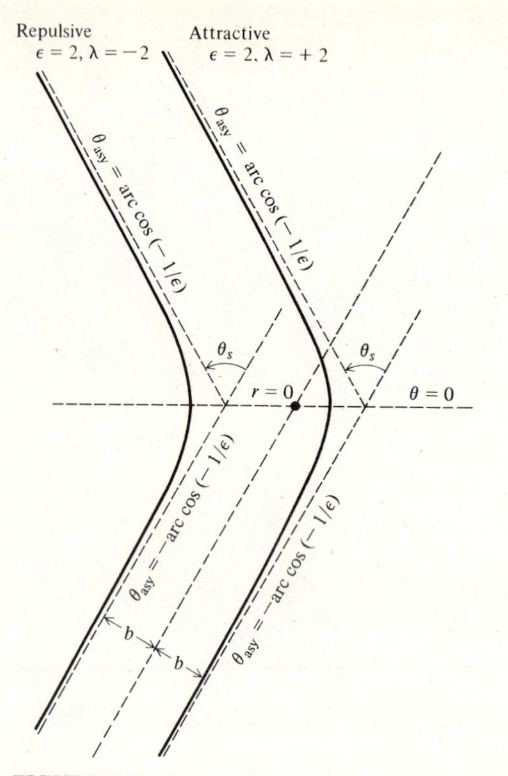

FIGURE 4-15 **Hyperbolic orbits for scattering from an attractive- or repulsive-force center at $r=0$.**

which relates the impact parameter and scattering angle of a single incident particle.

Under normal experimental conditions the incident beam consists of numerous particles with differing impact parameters. Since the scattering center is of atomic dimensions and is usually part of a solid, liquid, or gas, it is not possible to make measurements of the impact parameter for individual scatters. Under these circumstances the concept of a differential cross section is needed, as already discussed for BB scattering in Sec. 3-6.

For a single scattering center the number of particles per unit time $dN(\theta_s)$ that are scattered into the angular range θ_s to $\theta_s + d\theta_s$ is the number of incident particles per unit time that pass through a differential annulus with radii b and $b - db$, as illustrated in Fig. 4-16. We assume that the incident particles are uniformly distributed with

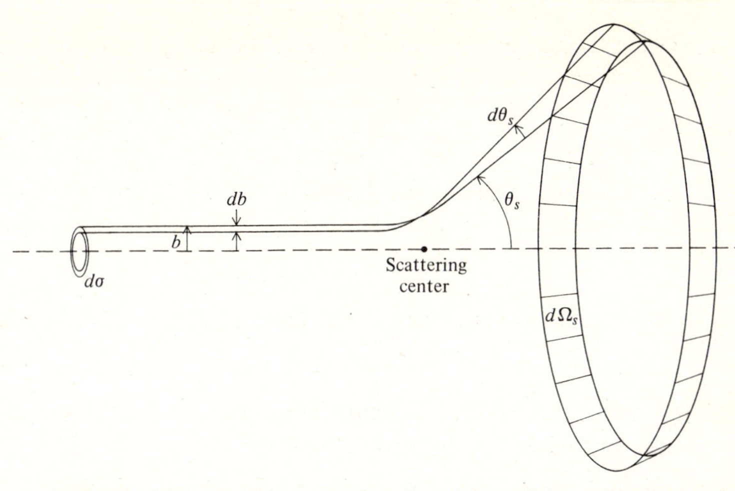

FIGURE 4-16 Repulsive scattering of particles with impact parameters between b and $b - db$.

a density N/A, where N is the number of incident particles which would strike a perpendicular area A. The number dN that pass through the annulus of area $d\sigma$ is

$$dN(\theta_s) = \frac{N}{A} d\sigma \tag{4-139}$$

where $d\sigma$ is the area of the annulus.

The quantity $d\sigma$ can be directly determined from experiment by Eq. (4-139).

$$d\sigma = \frac{dN}{N/A} \tag{4-140}$$

where dN is just the measured number of scatters into the solid angle Ω_s to $\Omega_s + d\Omega_s$. The incident flux

$$F = \frac{N}{A} \tag{4-141}$$

is the number of particles incident on a unit target area. The flux F is a property of the experimental conditions, and not of the force law between the particles. For this reason comparisons of theory and experiment are made for $d\sigma$ rather than dN.

In the scattering of nuclear particles, a thin-foil target has n nuclear-scattering centers per unit area. The effective scattering area is $nA\ d\sigma$, where A is the total area of the target. The number of scatters into θ_s to $\theta_s + d\theta_s$ is

$$dN(\theta_s) = N\frac{nA\,d\sigma}{A} \tag{4-142}$$

and the experimental value for $d\sigma$ is

$$d\sigma = \left(\frac{dN}{N}\right)\frac{1}{n} \tag{4-143}$$

The value of $d\sigma$ is given by the differential area of the annulus.

$$d\sigma = 2\pi b\,|db| \tag{4-144}$$

From Eq. (4-138) the differential of the cross section $d\sigma$ for scattering into the angular range θ_s to $\theta_s + d\theta_s$ is

$$d\sigma = \pi\left(\frac{\alpha}{mv_0{}^2}\right)^2 \frac{\cos(\theta_s/2)}{\sin^3(\theta_s/2)}\,d\theta_s \tag{4-145}$$

This result can be written in terms of the solid-angle element

$$d\Omega_s = 2\pi \sin\theta_s\,d\theta_s \tag{4-146}$$

as

$$\frac{d\sigma}{d\Omega_s} = \left(\frac{\alpha}{2mv_0{}^2}\right)^2 \frac{1}{\sin^4(\theta_s/2)} \tag{4-147}$$

This result was derived by Rutherford to explain the experimental results of Geiger and Marsden on the scattering of α particles by heavy nuclei. In the derivation of the Rutherford formula, the assumption was made that no incident particle interacted with more than one target nucleus, which is valid if the scattering angle is not too small.

In the repulsive scattering of α particles ($e_1 = 2e$) by atomic nuclei ($e_2 = Ze$), the distance of closest approach is found from Eq. (4-46) to be

$$r_{\min} = \lambda\left(\frac{1+\varepsilon}{1-\varepsilon}\right) = \frac{\lambda(1+\varepsilon)^2}{1-\varepsilon^2} \tag{4-148}$$

From Eqs. (4-44) and (4-47) we note that

$$1 - \varepsilon^2 = -\frac{2E|\mathbf{L}|^2}{m\alpha^2} = -\frac{2E\lambda(1+\varepsilon)}{\alpha}$$

so that

$$r_{\min} = \left(-\frac{\alpha}{2E}\right)(1+\varepsilon)$$

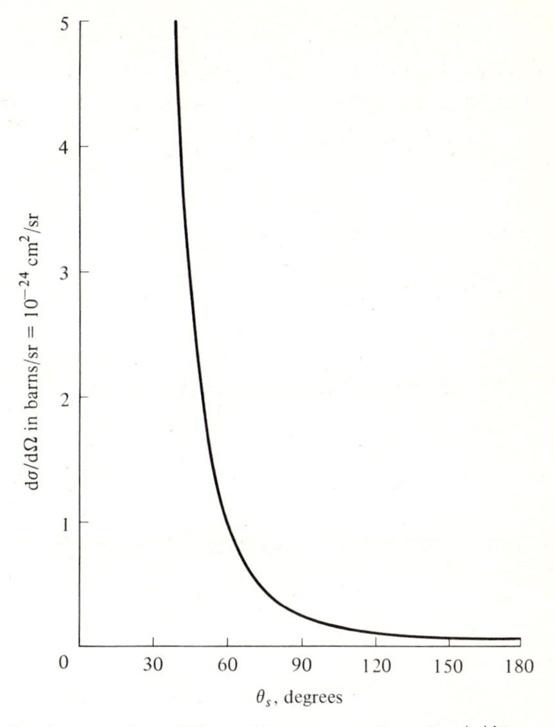

FIGURE 4-17 Rutherford scattering differential cross section for $|\alpha|/E = 10^{-12}$ cm.

The distance of closest approach can be expressed in terms of the scattering angle θ_s by use of Eq. (4-136).

$$r_{\min} = \frac{|\alpha|}{2E}\left(1 + \frac{1}{\sin(\theta_s/2)}\right) \tag{4-149}$$

The scatterer probes closest to the nucleus in the large-angle events. At $\theta_s = \pi$, the region down to $r_{\min} = |\alpha|/E$ is probed. By study of backward scattering events, Rutherford found that the Coulomb potential result in Eq. (4-147) held only for energies with $|\alpha|/E > 10^{-12}$ cm. This established the size of the typical atomic nucleus to be 10^{-12} cm, instead of 10^{-8} cm as was previously believed.

The integrated Rutherford scattering cross section

$$\sigma = \int \frac{d\sigma}{d\Omega_s}\, d\Omega_s = 2\pi\left(\frac{\alpha}{2mv_0^2}\right)^2 \int_0^\pi \frac{\sin\theta_s\, d\theta_s}{\sin^4(\theta_s/2)} \tag{4-150}$$

is infinite due to the divergence at $\theta_s = 0$. This is a consequence of the infinite range of the Coulomb force.

In scattering off a nucleus in the target, the momentum

$$\mathbf{q} = \mathbf{p}_f - \mathbf{p}_0 \tag{4-151}$$

is transferred to the α particle. The magnitude of $\mathbf{q}$ is related to the scattering angle by

$$q^2 = (\mathbf{p}_f - \mathbf{p}_0)^2 = p_f^2 + p_0^2 - 2\mathbf{p}_f \cdot \mathbf{p}_0 = p_f^2 + p_0^2 - 2p_f p_0 \cos\theta \tag{4-152}$$

For our idealization of an infinitely heavy nucleus ($M_N \to \infty$), the energy-conservation condition

$$\frac{p_0^2}{2m} = \frac{p_f^2}{2m} + \frac{q^2}{2M_N} \tag{4-153}$$

gives

$$p_f = p_0 = mv_0$$

The square of the momentum transfer in Eq. (4-152) then reduces to

$$q^2 = 2p_0^2(1 - \cos\theta_s) = 4p_0^2 \sin^2\frac{\theta_s}{2} \tag{4-154}$$

In terms of this variable the expression for the Rutherford differential cross section in Eq. (4-147) simplifies to

$$\frac{d\sigma}{d\Omega_s} = \left(\frac{2\alpha m}{q^2}\right)^2 \tag{4-155}$$

The calculation of Rutherford scattering in quantum mechanics accidentally gives the same result, though the physical principles are radically different.

PROBLEMS

1. A particle of mass m is subject to two forces, a central force $\mathbf{f}_1$ and a frictional force $\mathbf{f}_2$, with

$$\mathbf{f}_1 = F(r)\hat{\mathbf{r}}$$
$$\mathbf{f}_2 = -c\mathbf{v} \qquad c > 0$$

If the particle initially has angular momentum $\mathbf{L}_0$ about $\mathbf{r} = 0$, find the angular momentum for all subsequent times.

2. A spherical asteroid of uniform density 3 g/cm^3 and radius

100 m is rotating once per minute. It gradually acquires meteoritic material of the same density until a few billion years later its radius has doubled. If, on the average, this matter has arrived radially, what is the final rate of rotation?

3. Find the condition for stable circular orbits for a central potential energy of the form

$$V(r) = -\frac{c}{r^\lambda}$$

where $\lambda < 2$ and $\lambda \neq 0$. Show that the angular frequency for small radial oscillations ω_r is related to the orbit angular frequency ω_θ by

$$\omega_r = \sqrt{2 - \lambda}\, \omega_\theta$$

This result implies that the orbit is closed and the motion is periodic only if $\sqrt{2 - \lambda}$ is a rational number. Sketch the orbits for $\lambda = 1$ (Coulomb potential energy), $\lambda = -2$ (harmonic oscillator), and $\lambda = -7$.

4. A particle of mass m moves under the influence of the force

$$\mathbf{F} = -c^2 \frac{\mathbf{r}}{r^{5/2}}$$

a. Calculate the potential energy.
b. By means of the effective potential energy discuss the motion.
c. Find the radius of any circular orbit in terms of the angular momentum and calculate the period for the orbit.
d. Derive the frequency for small radial oscillations about the circular orbit of part *c*.

5. Find the force law for a central force which allows a particle to move in a spiral orbit given by $r = C\theta^2$, where C is a constant. [*Hint:* Use Eqs. (4-40) to find $V(r)$.]

6. A particle moves in a circular orbit in a force given by

$$\mathbf{F}(\mathbf{r}) = -\frac{\alpha}{r^2}\hat{\mathbf{r}}$$

If α suddenly decreases to half its original value, show that the particle's orbit becomes parabolic.

7. *a.* Calculate the orbital speed and period of revolution of the moon about the earth-moon center of mass. The earth-moon distance is approximately 384,000 km.
   b. Compare the orbital velocity of a satellite in a circular orbit 200 km above the surface of the earth, with the orbital velocity in a circular orbit at a similar distance from the surface of the moon. The ratio of lunar to earth mass is $M_L/M_e \approx 1/81.6$. The radii are $R_L = 1{,}741$ km and $R_e = 6{,}371$ km.

8. An astronaut can high-jump 2 m on earth. Is he in danger of not returning if he jumps while exploring Diemos, one of the moons of Mars? Can he by his own exertions put himself into any orbit? Assume Diemos has roughly the earth's density and has a diameter of 7 km.

9. Consider the motion of a particle in the central force.

$$\mathbf{F} = -k\mathbf{r}$$

Show that
   a. The orbit is an ellipse with the force center at the center of the ellipse.
   b. The radius vector sweeps out equal areas in equal time.
   c. The period is independent of the orbit parameters.

10. Give a simple derivation of Kepler's third law for a circular orbit.

11. An orbiting space station always remains vertically above an observer on the earth's surface. Where on the earth must the observer be located? What direction does the spaceship move? What is the radius of the orbit?

12. For a short rocket blast show that the most efficient way to change the energy of an orbit is to fire the rocket parallel (opposite) to the motion at perigee if the energy is to be increased (decreased).

13. Show that the close-circular-orbit period of a pebble around a boulder is roughly the same as the period of a low-altitude earth satellite (i.e., show that the close-orbit period depends only on the density of the large body).

14. Sputnik I had a perigee (point of closest approach to the earth) of 227 km above the earth's surface. At perigee its speed was 8 km/s.

Find its apogee (maximum distance from the earth's surface) and its period of revolution.

15. A rocket launching an earth satellite burns out at an altitude of 175 km. What must be the velocity and direction of motion at burnout to achieve a circular orbit? Where will the perigee and apogee of the orbit be if at burnout (*a*) the direction is correct for a circular orbit but the velocity is too high or (*b*) the speed is correct but the angle of elevation is too high by a small amount? Compute the required accuracy in velocity or angle in parts *a* and *b* needed to achieve equality within 10 percent of the apogee and perigee distances.

16. The most efficient way to transfer a spacecraft from a circular orbit with radius r_1 and period τ_1 to a new circular orbit with radius r_2 and period τ_2 is to insert the spacecraft in an intermediate elliptical orbit with extreme radii r_1 and r_2 from a focus. Find the transfer time on this inscribed ellipse and calculate the total velocity increment required for the transfer.

17. For a spacecraft launched from the surface of the earth, find the minimum velocity needed for escape from the solar system. Take into account the gravitational attraction of the earth. Compare your answer with the minimum velocity required for a trip to the moon.

18. A parallel beam of small projectiles is fired from space toward the moon with initial velocity v_0. What is the collision cross section σ for the projectiles to hit the moon? Express σ in terms of the moon's radius R_L, the escape velocity from the moon v^L_{esc}, and v_0. Neglect the motion of the moon.

19. The interaction between an atom and an ion at distances greater than contact is given by the potential energy $V(r) = -C/r^4$. [$C = (e^2/2)P_a{}^2$, where e is the ion charge and P_a is the polarizability of the atom.] Make a sketch of the effective potential energy vs. the radial coordinate. Note that if the total energy of the ion exceeds the maximum value of the effective potential energy, the ion spirals inward to the atom. Find the cross section for an ion of velocity v_0 to strike an atom. Assume that the ion is much lighter than the atom.

CHAPTER 5

Particle Systems and Rigid Bodies

For the treatment of rigid-body motion or the motion of a system of particles, it is advantageous to use the concept of center of mass (often abbreviated CM). The center of mass is that point in a rigid body or system of particles at which the total mass can be concentrated for purposes of calculating the translational motion.

5-1 CENTER OF MASS AND THE TWO-BODY PROBLEM

We begin with a derivation of the CM equation of motion for a two-particle system of masses m_1 and m_2 at positions $\mathbf{r}_1$ and $\mathbf{r}_2$. The equations of motion for the particles are

$$\mathbf{F}_1^{\text{ext}} + \mathbf{F}_1^{\text{int}} = m_1\ddot{\mathbf{r}}_1 \tag{5-1}$$

$$\mathbf{F}_2^{\text{ext}} + \mathbf{F}_2^{\text{int}} = m_2\ddot{\mathbf{r}}_2 \tag{5-2}$$

where $\mathbf{F}_i^{\text{ext}}$ represents the external force on particle i and $\mathbf{F}_i^{\text{int}}$ denotes the internal force on particle i exerted by the other particle in the system. If the internal forces satisfy Newton's third law,

$$\mathbf{F}_1^{\text{int}} = -\mathbf{F}_2^{\text{int}} \tag{5-3}$$

we obtain by addition of Eqs. (5-1) and (5-2) that

$$\mathbf{F}_1^{\text{ext}} + \mathbf{F}_2^{\text{ext}} = m_1\ddot{\mathbf{r}}_1 + m_2\ddot{\mathbf{r}}_2 \tag{5-4}$$

The total external force acting on the two-particle system is

$$\mathbf{F}^{\text{ext}} = \mathbf{F}_1^{\text{ext}} + \mathbf{F}_1^{\text{ext}} \tag{5-5}$$

If we introduce the center-of-mass (CM) vector,

$$\mathbf{R} \equiv \frac{m_1\mathbf{r}_1 + m_2\mathbf{r}_2}{m_1 + m_2} \tag{5-6}$$

then Eq. (5-4) can be written in the form

$$\mathbf{F}^{\text{ext}} = M\ddot{\mathbf{R}} \tag{5-7}$$

where $M = m_1 + m_2$ is the total mass. Equation (5-7) is the equation of motion of the CM; it has exactly the form of the equation of motion for a particle of mass M and position $\mathbf{R}$, under the influence of an external force $\mathbf{F}^{\text{ext}}$. If the total external force is zero, the CM point moves with constant velocity.

For a system composed of any number of particles, we can carry through the preceding argument unchanged. We sum over the equations of motion for all the particles,

$$\sum_i (\mathbf{F}_i^{\text{ext}} + \mathbf{F}_i^{\text{int}}) = \sum_i m_i\ddot{\mathbf{r}}_i$$

and observe that the sum $\sum_i \mathbf{F}_i^{\text{int}}$ over the internal forces vanishes by the third law. We obtain Eq. (5-7) again,

$$\mathbf{F}^{\text{ext}} = M\ddot{\mathbf{R}}$$

with

$$\mathbf{F}^{\text{ext}} = \sum_i \mathbf{F}_i^{\text{ext}}$$

$$M = \sum_i m_i \tag{5-8}$$

$$\mathbf{R} = \frac{\sum_i m_i \mathbf{r}_i}{\sum_i m_i}$$

Consequently, the second law of motion holds, not just for a particle, but for an arbitrary body, if the position of the body is interpreted to mean the position of its center of mass. In continuous systems we replace the mass elements m_i by $dm = \rho(r)\, dV$, where $\rho(r)$ is the density and dV is the differential volume element. Then the CM vector is given by

$$\mathbf{R} = \frac{\int \mathbf{r}\rho(\mathbf{r})\, dV}{\int \rho(\mathbf{r})\, dV} \tag{5-9}$$

We now return to the two-particle system and treat the part of the motion of the system which is not described by Eq. (5-7). In the most important physical applications, the force of one particle on the other depends only on the relative-position coordinate,

$$\mathbf{r} \equiv \mathbf{r}_1 - \mathbf{r}_2 \tag{5-10}$$

of the two particles, as illustrated in Fig. 5-1. Hence we want to form a combination of Eqs. (5-1) and (5-2), which is an equation of motion for $\mathbf{r}$. This is achieved by dividing Eq. (5-1) by m_1 and Eq. (5-2) by m_2 and then subtracting the two equations. Using Eq. (5-3), we find

$$\ddot{\mathbf{r}} = \ddot{\mathbf{r}}_1 - \ddot{\mathbf{r}}_2 = \left(\frac{1}{m_1} + \frac{1}{m_2}\right)\mathbf{F}_1^{\text{int}} + \left(\frac{\mathbf{F}_1^{\text{ext}}}{m_1} - \frac{\mathbf{F}_2^{\text{ext}}}{m_2}\right)$$

This equation can be rewritten in the form

$$\mu\ddot{\mathbf{r}} = \mathbf{F}_1^{\text{int}} + \mu\left(\frac{\mathbf{F}_1^{\text{ext}}}{m_1} - \frac{\mathbf{F}_2^{\text{ext}}}{m_2}\right) \tag{5-11}$$

where μ is the "reduced mass."

$$\frac{1}{\mu} = \frac{1}{m_1} + \frac{1}{m_2}$$

$$\mu = \frac{m_1 m_2}{m_1 + m_2} \tag{5-12}$$

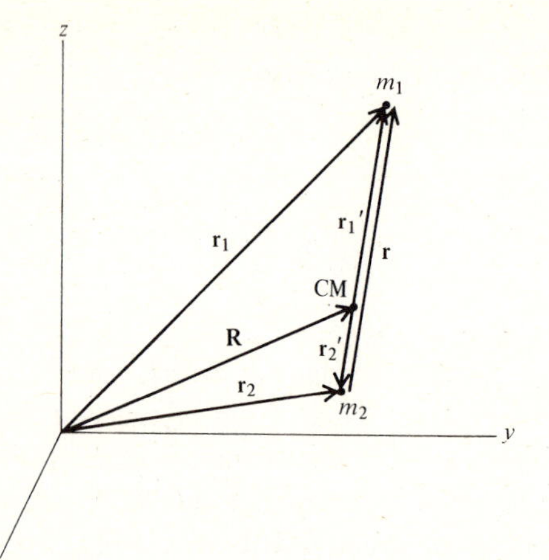

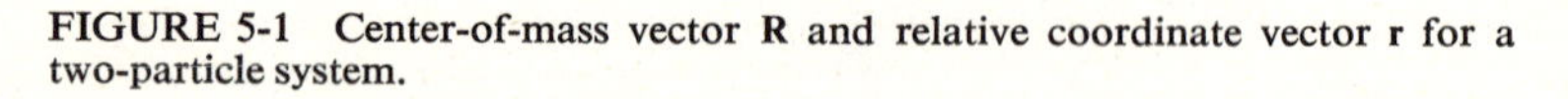

FIGURE 5-1 Center-of-mass vector **R** and relative coordinate vector **r** for a two-particle system.

Due to the presence of the external-force term on the right-hand side of Eq. (5-11), the motion of **r** is not independent of **R** in general, but that term vanishes in two important cases:

1. There is no external force on either particle

$$\mathbf{F}_1^{\text{ext}} = \mathbf{F}_2^{\text{ext}} = 0$$

2. The external forces are gravitational and due to distant sources, so that

$$\frac{\mathbf{F}_1^{\text{ext}}}{m_1} \approx \frac{\mathbf{F}_2^{\text{ext}}}{m_2}$$

If $\mathbf{F}^{\text{int}}$ depends only on $\mathbf{r} = \mathbf{r}_1 - \mathbf{r}_2$, the equation of motion for the relative coordinate **r** simplifies in these special cases to

$$\mu\ddot{\mathbf{r}} = \mathbf{F}_1^{\text{int}}(\mathbf{r}) \tag{5-13}$$

which has the same form as that for a single particle of mass μ in the influence of a force center at $\mathbf{r} = 0$. For a central force of interaction,

$$\mathbf{F}^{\text{int}} = F(r)\hat{\mathbf{r}}$$

we can therefore directly apply the methods of Chap. 4 to the solution of the two-body problem. For example, the exact form of Kepler's law for the period of revolution of two bodies of masses m_1, m_2 going around one another is immediately found from Eq. (4-72) by replacing m by $\mu = m_1 m_2/(m_1 + m_2)$, giving

$$\tau^2 = \left(\frac{4\pi^2\mu}{\alpha}\right)a^3 = \frac{4\pi^2}{G(m_1 + m_2)}\,a^3 \tag{5-14}$$

The total momentum of the two-body system,

$$\mathbf{P} = m_1\dot{\mathbf{r}}_1 + m_2\dot{\mathbf{r}}_2 \tag{5-15}$$

is just the momentum of the center of mass,

$$\mathbf{P} = M\dot{\mathbf{R}} \tag{5-16}$$

The kinetic energy and angular momentum of the system can be separated into parts associated with the center-of-mass motion and the relative motion by use of Eqs. (5-6) and (5-10).

$$\begin{aligned} T &= \tfrac{1}{2}m\dot{\mathbf{r}}_1{}^2 + \tfrac{1}{2}m\dot{\mathbf{r}}_2{}^2 \\ &= \tfrac{1}{2}M\dot{\mathbf{R}}^2 + \tfrac{1}{2}\mu\dot{\mathbf{r}}^2 \end{aligned} \tag{5-17}$$

$$\begin{aligned} \mathbf{L} &= m_1(\mathbf{r}_1 \times \dot{\mathbf{r}}_1) + m_2(\mathbf{r}_2 \times \dot{\mathbf{r}}_2) \\ &= M(\mathbf{R} \times \dot{\mathbf{R}}) + \mu(\mathbf{r} \times \dot{\mathbf{r}}) \end{aligned} \tag{5-18}$$

Although in the many-particle case it is always possible to separate out the CM motion, as in Eqs. (5-8), the remaining coordinates cannot in general be further separated. A complex system of coupled equations usually remains which frequently is not soluble by analytic techniques, so that one must make approximations or resort to a numerical treatment. Only in the two-body problem is the relative motion simple after the CM motion is separated out.

In the center-of-mass coordinate system the particles are measured from the CM point.

$$\begin{aligned} \mathbf{r}'_1 &= \mathbf{r}_1 - \mathbf{R} \\ \mathbf{r}'_2 &= \mathbf{r}_2 - \mathbf{R} \end{aligned} \tag{5-19}$$

From Eq. (5-6) we see that $\mathbf{r}'_1 \times \mathbf{r}'_2 = 0$, and hence the vectors $\mathbf{r}'_1$ and $\mathbf{r}'_2$ are collinear, as indicated in Fig. 5-1. The total momentum in the CM system is

$$\begin{aligned} \mathbf{P}' &= m_1\dot{\mathbf{r}}'_1 + m_2\dot{\mathbf{r}}'_2 = m_1\dot{\mathbf{r}}_1 + m_2\dot{\mathbf{r}}_2 - (m_1 + m_2)\dot{\mathbf{R}} \\ &= 0 \end{aligned} \tag{5-20}$$

in agreement with Eq. (3-40). In the CM system the scattering angle (angle between initial and final directions of particle 1) is given by

$$\cos\theta' = \hat{\mathbf{r}}'_{1i} \cdot \hat{\mathbf{r}}'_{1f} \tag{5-21}$$

From Eqs. (5-6) and (5-19), we can show that $\mathbf{r}_1'$ always points in the same direction as the relative coordinate $\mathbf{r} = \mathbf{r}_1 - \mathbf{r}_2$.

$$\mathbf{r}_1' = \mathbf{r}_1 - \mathbf{R} = \mathbf{r}_1 - \frac{m_1\mathbf{r}_1 + m_2\mathbf{r}_2}{m_1 + m_2}$$

$$= \frac{m_2}{m_1 + m_2}\mathbf{r} \tag{5-22}$$

Thus the CM scattering angle is the same as the scattering angle of the relative motion. Consequently, the CM scattering angle can be directly deduced from the equivalent particle scattering angle found in the solution to Eq. (5-13). The scattering angle in the laboratory system can in turn be determined from its relation to the CM angle given in Eq. (3-48).

We can apply these techniques for solving the two-body problem to Rutherford scattering on a target particle of finite mass. The projectile mass m in Eqs. (4-134) gets replaced by the reduced mass μ of the equivalent two-body problem. The scattering angle in the laboratory system in Eq. (4-135) is replaced by the scattering angle in relative coordinates θ' (which is also the CM scattering angle). The initial relative velocity v_0 is the same as the initial lab velocity. With these changes in Eq. (4-147) the CM differential cross section is given by

$$\frac{d\sigma}{d\Omega'} = \left(\frac{\alpha}{2\mu v_0^2}\right)^2 \frac{1}{\sin^4(\theta'/2)} \tag{5-23}$$

We can use Eq. (3-48) to write $d\sigma$ in terms of the laboratory angle. For equal masses for projectile and target, the conversion is particularly simple.

$$\theta' = 2\theta$$
$$d\Omega' = 4\cos\theta\, d\Omega \tag{5-24}$$

The resulting differential cross-section expression in the lab system is

$$\frac{d\sigma}{d\Omega} = \left(\frac{2\alpha}{mv_0^2}\right)^2 \frac{\cos\theta}{\sin^4\theta} \tag{5-25}$$

5-2 ROTATIONAL EQUATION OF MOTION

We now return to the general case of a system composed of many particles and derive an equation to describe the rotational motion analogous to Eq. (4-5) for a single particle. In analogy to the total momentum

$$\mathbf{P} = \sum \mathbf{p}_i = \sum_i m_i \dot{\mathbf{r}}_i \tag{5-26}$$

the total angular momentum about a point p (with coordinate $\mathbf{r}_p$) is the sum of the angular momenta about p of the particles in the system.

$$\mathbf{L} = \sum_i (\mathbf{r}_i - \mathbf{r}_p) \times m_i \dot{\mathbf{r}}_i \tag{5-27}$$

From this, we compute the time derivative of $\mathbf{L}$ to be

$$\dot{\mathbf{L}} = \sum_i (\mathbf{r}_i - \mathbf{r}_p) \times m_i \ddot{\mathbf{r}}_i - \dot{\mathbf{r}}_p \times \sum_i m \dot{\mathbf{r}}_i \tag{5-28}$$

We have allowed for the possibility that $\mathbf{r}_p$ is not a fixed point. If we now use the equations of motion

$$m_i \ddot{\mathbf{r}}_i = \mathbf{F}_i^{\text{ext}} + \mathbf{F}_i^{\text{int}}$$

in the first term of Eq. (5-28), we get

$$\dot{\mathbf{L}} = \mathbf{N}^{\text{ext}} + \mathbf{N}^{\text{int}} - \dot{\mathbf{r}}_p \times \mathbf{P} \tag{5-29}$$

where

$$\begin{aligned} \mathbf{N}^{\text{ext}} &= \sum_i (\mathbf{r}_i - \mathbf{r}_p) \times \mathbf{F}_i^{\text{ext}} \\ \mathbf{N}^{\text{int}} &= \sum_i (\mathbf{r}_i - \mathbf{r}_p) \times \mathbf{F}_i^{\text{int}} \end{aligned} \tag{5-30}$$

are the total external and internal torques, respectively, about p. The part $\mathbf{N}^{\text{int}}$ vanishes if the "extended third law" holds; namely, action equals reaction *and* is directed along a line between the particles. To show that $\mathbf{N}^{\text{int}} = 0$ under these circumstances, we consider the contribution to $\mathbf{N}^{\text{int}}$ from two particles a and b.

$$\mathbf{N}^{[a,\,b]} = (\mathbf{r}_a - \mathbf{r}_p) \times \mathbf{F}_a^{[b]} + (\mathbf{r}_b - \mathbf{r}_p) \times \mathbf{F}_b^{[a]} \tag{5-31}$$

Here $\mathbf{F}_a^{[b]}$ denotes the force on particle a due to b. According to the "extended third law,"

$$\mathbf{F}_b^{[a]} = -\mathbf{F}_a^{[b]} \tag{5-32}$$

and $\mathbf{F}_a^{[b]}$ is parallel to $(\mathbf{r}_a - r_b)$. As a consequence we find

$$\mathbf{N}^{[a,\,b]} = 2(\mathbf{r}_a - \mathbf{r}_b) \times \mathbf{F}_a^{[b]} = 0$$

Since the contribution to $\mathbf{N}^{\text{int}}$ from every pair of particles vanishes, $\mathbf{N}^{\text{int}}$ vanishes. Equation (5-29) simplifies to

$$\dot{\mathbf{L}} = \mathbf{N}^{\text{ext}} - \dot{\mathbf{r}}_p \times \mathbf{P} \tag{5-33}$$

The last term in Eq. (5-33) vanishes if $\dot{\mathbf{r}}_p = 0$ (i.e., if $\mathbf{r}_p$ is a fixed point) or if $\mathbf{r}_p = \mathbf{R}$ (because $\dot{\mathbf{R}} \times \mathbf{P} = M\dot{\mathbf{R}} \times \dot{\mathbf{R}} = 0$). In either of these cases the rotational equation of motion is

$$\frac{d\mathbf{L}}{dt} = \mathbf{N}^{\text{ext}} \qquad \dot{\mathbf{r}}_p = 0 \quad \text{or} \quad \mathbf{r}_p = \mathbf{R} \tag{5-34}$$

The equation $\dot{\mathbf{L}} = \mathbf{N}^{\text{ext}}$ bears a close resemblance to $\dot{\mathbf{P}} = \mathbf{F}^{\text{ext}}$. For an isolated system both $\mathbf{P}$ and $\mathbf{L}$ are constants, since $\mathbf{F}^{\text{ext}} = 0$ and $\mathbf{N}^{\text{ext}} = 0$. Even though the two equilibrium conditions $\mathbf{F}_{\text{ext}} = 0$ and $\mathbf{N}^{\text{ext}} = 0$ appear similar, they exhibit some interesting differences for systems in which internal motion is possible. The vanishing of the net force assures that the center of mass once fixed will remain so, no matter what the internal forces or internal motion. If the net external torque is zero, the total angular momentum is constant, and if initially zero, will remain zero. However, $\mathbf{L} = 0$ does not preclude changes in orientation of the system by exclusive use of internal forces. This can be demonstrated by a person sitting on a piano stool that is free to rotate, with a dumbbell in each hand. The dumbbells are initially held close to his body. When the dumbbells are extended radially outward at arm's length, the stool remains stationary. Then, when the dumbbells are moved by the person in a circular arc parallel to the floor, the person is rotated on the stool in the opposite direction to the motion of the dumbbells. The angular momentum from the rotation of the person cancels the angular momentum due to the dumbbell motion, so that the total angular momentum remains zero. When the dumbbells are then drawn radially back to the body, a net rotation has been achieved, with no change in the system or angular momentum. Repetition of the process enables the person to face in any arbitrary direction.

A dramatic illustration of the possibility of a change in orientation by exclusive use of internal forces is the ability of a cat to turn itself in midair and land upright on its feet, even when dropped vertically from an upside-down position. By contrast, the cat can do nothing whatever to alter its fall, that is, to change the motion of its CM.

For an isolated system the condition that

$$\mathbf{P} = M\dot{\mathbf{R}} = \text{constant}$$

leads to uniform motion of the CM.

$$\mathbf{R} = \mathbf{R}_0 + \frac{\mathbf{P}}{M}t$$

This can be stated as a conservation law:

$$M\mathbf{R} - \mathbf{P}t = \text{constant}$$

No corresponding result can be derived from $\mathbf{L} = \text{constant}$, in general, because there is no "rotational coordinate" analogous to $\mathbf{R}$.

5-3 RIGID BODIES: STATIC EQUILIBRIUM

Applications of mechanics to many particle systems commonly deal with the motion of rigid bodies. A rigid body is a system of particles whose distances from one another are fixed. The position of every particle in a rigid body is determined by the position of any one point of the body (such as the CM) plus the orientation of the body about that point. A total of six coordinates is needed to specify the motion of a rigid body. The position of one particle in the body requires the specification of three coordinates. The position of a second particle can be specified by two angular coordinates, since it lies at a fixed distance from the first. The position of a third particle is determined by only one coordinate because its distances from the first and second particles are fixed. The positions of any other particles in the rigid body are completely fixed by their distances from the first three particles. Thus $3 + 2 + 1 = 6$ coordinates determine the positions of the particles in a rigid body. Consequently, the motion of a rigid body is controlled by only six equations of motion. The translational motion of the CM is determined by

$$\dot{\mathbf{P}} = \mathbf{F}^{\text{ext}} \tag{5-35}$$

and the rotational motion about the CM, or a fixed point is determined by

$$\dot{\mathbf{L}} = \mathbf{N}^{\text{ext}} \tag{5-36}$$

These six equations, which hold for any system of particles, completely control the motion of a rigid body.

The conditions under which a rigid body remains in equilibrium under the action of a set of forces are of great practical importance in the design of permanent structures. From Eqs. (5-35) and (5-36) the six conditions for complete equilibrium of a rigid body are

$$\mathbf{F}^{\text{ext}} = \sum_i \mathbf{F}_i^{\text{ext}} = 0 \tag{5-37}$$

$$\mathbf{N}_{\text{CM}}^{\text{ext}} = \sum_i (\mathbf{r}_i - \mathbf{R}) \times \mathbf{F}_i^{\text{ext}} = 0 \tag{5-38}$$

The net external force must vanish in order that the CM move with constant velocity. The torque about the CM point $\mathbf{R}$ must vanish in order that the angular momentum about the CM (and thereby the rotational motion) does not change. For static equilibrium, the CM must be initially at rest and the total angular momentum about the CM must initially be zero.

The torque about an arbitrary point p can be easily related to the torque about the CM. If the vector distance between $\mathbf{r}_p$ and $\mathbf{R}$ is

$$\mathbf{a} = \mathbf{r}_p - \mathbf{R} \tag{5-39}$$

then the torque about p is

$$\mathbf{N}_p^{\text{ext}} = \sum_i (\mathbf{r}_i - \mathbf{r}_p) \times \mathbf{F}_i^{\text{ext}} = \sum_i (\mathbf{r}_i - \mathbf{R}) \times \mathbf{F}_i^{\text{ext}} - \mathbf{a} \times \sum_i \mathbf{F}_i^{\text{ext}}$$

or

$$\mathbf{N}_p^{\text{ext}} = \mathbf{N}_{\text{CM}}^{\text{ext}} - \mathbf{a} \times \mathbf{F}^{\text{ext}} \tag{5-40}$$

For a body in complete equilibrium, $\mathbf{F}^{\text{ext}} = 0$ and $\mathbf{N}_{\text{CM}}^{\text{ext}} = 0$ from Eqs. (5-37) and (5-38). In this case we find from Eq. (5-40) that the external torque about an arbitrary point p also vanishes.

$$\mathbf{N}_p^{\text{ext}} = 0 \tag{5-41}$$

Often, however, equilibrium is desired only for a subset of the six independent directions of motion. The drag-strip racer of Sec. 1-3 is such an example. The external force in the direction of racer's motion was nonzero, but equilibrium was to be maintained in all other directions. For rotational equilibrium, the torque about the CM of the racer vanishes according to Eq. (5-38). However, by Eq. (5-40) the torque about another point p may not be zero.

A set of two antiparallel forces with equal magnitudes $\mathbf{F}$ and $-\mathbf{F}$, separated by a vector $\mathbf{c}$ as in Fig. 5-2, is called a *couple*. Since the net external force of a couple is zero, by Eq. (5-40) the torque produced is independent of the point of reference. The torque of the couple is

$$\mathbf{N}_{\text{couple}} = \mathbf{c} \times \mathbf{F} \tag{5-42}$$

It is sometimes convenient to represent a system of forces by a single total-force vector on a given point plus a couple with the equivalent torque of the original forces about that point.

As a simple application, we discuss the motion which results when a light bar magnet is rigidly attached near one end of a rectangular board floating in a pan of water. A couple acts on the magnet, tending to align it in a north-south direction. Since there is no net external force on the board, the CM of the board must remain stationary. The torque produced by the couple therefore acts to rotate the board about its CM point (not about the end of the board where the magnet is placed) until the magnet reaches a north-south alignment.

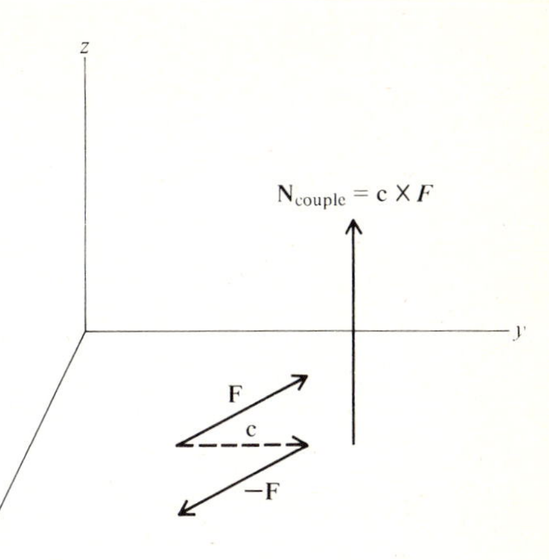

FIGURE 5-2 A vector 'couple.'

5-4 ROTATIONS OF RIGID BODIES

A *rotation* can be defined as the motion of a point p about a line such that the distance from p to each point on the line is constant. In an infinitesimal rotation, the rotational displacement $d\mathbf{r}$ must be perpendicular to a vector $(\mathbf{r} - \mathbf{r}_0)$ from a point O on the line to the point p. Furthermore, $d\mathbf{r}$ must be perpendicular to a unit vector $\hat{\mathbf{n}}$ directed along the line, as illustrated in Fig. 5-3. These two conditions can be mathematically expressed as

$$d\mathbf{r} \equiv \hat{\mathbf{n}} \times (\mathbf{r} - \mathbf{r}_0)\, d\phi \tag{5-43}$$

Since $|\hat{\mathbf{n}} \times (\mathbf{r} - \mathbf{r}_0)| = |\mathbf{r} - \mathbf{r}_0| \sin\theta$ is the perpendicular radius of rotation about the line, and $|d\mathbf{r}| = |\hat{\mathbf{n}} \times (\mathbf{r} - \mathbf{r}_0)|\, d\phi$, the quantity $d\phi$ is the infinitesimal angle of rotation about the axis $\hat{\mathbf{n}}$. The velocity of the point p relative to the point O due to rotation is

$$\mathbf{v} = \frac{d\mathbf{r}}{dt} = \hat{\mathbf{n}} \times (\mathbf{r} - \mathbf{r}_0)\frac{d\phi}{dt} \tag{5-44}$$

If we introduce an angular velocity vector

$$\boldsymbol{\omega} = \hat{\mathbf{n}}\omega \equiv \hat{\mathbf{n}}(t)\frac{d\phi}{dt} \tag{5-45}$$

for the rotation about $\hat{\mathbf{n}}$, the rotational velocity of p can be expressed as

$$\mathbf{v} = \boldsymbol{\omega} \times (\mathbf{r} - \mathbf{r}_0) \tag{5-46}$$

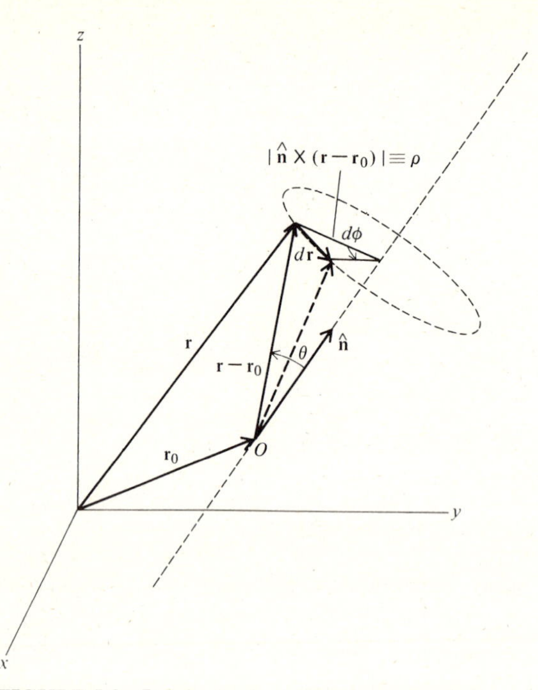

FIGURE 5-3 Infinitesimal rotation about an axis n̂.

If the point O is moving with a translational velocity v_0 relative to a fixed reference frame, the velocity of the point p in the fixed frame is the vector sum of the translational velocity $\mathbf{v}_0$ of the point O and the rotational velocity about O.

$$\mathbf{v}(\mathbf{r}) = v_0 + \boldsymbol{\omega} \times (\mathbf{r} - \mathbf{r}_0) \tag{5-47}$$

Over a period of time both the direction and magnitude of $\boldsymbol{\omega}$ may change as the body rotates.

In a rigid body the distance between any two points $\mathbf{r}_A$, $\mathbf{r}_B$ stays constant.

$$\frac{d}{dt} |\mathbf{r}_A - \mathbf{r}_B|^2 = 0 \tag{5-48}$$

Since

$$\frac{d}{dt} |\mathbf{r}_A - \mathbf{r}_B|^2 = 2(\mathbf{r}_A - \mathbf{r}_B) \cdot (\mathbf{v}_A - \mathbf{v}_B)$$

the rigid-body constraint in Eq. (5-48) is satisfied by the velocity field in Eq. (5-47), provided that $\boldsymbol{\omega}$ is the same for all points in the body.

$$(\mathbf{r}_A - \mathbf{r}_B) \cdot (\mathbf{v}_A - \mathbf{v}_B) = (\mathbf{r}_A - \mathbf{r}_B) \cdot \boldsymbol{\omega} \times (\mathbf{r}_A - \mathbf{r}_B) = 0$$

Hence $\mathbf{v}(\mathbf{r})$ as given by Eq. (5-47) is the velocity field of a rigid body. The velocities of all particles in a rigid body can be specified by six independent numbers, the components of $\mathbf{v}_0$ and $\boldsymbol{\omega}$.

As an example, we consider a wheel of radius R which rolls without slipping on a level surface, as illustrated in Fig. 5-4*a*. For no slipping we have

$$dx = R\,d\theta$$

where dx and $d\theta$ are infinitesimal horizontal and angular displacements, respectively. Dividing both sides by the time interval dt, we get

$$\mathbf{v}_0 = R\omega\hat{\mathbf{x}}$$

where $\mathbf{v}_0$ is the velocity of the center of mass. The velocity of a point on the wheel relative to the CM is

$$\mathbf{v}_{\text{rot}} = \boldsymbol{\omega} \times \mathbf{r} = v_0\hat{\boldsymbol{\omega}} \times \left(\frac{\mathbf{r}}{R}\right)$$

as illustrated in Fig. 5-4*b*. With respect to a reference frame at rest on the level surface, the velocity of the point on the wheel is given by

$$\mathbf{v} = \mathbf{v}_0 + \mathbf{v}_{\text{rot}} = \mathbf{v}_0 + \mathbf{v}_0\hat{\boldsymbol{\omega}} \times \left(\frac{\mathbf{r}}{R}\right)$$

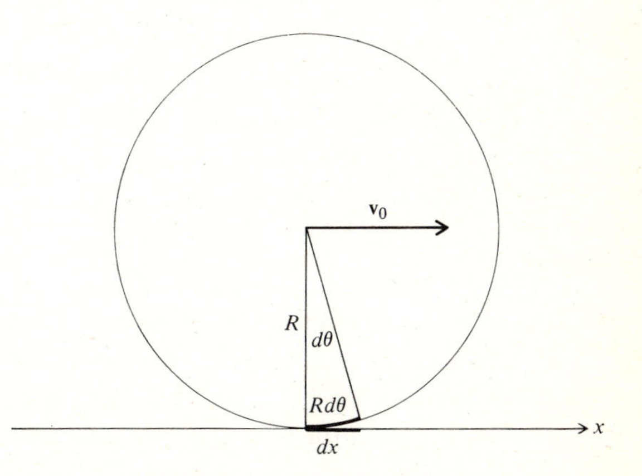

FIGURE 5-4*a* **Wheel rolling without slipping on a level surface.**

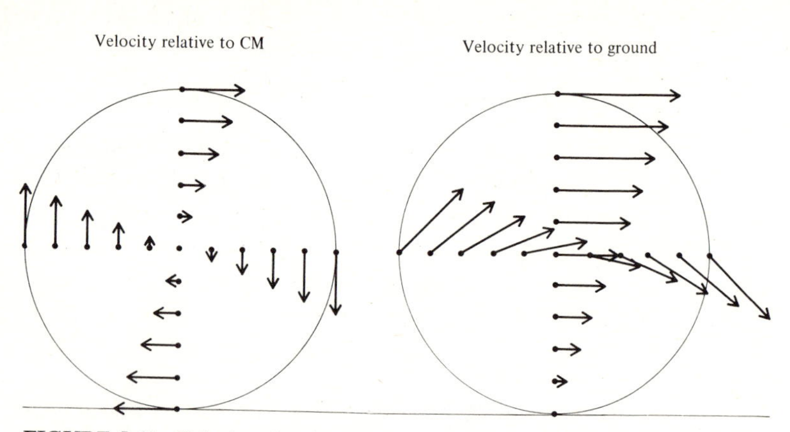

FIGURE 5-4b Velocity of points on a wheel relative to the center of mass and relative to the ground.

We note that $\mathbf{v} = 0$ at the point of contact. No slipping means that there is no relative motion of the wheel and surface at the contact point.

The velocity $\mathbf{v}_0$ in Eq. (5-47) is the time derivative of the coordinate $\mathbf{r}_0$. In general, the angular velocity $\boldsymbol{\omega}$ is not the time derivative of any coordinate ϕ. As a consequence, angular displacements are quite different in nature from translational displacements. Only in the special case of a fixed axis of rotation is it possible to express $\boldsymbol{\omega}$ as the time derivative of a coordinate. We can show that it is not possible to write $\boldsymbol{\omega} = \dot{\boldsymbol{\phi}}$ in general by the following simple example. If such a representation were possible, we could compute the coordinates ϕ_x, ϕ_y, ϕ_z describing the orientation of the rigid body by integrating over the components of $\boldsymbol{\omega}$. For example,

$$\phi_x(t) - \phi_x(0) = \int_0^t \omega_x(t')\,dt'$$

This would say that the change in orientation $\boldsymbol{\phi}(t) - \boldsymbol{\phi}(0)$ resulting from a motion $\boldsymbol{\omega}(t)$ depends only on the three numbers $\int_0^t \omega_i(t')\,dt'$. That this is not true can be seen from the following demonstration. Take a book and choose fixed axes $\hat{\mathbf{x}}$, $\hat{\mathbf{y}}$, $\hat{\mathbf{z}}$. First rotate the book by 90° around the $\hat{\mathbf{x}}$ axis and then by 90° around the $\hat{\mathbf{y}}$ axis. Then start again from the original orientation and make the same rotations in the opposite order. In the two cases the resulting orientations of the book are different, but the integral $\int \boldsymbol{\omega}\, dt$ is the same, namely,

$$\int \omega_x\, dt = \frac{\pi}{2} \qquad \int \omega_y\, dt = \frac{\pi}{2} \qquad \int \omega_z\, dt = 0$$

5-5 GYROSCOPE EFFECT

As an interesting application of the rotational equation of motion (5-34) we will discuss the gyroscope effect experienced by a wheel spinning in a vertical plane, as illustrated in Fig. 5-5. With a counterclockwise spin, the angular-momentum vector points along the positive x axis. When a torque which tends to turn the wheel in a counterclockwise sense about the positive y axis is applied, the wheel is observed to precess about the z axis. We can predict this precession from Eq. (5-34) and derive an expression for the precession frequency. According to Eq. (5-34), the change in angular momentum in an infinitesimal time interval dt is

$$d\mathbf{L} = \mathbf{N}\, dt \tag{5-49}$$

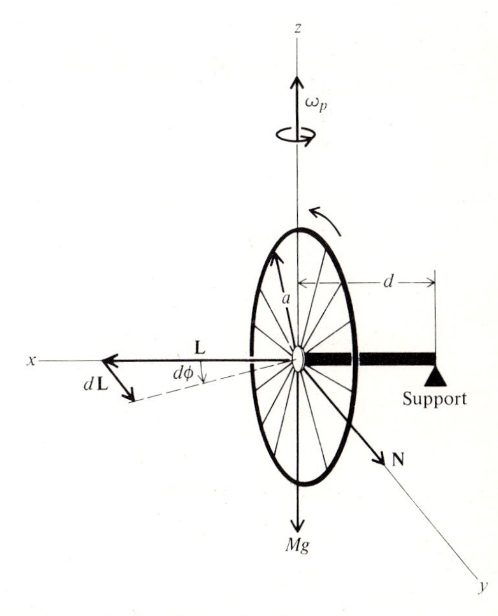

FIGURE 5-5 Gyroscope effect for a wheel with massive rim.

The increment $d\mathbf{L}$ is parallel to $\mathbf{N}$ and perpendicular to $\mathbf{L}$, as shown in Fig. 5-5. Since $\mathbf{L}$ and $\mathbf{N}$ are perpendicular, the length of $\mathbf{L}$ is unchanged to first order dt. The direction of $\mathbf{L}$ is rotated counterclockwise in the xy plane through an angle $d\phi$ given by

$$d\phi = \frac{dL}{L} = \frac{N}{L}\,dt \tag{5-50}$$

If $\mathbf{N}$ remains perpendicular to $\mathbf{L}$ and in the xy plane, the angular velocity of precession about the z axis is

$$\omega_P = \frac{d\phi}{dt} = \frac{N}{L} \tag{5-51}$$

This result is known as *simple precession*, inasmuch as we neglected the angular momentum associated with the precession motion. Whenever the applied torque is small or the spin large, simple precession is a good approximation. When the precession angular momentum is taken into account in the description of the motion, an oscillation called *nutation* about the xy plane may be present, in addition to the precessional motion.

A popular lecture demonstration experiment that illustrates simple precession uses a bicycle rim loaded with lead. The wheel is oriented in a vertical plane as in Fig. 5-5. The suspension point is located a distance d from the plane of the wheel along the wheel axis. The weight of the wheel then supplies the torque.

$$\mathbf{N} = Mgd\hat{\mathbf{y}} \tag{5-52}$$

If the wheel has radius a and mass M and spins with angular velocity ω, the angular momentum is

$$\mathbf{L} = M\omega a^2\hat{\mathbf{x}} \tag{5-53}$$

The resulting angular velocity of precession about the z axis is

$$\omega_P = \frac{gd}{\omega a^2} \tag{5-54}$$

where we have used Eqs. (5-51) to (5-53). For $a = d = 0.3$ m and a spin rate of $\omega/2\pi = 200$ r/min, we find a precession rate of

$$\frac{\omega_P}{2\pi} = \frac{9.8 \times 3{,}600}{200(2\pi)^2(0.3)} = 15 \text{ r/min} \tag{5-55}$$

5-6 THE BOOMERANG

An explanation of why a boomerang returns can be given in terms of the gyroscope effect. The boomerang can take on a variety of shapes.[1] In its most common form it appears as two airfoil-shaped blades meeting at an angle near 90°, as illustrated in Fig. 5-6. However, the characteristic bananalike shape of most boomerangs has little to do with their ability to return. Another version consists of two crossed blades, as shown in Fig. 5-7. The boomerang is thrown overhand in a vertical plane in the manner of Fig. 5-8. As it leaves the hand, the blades are rapidly rotating about the CM, and the CM is moving parallel to the ground. Due to its spin, the boomerang has an angular momentum about the CM that is initially directed to the left, as shown in Fig. 5-9.

The aerodynamic "lift" forces on the airfoils act perpendicular to the plane of rotation, as indicated in Fig. 5-10. The total aerodynamic force on the boomerang accelerates it perpendicular to the plane of rotation, in the direction of **L**. An upper blade of the boomerang moves more rapidly through the air than a lower blade because the rotation and translation velocities add on the upper blade and subtract on the lower blade. Since the aerodynamic force is larger for higher blade velocities, the upper blade experiences a greater force, and an external torque about the CM is generated by the forces on the airfoils. The torque points opposite the CM velocity direction. Thus the initial directions of **N** and **L** are identical with those for the wheel in Fig. 5-5. From the gyroscope effect discussed in Sec. 5-5 we

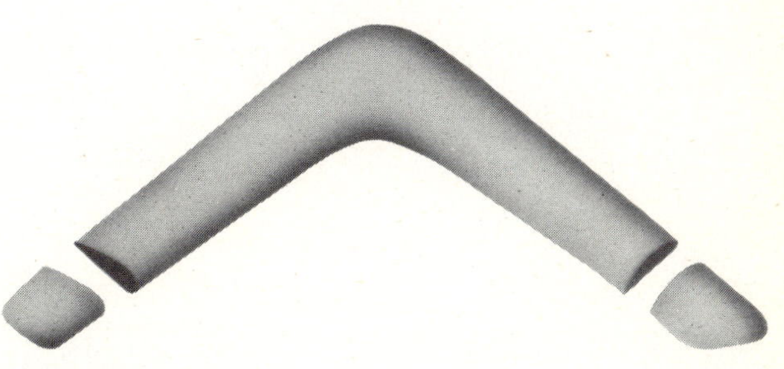

FIGURE 5-6 **Common boomerang.**

[1] See, for example, Felix Hess, *Sci. Amer.*, November 1969.

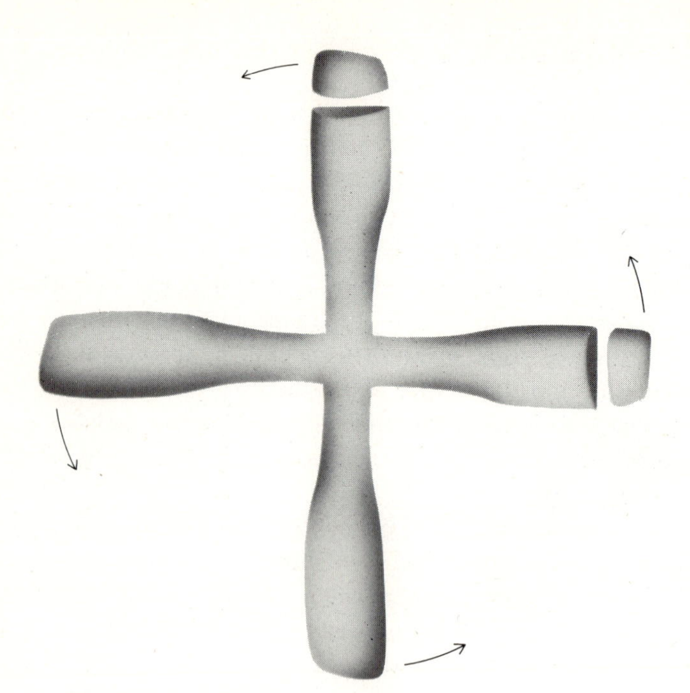

FIGURE 5-7 Cross-blade boomerang.

FIGURE 5-8 Proper method of throwing a boomerang.

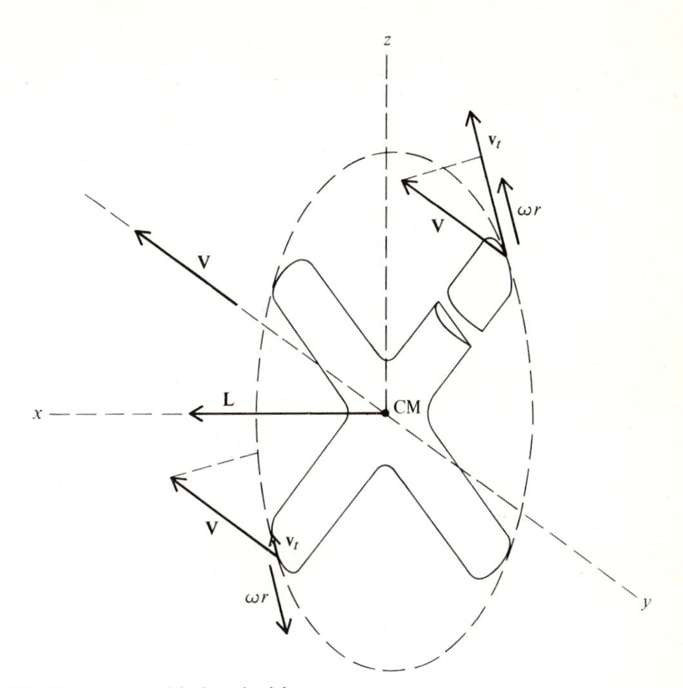

FIGURE 5-9 Boomerang-blade velocities.

predict that the plane of rotation precesses counterclockwise about the vertical axis. This precession of the rotational plane accompanied by translational acceleration perpendicular to the rotational plane allows the boomerang to travel in a circular orbit and return to the thrower.

To discuss the flight of the boomerang in a more quantitative fashion, we consider the crossed-blade boomerang of Fig. 5-7. In this case the CM lies at the blade hub, which simplifies the analysis considerably. We choose the origin of our coordinate system at the hub. Initially, we take the CM motion along the negative y axis with velocity $\mathbf{V} = -V\hat{\mathbf{j}}$. Rotation occurs around the x axis in a counterclockwise sense with angular velocity ω. One of the blades with length l, mass $M/4$, and linear mass density $\delta = \frac{1}{4}(M/l)$ is depicted in Fig. 5-11. A point on the blade at a distance r from the CM is specified as a function of time by

$$\mathbf{r} = r(\hat{\mathbf{j}} \sin \omega t + \hat{\mathbf{k}} \cos \omega t) \tag{5-56}$$

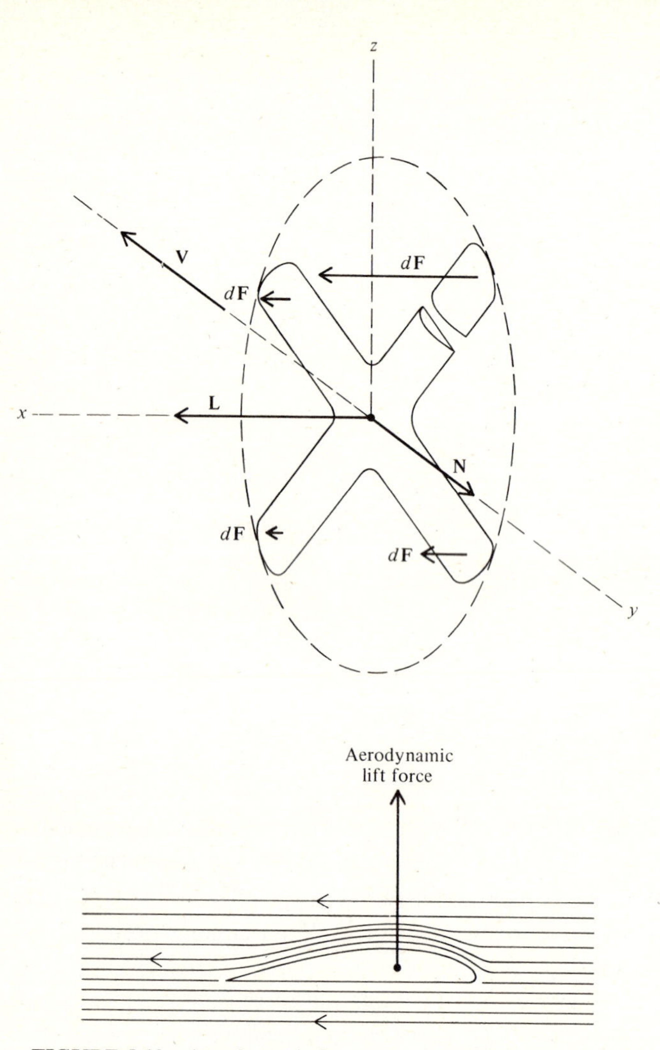

FIGURE 5-10 Aerodynamic forces on a cross-blade boomerang.

The aerodynamic force is dependent on the transverse component v_t of the air velocity over the airfoil in a direction perpendicular to the long edge of the blade. The force will be approximated by a quadratic dependence on v_t. The force on an element of blade at a distance r is

$$d\mathbf{F} = \hat{\mathbf{i}} c v_t^2 \, dr \tag{5-57}$$

The perpendicular air-velocity component v_t is due to the rotational motion of the blade and to the translation motion of the boomerang

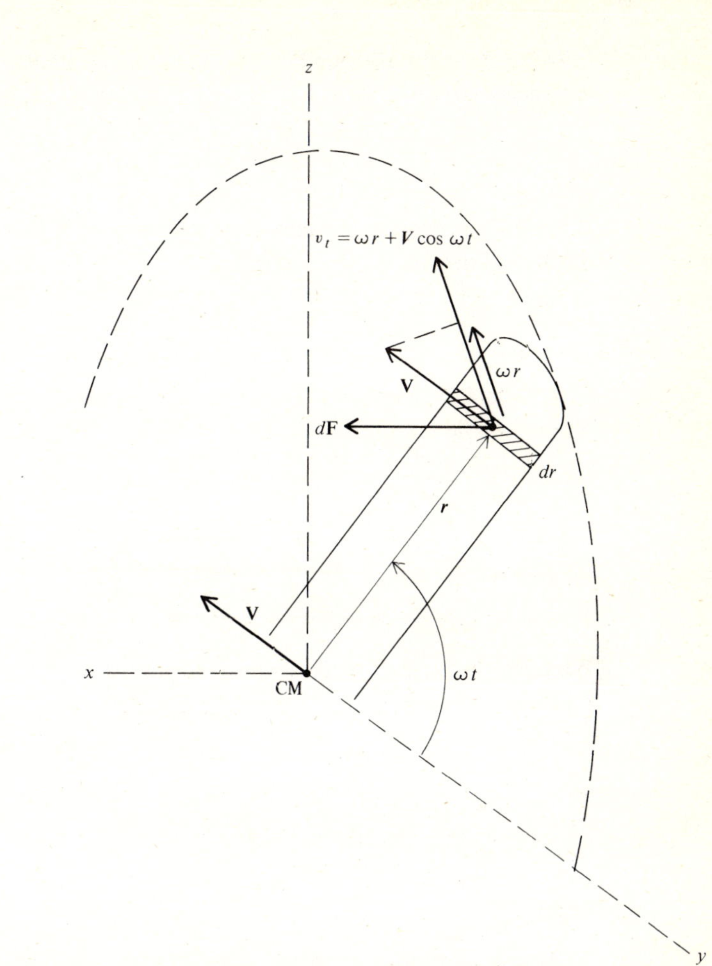

FIGURE 5-11 **Diagram of one blade in a four-blade boomerang.**

CM at velocity V. From Eq. (2-70) the $\hat{\mathbf{I}}$ component, which is v_t, is given by

$$v_t = \omega r + V \cos \omega t \tag{5-58}$$

Using Eqs. (5-57) and (5-58), we find

$$d\mathbf{F}(r, t) = \hat{\mathbf{i}}c(\omega^2 r^2 + 2\omega Vr \cos \omega t + V^2 \cos^2 \omega t)\, dr \tag{5-59}$$

As the blade rotates, $d\mathbf{F}$ varies in magnitude. The three remaining blades contribute forces similar to Eq. (5-59), but with ωt replaced by $\omega t + \pi/2$, $\omega t + \pi$, and $\omega t + 3\pi/2$, respectively. The net force on the boomerang from all four blades due to the elements of length dr at distance r is

$$d\mathbf{F}(r) = 4\hat{\mathbf{i}}c\left(\omega^2 r^2 + \frac{V^2}{2}\right) \tag{5-60}$$

Adding the elements by integration over r, we find the total force normal to the plane of rotation is

$$\mathbf{F} = 4\hat{\mathbf{i}}cl\left(\frac{\omega^2 l^2}{3} + \frac{V^2}{2}\right) \tag{5-61}$$

From Eqs. (5-56) and (5-57) the torque about the CM from an element on one of the blades is

$$\begin{aligned} d\mathbf{N} = \mathbf{r} \times d\mathbf{F} &= r(\hat{\mathbf{j}} \sin \omega t + \hat{\mathbf{k}} \cos \omega t) \times \hat{\mathbf{i}} c v_t^2 \, dr \\ &= c r v_t^2 (\hat{\mathbf{j}} \cos \omega t - \hat{\mathbf{k}} \sin \omega t) \, dr \end{aligned}$$

Expansion of this result, using Eq. (5-58), gives

$$d\mathbf{N} = cr(\omega^2 r^2 + 2\omega V r \cos \omega t + V^2(\cos \omega t)^2)(\hat{\mathbf{j}} \cos \omega t - \hat{\mathbf{k}} \sin \omega t) \, dr$$

We add to this the torques from the other three blades' elements to find the net torque

$$d\mathbf{N} = 4c\omega V r^2 \hat{\mathbf{j}} \, dr \tag{5-62}$$

The torque due to all elements is then obtained by integrating Eq. (5-62) over the length of a blade.

$$\mathbf{N} = \tfrac{4}{3} c\omega V l^3 \hat{\mathbf{j}} \tag{5-63}$$

The angular momentum $\mathbf{L}$ about the CM of the boomerang can be computed as follows. A blade element at distance r has mass $dm = \delta \, dr = \frac{1}{4}(M/l) \, dr$. As the blade rotates with angular velocity ω, the angular momentum of the element is $d\mathbf{L} = dm \, r^2 \boldsymbol{\omega}$, where $\boldsymbol{\omega} = \hat{\mathbf{i}}\omega$. The angular momentum of the whole blade is then

$$\mathbf{L} = \frac{M}{4l} \boldsymbol{\omega} \int_0^l r^2 \, dr = \tfrac{1}{12} M l^2 \boldsymbol{\omega} \tag{5-64}$$

The complete boomerang has angular momentum

$$\mathbf{L} = \tfrac{1}{3} M l^2 \boldsymbol{\omega} \tag{5-65}$$

The constant of proportionality between $\mathbf{L}$ and $\boldsymbol{\omega}$ is called the *moment of inertia*.

$$I = \tfrac{1}{3} M l^2$$

The torque in the $\hat{\mathbf{j}}$ direction induces a precession of the **L** vector as given by Eq. (5-51). The precession angular velocity is

$$\omega_P = \frac{N}{L} = \frac{\frac{4}{3}c\omega V l^3}{\frac{1}{3}Ml^2\omega} = \frac{4cVl}{M} \tag{5-66}$$

The motion of the CM is influenced by the aerodynamic force normal to the plane of rotation, gravity, and a drag force due to air resistance. For boomerangs with a small path circumference and quick return time, the last two effects are small in comparison with the aerodynamic force. Since a boomerang is supposed to return to the thrower, we investigate under what conditions the CM will travel in a circular orbit of radius R with angular velocity $\boldsymbol{\omega}_0$. In a circular orbit the aerodynamic force exactly balances the centrifugal force. In order for the aerodynamic force to be always radial inward toward the center of a circle, the orbital angular velocity $\boldsymbol{\omega}_0$ must match the precession rate $\boldsymbol{\omega}_P$.

$$\omega_0 = \frac{V}{R} = \omega_P \tag{5-67}$$

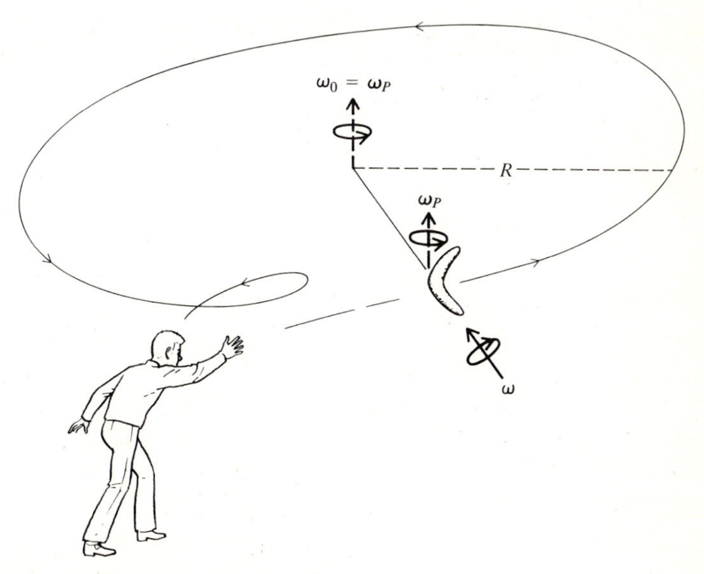

FIGURE 5-12 Typical boomerang orbit.

From Eqs. (5-66) and (5-67) we can determine the radius of the orbit as

$$R = \frac{M}{4lc} = \frac{\delta}{c} \tag{5-68}$$

where δ is the linear mass density and c is the lift constant determined by the airfoil shape and air properties. By equating the magnitude of the lift force in Eq. (5-61) to the centrifugal force MV^2/R,

$$F = 4cl\left(\frac{\omega^2 l^2}{3} + \frac{V^2}{2}\right) = \frac{MV^2}{R} \tag{5-69}$$

we obtain

$$V = \sqrt{\tfrac{2}{3}}\,\omega l \tag{5-70}$$

For a simple circular return flight the CM velocity V and spin ω must be related, as in Eq. (5-70).

From Eq. (5-68) we see that the boomerang has a flight radius which is independent of how hard it is thrown. Of course, if it is thrown very slowly, the effects of gravity will become important, and our theory breaks down. In fact, a boomerang is usually aimed upward at a slight angle, to compensate for the gravitational fall. If an indoor boomerang is desired, it should have an exaggerated airfoil shape to obtain a small orbit radius in Eq. (5-68). For long flights a boomerang made of dense material is needed, and of course the design should minimize drag. It is said that some native Australians can throw the boomerang 90 m and have it return to their feet. Such a record-setting boomerang would be useless to someone without a very strong arm since it could not be thrown with a smaller radius of orbit. A typical outdoor boomerang orbit may have a diameter of about 25 m. The boomerang starts its flight with a CM velocity of about 25 m/s and a rotation rate of about 10 r/s. It stays in the air for about 8 s.

5-7 MOMENTS AND PRODUCTS OF INERTIA

The dynamics of rigid-body rotations are contained in Eq. (5-34), which relates the time rate of change of the total angular momentum to the external torque. The angular momentum $\mathbf{L}$ about a point O can be computed in terms of the angular velocity $\boldsymbol{\omega}$ in Eq. (5-46). We denote the location relative to O of a point mass m_i in the rigid body by $\mathbf{r}_i$. Then from Eq. (5-46) the velocity of the mass m_i relative to O is

$$\mathbf{v}_i = \boldsymbol{\omega} \times \mathbf{r}_i \tag{5-71}$$

The angular momentum about O is[1]

$$\mathbf{L} = \sum_i m_i(\mathbf{r}_i \times \mathbf{v}_i) = \sum_i m_i \mathbf{r}_i \times (\boldsymbol{\omega} \times \mathbf{r}_i) \tag{5-72}$$

The summation is over all mass points in the body. Using Eq. (2-44*a*) to expand the triple cross product in Eq. (5-72), we obtain

$$\mathbf{L} = \boldsymbol{\omega}\left(\sum_i m_i |\mathbf{r}_i|^2\right) - \sum_i m_i \mathbf{r}_i(\mathbf{r}_i \cdot \boldsymbol{\omega}) \tag{5-73}$$

We observe that the angular-momentum vector **L** will not necessarily be parallel to the angular-velocity vector **ω**. We can write Eq. (5-73) in cartesian components as

$$\begin{aligned}
L_x &= \omega_x \sum_i m_i(y_i^2 + z_i^2) - \omega_y \sum_i m_i x_i y_i - \omega_z \sum_i m_i x_i z_i \\
L_y &= -\omega_x \sum_i m_i y_i x_i + \omega_y \sum_i m_i(x_i^2 + z_i^2) - \omega_z \sum_i m_i y_i z_i \\
L_z &= -\omega_x \sum_i m_i z_i x_i - \omega_y \sum_i m_i z_i y_i + \omega_z \sum_i m_i(x_i^2 + y_i^2)
\end{aligned} \tag{5-74}$$

For the coefficients of the angular-velocity components in Eqs. (5-74) we introduce the notation

$$\begin{aligned}
I_{xx} &= \sum_i m_i(y_i^2 + z_i^2) & I_{xy} &= -\sum_i m_i x_i y_i & I_{xz} &= -\sum_i m_i x_i z_i \\
I_{yx} &= -\sum_i m_i y_i x_i & I_{yy} &= \sum_i m_i(x_i^2 + z_i^2) & I_{yz} &= -\sum_i m_i y_i z_i \\
I_{zx} &= -\sum_i m_i z_i x_i & I_{zy} &= -\sum_i m_i z_i y_i & I_{zz} &= \sum_i m_i(x_i^2 + y_i^2)
\end{aligned} \tag{5-75}$$

In terms of these quantities we have

$$\begin{aligned}
L_x &= I_{xx}\omega_x + I_{xy}\omega_y + I_{xz}\omega_z \\
L_y &= I_{yx}\omega_x + I_{yy}\omega_y + I_{yz}\omega_z \\
L_z &= I_{zx}\omega_x + I_{zy}\omega_y + I_{zz}\omega_z
\end{aligned} \tag{5-76}$$

The components of angular momentum in Eqs. (5-76) are then compactly written

$$L_j = \sum_k I_{jk}\omega_k \tag{5-77}$$

[1] This definition for **L** is equivalent to Eq. (5-27) when O is a fixed point or the CM point.

where the subscripts j, k take on the values x, y, z. The three quantities I_{jj} are known as *moments of inertia*, and the six I_{jk} with $j \neq k$ are called *products of inertia*. From Eqs. (5-75) we note that the products of inertia are symmetric.

$$I_{jk} = I_{kj} \tag{5-78}$$

The nine quantities I_{jk} form a *symmetric tensor* and can be written as a 3 × 3 matrix.

$$\mathbf{I} \equiv \begin{pmatrix} I_{xx} & I_{xy} & I_{xz} \\ I_{yx} & I_{yy} & I_{yz} \\ I_{zx} & I_{zy} & I_{zz} \end{pmatrix} \tag{5-79}$$

In vector notation Eq. (5-77) can be written

$$\mathbf{L} = \mathbf{I} \cdot \boldsymbol{\omega}$$

From Eqs. (5-34) and (5-79) the equation of motion for general rotations of a rigid body about its CM point or about a fixed point in space is

$$N_j = \frac{dL_j}{dt} = \sum_k \frac{d}{dt}(I_{jk}\,\omega_k) \tag{5-80}$$

or more compactly, in vector notation,

$$\mathbf{N} = \frac{d\mathbf{L}}{dt} = \frac{d}{dt}(\mathbf{I} \cdot \boldsymbol{\omega})$$

The kinetic energy of a rigid body can likewise be expressed in terms of the moments and products of inertia. The kinetic energy is given by

$$T = \tfrac{1}{2}\sum_i m_i \mathbf{v}_i \cdot \mathbf{v}_i = \tfrac{1}{2}\sum_i m_i(\boldsymbol{\omega} \times \mathbf{r}_i) \cdot (\boldsymbol{\omega} \times \mathbf{r}_i) = \tfrac{1}{2}\boldsymbol{\omega} \cdot [\sum_i m_i \mathbf{r}_i \times (\boldsymbol{\omega} \times \mathbf{r}_i)] \tag{5-81}$$

where we have interchanged dot and cross products in the final step. Using Eqs. (5-72) and (5-77), we find

$$T = \tfrac{1}{2}\boldsymbol{\omega} \cdot \mathbf{L} = \tfrac{1}{2}\sum_{jk} I_{jk}\,\omega_j\,\omega_k \tag{5-82}$$

In vector notation T can be written

$$T = \tfrac{1}{2}\boldsymbol{\omega} \cdot \mathbf{I} \cdot \boldsymbol{\omega}$$

5-8 SINGLE-AXIS ROTATIONS

The equation of motion in Eq. (5-80) simplifies considerably for the case of rotation about a single fixed axis. For definiteness we choose the z axis as the axis of rotation, $\boldsymbol{\omega} = \omega_z \hat{\mathbf{z}}$. The components of the angular momentum in Eqs. (5-76) are then

$$\begin{aligned} L_x &= I_{xz}\,\omega_z \\ L_y &= I_{yz}\,\omega_z \\ L_z &= I_{zz}\,\omega_z \end{aligned} \tag{5-83}$$

The equations of motion from Eq. (5-80) are

$$\begin{aligned} N_x &= \dot{L}_x = \frac{d}{dt}(I_{xz}\,\omega_z) \\ N_y &= \dot{L}_y = \frac{d}{dt}(I_{yz}\,\omega_z) \\ N_z &= \dot{L}_z = \frac{d}{dt}(I_{zz}\,\omega_z) \end{aligned} \tag{5-84}$$

For a rigid body that is symmetrical about the z axis, we find from Eqs. (5-75) that

$$I_{xz} = I_{yz} = 0 \tag{5-85}$$

In this case the torques N_x and N_y in Eqs. (5-84) are zero. On the other hand, if one of the products of inertia, I_{xz} or I_{yz}, is nonzero, the body is unbalanced and the bearings must provide the torques N_x or N_y to keep the axis of rotation from moving.

The rotational motion about the z axis is accelerated by the external torque N_z. From Eqs. (5-75) we observe that the moment of inertia

$$I_{zz} = \sum_i m_i(x_i^2 + y_i^2) \tag{5-86}$$

is time-independent, since the perpendicular distance $\sqrt{x_i^2 + y_i^2}$ from the rotation axis of the mass point m_i is fixed in a rigid body. Thus the equation of motion for the z component in Eqs. (5-84) is

$$N_z = \dot{L}_z = I_{zz}\,\dot{\omega}_z \tag{5-87}$$

The kinetic energy for rigid-body rotation about the fixed z axis is found from Eq. (5-82) to be given by

$$T = \tfrac{1}{2} I_{zz}\,\omega_z^2 \tag{5-88}$$

The equation for rotational motion about a fixed axis in Eq. (5-87) has the same mathematical structure as the equation for linear motion in one direction.

$$F_z = M\dot{v}_z \tag{5-89}$$

In fact, the following direct correspondence can be made between the physical quantities of angular and linear motion:

Angular motion	Linear motion
Moment of inertia, I_{zz}	Mass, M
Angular acceleration, $\alpha_z = \dot{\omega}_z = \dfrac{d^2\phi}{dt^2}$	Linear acceleration, $a_z = \dot{v}_z = \dfrac{d^2z}{dt^2}$
Torque, N_z	Force, F_z
Angular velocity, $\omega_z = \dfrac{d\phi}{dt}$	Linear velocity, $v_z = \dfrac{dz}{dt}$
Angular position, ϕ	Linear position, z
Angular momentum, $L_z = I_{zz}\omega_z$	Linear momentum, $P_z = Mv_z$
Kinetic energy, $T = \frac{1}{2}I_{zz}\,\omega_z^{\,2}$	Kinetic energy, $T = \frac{1}{2}Mv_z^{\,2}$

As a consequence, we can directly apply the techniques for solving one-dimensional problems in Chaps. 1 and 2 to solve Eq. (5-87) for rotations about a single axis. For example, if the torque is conservative (function only of the angle ϕ), we can define a potential energy

$$V(\phi) = -\int_{\phi_s}^{\phi} N_z(\phi)\, d\phi \tag{5-90}$$

in correspondence with Eq. (2-5).

5-9 MOMENTS-OF-INERTIA CALCULATIONS

The moment of inertia I_0 of a rigid body about a given axis through the point O is related to the moment of inertia I_{CM} about a parallel axis which passes the center of mass by the parallel-axis rule

$$I_0 = I_{CM} + Md^2 \tag{5-91}$$

where d is the perpendicular distance between the two axes. In practice I_{CM} is often easier to compute than I_0, making it advantageous to use this parallel-axis rule. To prove Eq. (5-91), we take I_0 to be the moment of inertia about the z axis through the origin O. For continuous systems we have

$$I_0 = \int (x^2 + y^2)\rho\, dV \tag{5-92}$$

The coordinates of a point (x, y, z) in the body can be expressed in terms of coordinates (x', y', z') relative to the CM point as

$$x = x' + X$$
$$y = y' + Y \tag{5-93}$$
$$z = z' + Z$$

(X, Y, Z) is the CM location. When this relation is plugged into Eq. (5-92), we get

$$I_0 = \int (x'^2 + y'^2)\rho\, dV + (X^2 + Y^2)\int \rho\, dV + 2X \int x'\rho\, dV + 2Y \int y'\rho\, dV$$

The last two terms vanish by the defining equation (5-9) for the CM point. We identify the first two terms as

$$I_{\text{CM}} = \int (x'^2 + y'^2)\rho\, dV$$
$$Md^2 = (X^2 + Y^2)\int \rho\, dV \tag{5-94}$$

which completes the proof of Eq. (5-91).

Another useful rule for moments of inertia applies to bodies whose mass is distributed in a single plane. For a body in the xy plane with mass density per unit area σ, the moments of inertia about the three axes are

$$I_{xx} = \int y^2\sigma\, dA$$
$$I_{yy} = \int x^2\sigma\, dA \tag{5-95}$$
$$I_{zz} = \int (x^2 + y^2)\sigma\, dA$$

From these we derive the perpendicular-axis rule

$$I_{xx} + I_{yy} = I_{zz} \tag{5-96}$$

When the mass distribution is symmetrical about the z axis, we obtain

$$I_{xx} = I_{yy} = \tfrac{1}{2}I_{zz} \tag{5-97}$$

In the applications to be considered in the following two sections, we shall need the moment of inertia of a spherical body about an axis through its center of mass. For a sphere of radius a the mass density is

$$\rho = \frac{M}{\frac{4}{3}\pi a^3} \tag{5-98}$$

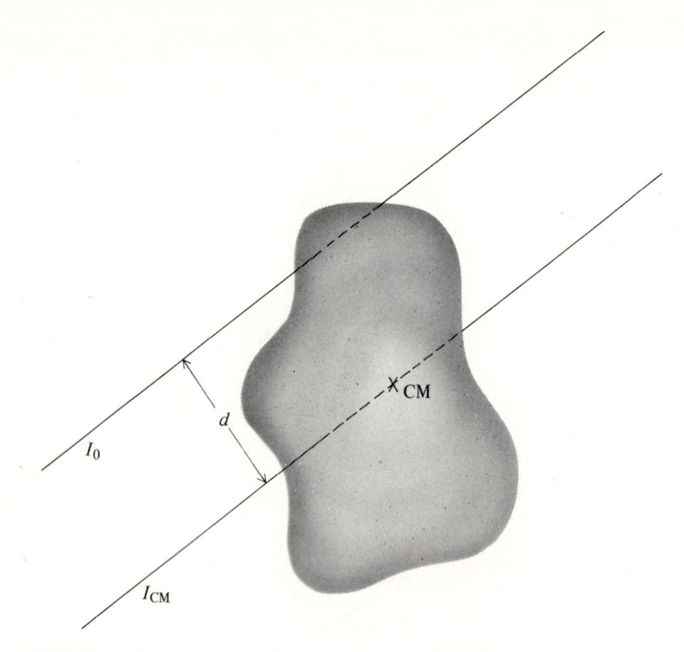

FIGURE 5-13 Parallel-axis rule for moments of inertia.

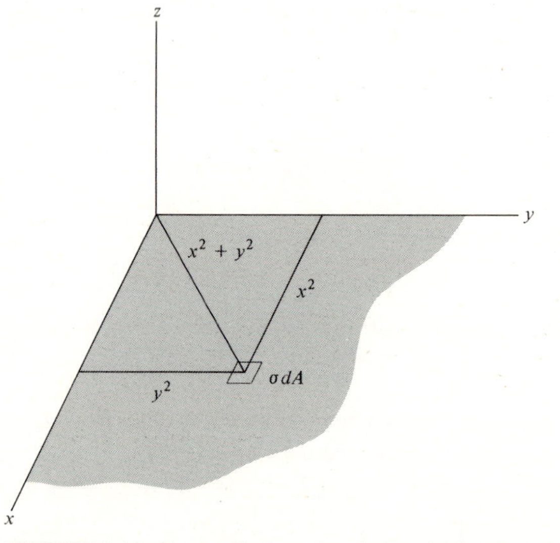

FIGURE 5-14 Perpendicular-axis rule for a body whose mass is distributed only in the xy plane.

where M is the total mass. The integral for the moment of inertia in Eqs. (5-94),

$$I_{\text{CM}} = \frac{M}{\frac{4}{3}\pi a^3} \int (x'^2 + y'^2)\, dV \tag{5-99}$$

can be carried out simply in cylindrical coordinates.

$$\begin{aligned} x' &= r' \cos \phi' \\ y' &= r' \sin \phi' \\ dV &= (r'\, d\phi')\, dr'\, dz' \end{aligned} \tag{5-100}$$

After the transformation to cylindrical variables, we find

$$\begin{aligned} I_{\text{CM}} &= \frac{M}{\frac{4}{3}\pi a^3} \int_{-a}^{+a} dz' \int_0^{\sqrt{a^2 - z'^2}} r'^3\, dr' \int_0^{2\pi} d\phi' = \frac{M}{\frac{4}{3}\pi a^3} \int_{-a}^{+a} dz' \frac{(a^2 - z'^2)^2}{4} (2\pi) \\ &= \tfrac{2}{5} M a^2 \end{aligned} \tag{5-101}$$

Some other frequently used moments of inertia of simple uniform bodies are tabulated in Table 5-1.

TABLE 5-1 MOMENTS OF INERTIA OF SOME SIMPLE BODIES

Body	Axis through CM	Moment of inertia
Rod, length l	Perpendicular to rod	$I_{\text{CM}} = \frac{1}{2} M l^2$
Rectangular plate, sides a, b	Parallel to side b	$I_{\text{CM}} = \frac{1}{12} M a^2$
	Perpendicular to plate	$I_{\text{CM}} = \frac{1}{12} M (a^2 + b^2)$
Cube, sides a	Perpendicular to face	$I_{\text{CM}} = \frac{1}{6} M a^2$
Hoop, radius a	Perpendicular to plane	$I_{\text{CM}} = M a^2$
Disk, radius a	Perpendicular to plane	$I_{\text{CM}} = \frac{1}{2} M a^2$
	Parallel to plane	$I_{\text{CM}} = \frac{1}{4} M a^2$
Solid cylinder:		
Radius a	Along cylinder axis	$I_{\text{CM}} = \frac{1}{2} M a^2$
Length l	Perpendicular to cylinder axis	$I_{\text{CM}} = \frac{1}{12} M (3a^2 + l^2)$
Spherical shell, radius a	Any axis	$I_{\text{CM}} = \frac{2}{3} M a^2$
Solid sphere, radius a	Any axis	$I_{\text{CM}} = \frac{2}{5} M a^2$
Solid ellipsoid of revolution:		
Semimajor axis a	Along axis a (symmetry axis)	$I_{\text{CM}} = \frac{2}{5} M b^2$
Semiminor axis b	Along axis b	$I_{\text{CM}} = \frac{1}{5} M (a^2 + b^2)$

5-10 IMPULSES AND BILLIARD SHOTS

For forces that act only during a very short time, it is convenient to use an integrated form of the laws of motion. The translational motion of the center-of-mass point is determined by

$$\dot{\mathbf{P}} = \mathbf{F}^{\text{ext}} \tag{5-102}$$

If we multiply both sides of this equation by dt and integrate over the short time interval $\Delta t = t_1 - t_0$, during which the force acts, we obtain

$$\Delta\mathbf{P} = \mathbf{P}^1 - \mathbf{P}^0 = \int_{t0}^{t1} \mathbf{F}^{\text{ext}}\, dt \tag{5-103}$$

The time integral of the force on the right is called the *impulse*. For angular motion the integrated form of the equation of motion in Eq. (5-34) is

$$\Delta\mathbf{L} = \mathbf{L}^1 - \mathbf{L}^0 = \int_{t0}^{t1} \mathbf{N}^{\text{ext}}\, dt \tag{5-104}$$

The time integral of the torque is called the *angular impulse*. For rigid-body rotations about a fixed z axis, we can use Eq. (5-87) to rewrite the angular-impulse equation (5-104) as

$$\Delta L_z = I_{zz}\, \Delta\omega_z = I_{zz}(\omega_z{}^1 - \omega_z{}^0) = \int_{t0}^{t1} N_z^{\text{ext}}\, dt \tag{5-105}$$

As an example of the usefulness of the impulse formulation of the equations of motion, we discuss the dynamics of billiard shots. For simplicity we consider only shots in which the cue hits the ball in its vertical median plane in a horizontal direction. In billiard jargon these are shots without "English."

The cue imparts an impulse to the stationary ball at a vertical distance h above the table, as illustrated in Fig. 5-15. The linear impulse from Eq. (5-103) is

$$M\,\Delta V_x = MV_x{}^1 = \int_{t0}^{t1} F_x\, dt \tag{5-106}$$

where $V_x{}^1$ is the velocity of the CM just after impact. The angular impulse of Eq. (5-105) about the z axis in Fig. 5-15 which passes through the CM of the ball is

$$\Delta L_z = I_{zz}\,\omega_z{}^1 = \int_{t0}^{t1} (h - a)F_x\, dt \tag{5-107}$$

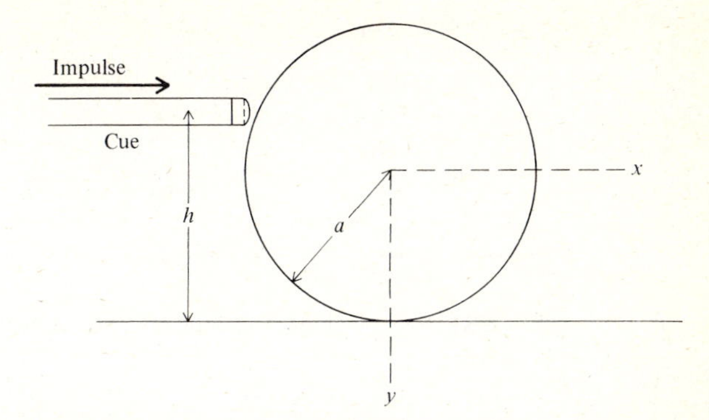

FIGURE 5-15 **Impulse imparted to a billiard ball by the cue stick.**

where a is the radius of the ball. By elimination of the force integral between Eqs. (5-106) and (5-107) and substitution of the moment of inertia from Eq. (5-101), we arrive at the following relation between the spin and velocity of the ball immediately after the impulse:

$$\omega_z{}^1 = \frac{5}{2}\left(\frac{h-a}{a^2}\right)V_x{}^1 \tag{5-108}$$

The velocity of the ball at the point of contact with the table is

$$V_c = V_x{}^1 - a\omega_z{}^1 = V_x{}^1\left(\frac{7a-5h}{2a}\right) \tag{5-109}$$

If the ball is to roll without slipping ($V_c = 0$), we find that

$$h = \tfrac{7}{5}a \tag{5-110}$$

Only if the ball is hit exactly at this height does pure rolling take place from the very start. For a "high shot" with $h > \frac{7}{5}a$, V_c is opposite in direction to $V_x{}^1$. Since the friction at the billiard cloth opposes V_c, it causes V_x to increase and $\omega_z{}^1$ to decrease until pure rolling sets in. For a "low shot" with $h < \frac{7}{5}a$, the contact velocity V_c is in the direction of $V_x{}^1$. In this case the friction force decreases $V_x{}^1$ and increases $\omega_z{}^1$ until rolling occurs. The diagram of Fig. 5-16 summarizes these results. For a ball which is rolling uniformly, the CM moves uniformly, and there is therefore no static frictional force in the direction of the motion.

When the moving cue ball makes a head-on collision with a target ball at rest, the CM of the cue ball momentarily stops, and the target

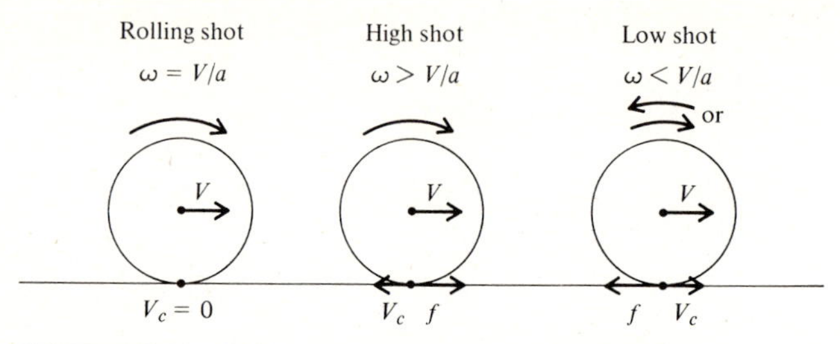

FIGURE 5-16 Rolling, high, and low shots in billiards.

ball moves forward with the CM velocity of the cue ball, as shown earlier, in Sec. 3-4. Since the balls are assumed smooth, the cue ball retains its spin $\omega_z{}^1$ in the collision. Consequently, the contacting point on the cue ball moves with velocity $V_c = -a\omega_z{}^1$ immediately after collision. If $\omega_z{}^1 > 0$ at the moment of collision, the friction force acting opposite to the direction of V_c accelerates the cue ball forward, as illustrated in Fig. 5-17. This is the so-called *follow shot*. If $\omega_z{}^1 < 0$, the friction force accelerates the cue ball backward until pure rolling motion sets in. This is the *draw shot*.

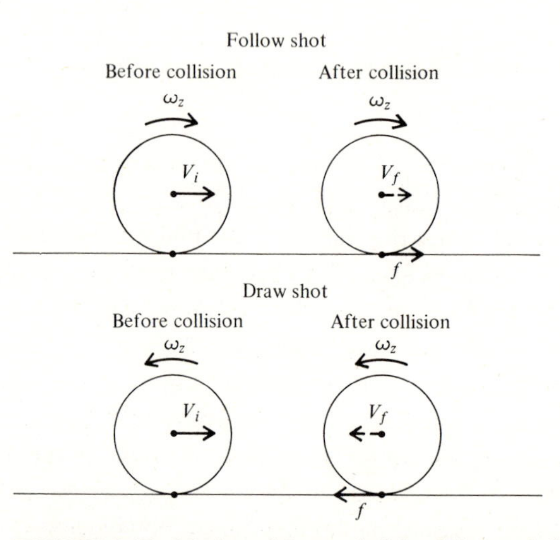

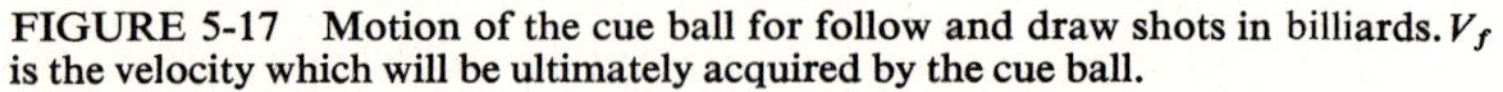
FIGURE 5-17 Motion of the cue ball for follow and draw shots in billiards. V_f is the velocity which will be ultimately acquired by the cue ball.

5-11 SUPER-BALL BOUNCES

The bizarre behavior observed in bounces of the Wham-O Super-Ball[1] can be predicted from the rigid-body equations of motion.[2] The Super-Ball is a hard spherical rubber ball. The bounces of a Super-Ball on a hard surface are almost elastic (i.e., energy-conserving) and essentially nonslip at the point of contact. As an idealization, we shall assume kinetic-energy conservation during a bounce. We shall also neglect gravity in our calculations, though its inclusion does not change the principal results.

We begin with an analysis of a single bounce from the floor. We denote the initial components of the CM velocity by $V_x{}^0$ and $V_y{}^0$ and the initial spin of the ball about the z axis through the CM by $\omega_z{}^0$, as pictured in Fig. 5-18. The frictional force f_x and the normal force f_y act on the ball only for a very short time duration, Δt. We can determine the changes in the velocities ΔV_x, ΔV_y from the linear-impulse equation (5-103), and the change in spin from the angular-impulse equation (5-105). We obtain

$$
\begin{aligned}
M\,\Delta V_x &= -\int_{t0}^{t1} f_x\,dt \\
M\,\Delta V_y &= -\int_{t0}^{t1} f_y\,dt \\
I_{zz}\,\Delta\omega_z &= a\int_{t0}^{t1} f_x\,dt
\end{aligned}
\tag{5-111}
$$

By elimination of the frictional force f_x from the first and third equations, we obtain a relation between ΔV_x and $\Delta\omega_z$ caused by f_x.

$$
M(V_x{}^1 - V_x{}^0) = -\frac{I_{zz}}{a}(\omega_z{}^1 - \omega_z{}^0) \tag{5-112}
$$

The assumption that f_x and f_y are independent (i.e., that the deformations of the superball result in stresses in the x and y directions, which are independent of one another) requires that the energies associated with the x and y motions be separately conserved. In other words,

[1] Registered trademark of Wham-O Corporation, San Gabriel, Calif.
[2] R. L. Garwin, *Amer. J. Phys.*, **37**: 88 (1969).

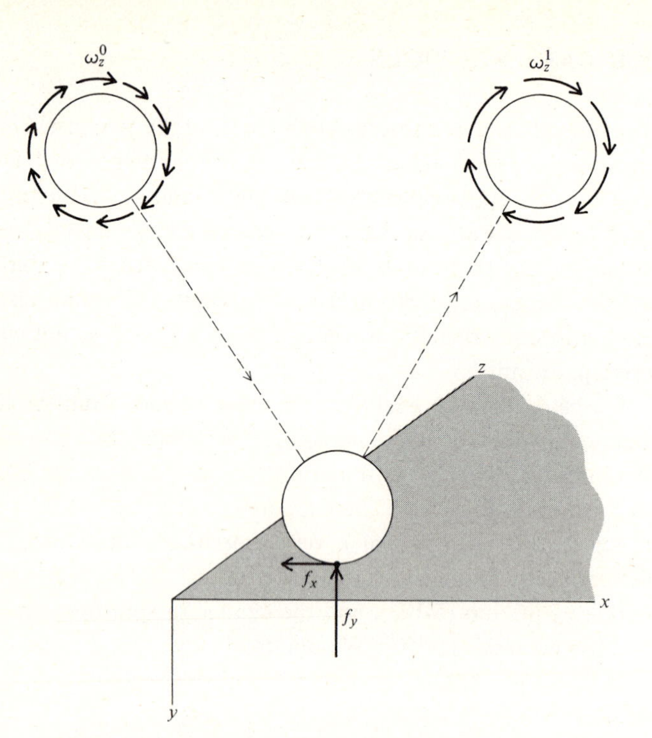

FIGURE 5-18 Super-Ball bounce from a hard surface.

both f_x and f_y are conservative forces. The conservative nature of f_y leads to

$$\tfrac{1}{2}M(V_y{}^1)^2 = \tfrac{1}{2}M(V_y{}^0)^2 \tag{5-113}$$

We conclude that the vertical component of velocity must be reversed by the action of the normal force f_y:

$$V_y{}^1 = -V_y{}^0 \tag{5-114}$$

The stipulation that f_x be energy-conserving (i.e., no slipping) yields the condition

$$\tfrac{1}{2}I_{zz}(\omega_z{}^1)^2 + \tfrac{1}{2}M(V_x{}^1)^2 = \tfrac{1}{2}I_{zz}(\omega_z{}^0)^2 + \tfrac{1}{2}M(V_x{}^0)^2 \tag{5-115}$$

Energy can be interchanged in the bounce between the rotational and horizontal translational modes via the coupling between the spin and horizontal velocity changes in Eqs. (5-111). We can rewrite the condition in Eq. (5-115) as

$$M(V_x{}^1 - V_x{}^0)(V_x{}^1 + V_x{}^0) = -I_{zz}(\omega_z{}^1 - \omega_z{}^0)(\omega_z{}^1 + \omega_z{}^0) \tag{5-116}$$

Eqs. (5-112), (5-113), and (5-116) govern the dynamics of a Super-Ball bounce. One possible solution to Eqs. (5-112) and (5-116) is

$$\begin{aligned} V_x{}^1 &= V_x{}^0 \\ \omega_z{}^1 &= \omega_z{}^0 \end{aligned} \tag{5-117}$$

This solution corresponds to zero frictional force in Eqs. (5-111) and is therefore relevant only for a smooth ball. The solution appropriate for a Super-Ball can be obtained by division of Eqs. (5-112) and (5-116). We find

$$V_x{}^1 + V_x{}^0 = a(\omega_z{}^1 + \omega_z{}^0) \tag{5-118}$$

or by rearrangement,

$$(V_x{}^1 - a\omega_z{}^1) = -(V_z{}^0 - a\omega_z{}^0) \tag{5-119}$$

The quantity $(V_x - a\omega_z)$ is just the horizontal velocity at the point on the ball that makes contact with the floor. Hence the velocity at the point of contact is exactly reversed by a bounce. From Eqs. (5-112) and (5-119) we can solve for the spin and horizontal velocity immediately after the bounce in terms of the initial spin and velocity. With the moment of inertia of the Super-Ball given by Eq. (5-101), we arrive at

$$\begin{aligned} V_x{}^1 &= \tfrac{3}{7}V_x{}^0 + \tfrac{4}{7}\omega_z{}^0 a \\ \omega_z{}^1 &= -\tfrac{3}{7}\omega_z{}^0 + \tfrac{10}{7}\frac{V_x{}^0}{a} \end{aligned} \tag{5-120}$$

as the general solution for a Super-Ball bounce.

As an example of the result in Eqs. (5-120), a Super-Ball which approaches the floor from a vertical direction ($V_x{}^0 = 0$) with initial spin $\omega_z{}^0$ will leave the floor with

$$\begin{aligned} V_x{}^1 &= \tfrac{4}{7}\omega_z{}^0 a \\ \omega_z{}^1 &= -\tfrac{3}{7}\omega_z{}^0 \\ V_y{}^1 &= -V_y{}^0 \end{aligned} \tag{5-121}$$

as illustrated in Fig. 5-19. A smooth ball with the same initial velocity and spin would bounce back in the vertical direction.

The unexpected behavior of a Super-Ball is even more dramatically exhibited in successive bounces. As indicated in Fig. 5-20, a Super-Ball thrown to the floor in such a way that it bounces from the underside of a table will return to the hand. We can show this quite simply from

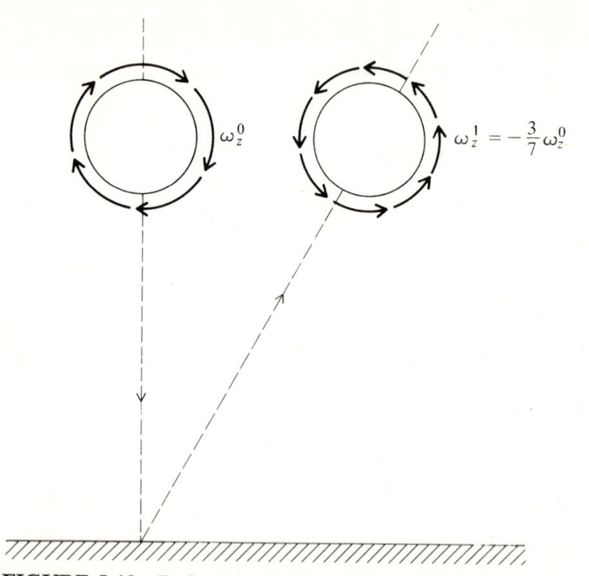

FIGURE 5-19 Deflection of a Super-Ball from a vertical bounce.

repeated applications of Eqs. (5-120). If the initial spin of the ball is zero, the velocity and spin after the first bounce from the floor are

$$V_x{}^1 = \tfrac{3}{7}V_x{}^0$$
$$\omega_z{}^1 = \tfrac{10}{7}\frac{V_x{}^0}{a} \qquad (5\text{-}122)$$

For the bounce off the underside of the table, the angular impulse is opposite in sign to the impulse in Eqs. (111). With the angular impulse reversed, the appropriate modifications of Eqs. (5-120) for the second bounce are

$$V_x{}^2 = \tfrac{3}{7}V_x{}^1 - \tfrac{4}{7}\omega_z{}^1 a$$
$$\omega_z{}^2 = -\tfrac{3}{7}\,\omega_z{}^1 - \tfrac{10}{7}\frac{V_x{}'}{a} \qquad (5\text{-}123)$$

When we substitute Eqs. (5-122) into Eqs. (5-123), we get

$$V_x{}^2 = -\tfrac{31}{49}V_x{}^0$$
$$\omega_z{}^2 = -\tfrac{60}{49}\frac{V_x{}^0}{a} \qquad (5\text{-}124)$$

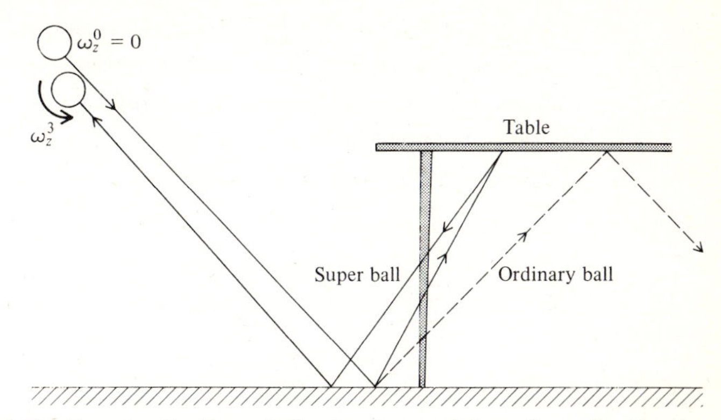

FIGURE 5-20 Return of a Super-Ball when bounced from the underside of a table.

Thus the horizontal direction of motion has been reversed. For the final bounce off the floor we again apply Eqs. (120), with Eqs. (5-124) as initial values. We find

$$\begin{aligned} V_x{}^3 &= -\tfrac{333}{343}\, V_x{}^0 \\ \omega_z{}^3 &= -\tfrac{130}{343}\, V_x{}^0 \end{aligned} \qquad (5\text{-}125)$$

Thus the Super-Ball returns after the three bounces with a slightly lower velocity than when it started, although the total kinetic energy remains the same. A smooth ball would not return, but would continue bouncing between the floor and the table, as indicated by the dashed line in Fig. 5-20.

PROBLEMS

1. Find the mean location of the center of mass of the earth-moon system.

2. For a two-particle system in a region of uniform gravitational acceleration g, show that the net gravitational torque about the CM point of the system is zero.

3. A boat of mass 60 kg and length 4 m is at rest in quiet water. If a man of mass 80 kg walks from the bow to the stern, what distance will the boat move? Neglect water resistance and inertia.

4. A circular tabletop of radius 1 m and mass 3 kg is supported by three equally spaced legs on the circumference. When a vase is placed on the table, the legs support 1, 2, and 3 kg, respectively. How heavy is the vase, and where is it located on the table? What is the lightest vase which might upset the table?

5. A cylindrical glass full of ice weighs four times as much as when empty. At what intermediate level of filling is the glass least likely to tip? Neglect the mass of the bottom of the glass.

6. Two particles on a line are mutually attracted by a force

$$F = -fr$$

where f is a constant and r is the distance of separation. At time $t = 0$, particle A of mass M is located at $x = 5$ cm, and particle B of mass $M/4$ is located at $x = 10$ cm. If the particles are at rest at time $t = 0$, at what value of x do they collide? What is the relative velocity of the two particles at the moment the collision occurs?

7. Compare the gravity forces on the moon due to the earth and sun. Show from Eq. (5-11) that the sun is not very important in determining the moon's motion relative to the earth.

8. The two atoms in a diatomic molecule (masses m_1 and m_2) interact through a potential energy

$$V(r) = \frac{a^2}{4r^4} - \frac{b^2}{3r^3}$$

where r is the separation of the atoms.

a. Find the equilibrium separation of the atoms and the frequency of small oscillations about equilibrium assuming that the molecule does not rotate. How much energy must be supplied to the molecule in equilibrium in order to break it up?

b. Determine the maximum angular momentum which the molecule can have without breaking up, assuming that the motion is in circular orbits. Find the particle separation at the breakup angular momentum.

c. Calculate the velocity of each particle in the laboratory system at breakup, assuming that the center of mass is at rest.

9. Two point masses are connected by a spring with spring constant k but are otherwise free to move in space. The equations of motion are

$$m_1 \frac{d^2\mathbf{r}_1}{dt^2} = -k(\mathbf{r}_1 - \mathbf{r}_2 - \mathbf{l}) \qquad m_2 \frac{d^2\mathbf{r}_2}{dt^2} = k(\mathbf{r}_1 - \mathbf{r}_2 - \mathbf{l})$$

where $\mathbf{l} = l(\mathbf{r}_1 - \mathbf{r}_2)/|\mathbf{r}_1 - \mathbf{r}_2|$.

a. Find the equilibrium separation of the masses and the frequency of oscillation of the masses about equilibrium assuming that the system does not rotate.

b. How will the equilibrium separation of the masses and the frequency of small oscillations about equilibrium change as the system rotates about an axis through the CM perpendicular to the axis of the oscillator?

c. Show that the total energy of this system is conserved.

10. Two particles with masses m_1 and m_2 collide head on. Particle 1 has an initial velocity v_1, and particle 2 is initially at rest in the laboratory system. The particles interact through a potential energy

$$V = V_0 \left(\frac{a}{r_{12}}\right)^2$$

where $V_0 > 0$ and $r_{12} = |\mathbf{r}_1 - \mathbf{r}_2|$.

a. Compute the total energy and angular momentum of the two particles in the CM system. Express the results in terms of m_1, m_2, and v_1.

b. Describe the motion qualitatively as it appears in the CM system and in the lab system.

c. Find the distance of closest approach (minimum separation between the particles).

d. Find the velocity of particle 2 in the lab system after the collision.

11. Two particles of masses m_1 and m_2 collide. The initial velocity of particle 1 in the lab system is $\mathbf{v}_1$, while particle 2 is initially at rest. The initial impact parameter is b_0, as shown. The particles interact through a potential $V = V_0/|\mathbf{r}_1 - \mathbf{r}_2|^4$.

a. Calculate the total energy and angular momentum in the CM system in terms of the particle masses and velocities.

b. Derive the equation of motion for the relative coordinate $\mathbf{r} = \mathbf{r}_1 - \mathbf{r}_2$. Find the distance of closest approach.

c. Show how the angle β between the final velocities of the particles in the lab can be calculated if the magnitude $|\mathbf{v}_{1f}|$ of the final lab velocity of particle 1 is measured.

12. A disk of radius R is oriented in a vertical plane and spinning about its axis with angular velocity ω. If the spinning disk is set down on a horizontal surface, with what translational CM velocity will it roll away?

13. A heavy axially symmetric gyroscope is supported at a pivot, as shown. The mass of the gyroscope is M, and the moment of inertia about its symmetry axis is I. The initial angular velocity about its symmetry axis is $\boldsymbol{\omega}$. Give a suitable approximate equation of motion for the system, assuming that ω is very large. Find the angular frequency of the gyroscopic precession. Show that the above approximation is justified for

$$\omega \gg \sqrt{\frac{g}{l}}$$

where all moments of inertia are taken to be roughly Ml^2.

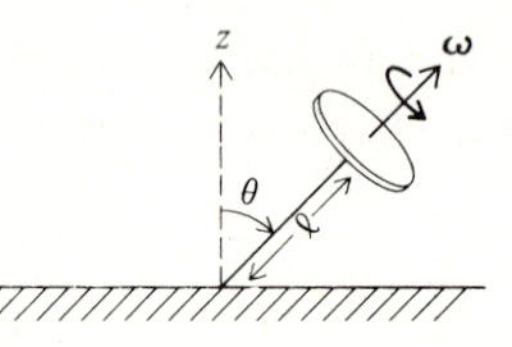

14. Show that if the aerodynamic force on a boomerang blade is proportional to $|v_t|$ (not v_t^2 as before), the ratio of spin to CM velocity must still be related as $V = \sqrt{\frac{2}{3}}\omega l$ for a successful return. Show that the radius of the orbit is now proportional to the velocity.

15. A thin, uniform rod of mass M is supported by two vertical strings, as shown. Find the tension in the remaining string immediately after one of the strings is severed.

16. A physical pendulum consists of a solid cylinder which is free to rotate about a transverse axis displaced by a distance l along the symmetry axis from the center of mass, as illustrated. Find the value of l for which the period is a minimum. Express the result in terms of the mass M and moment of inertia I about a transverse axis through the CM.

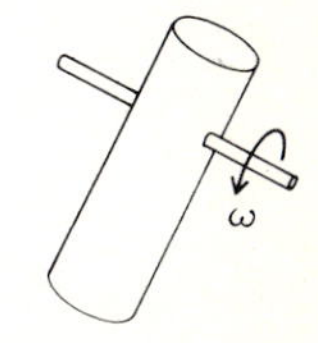

17. For rotations about the z axis derive a "parallel-axis theorem" for the products of inertia.

18. A pendulum consists of two masses connected by a very light rigid rod, as shown. The pendulum is free to oscillate in the vertical plane about a horizontal axis located a distance a from m_a at a distance b from m_b.

a. Calculate the moment of inertia of the system about 0. Find the location of the center of mass.

b. Set up the equation of motion for the system and derive the potential-energy function.

c. Take $b > a$ and determine the frequency of oscillation for small angles of displacement from the vertical.

d. Derive an exact expression for the period of the pendulum ($|\theta_{\max}| < \pi$).

e. Find the minimum angular velocity which must be given to the system (starting at equilibrium) if it is to continue in rotation instead of oscillating.

19. A Yo-Yo of mass M is composed of two disks of radius R separated by a distance t by a shaft of radius r. A massless string is wound on the shaft, and the loose end is held in the hand. Upon release the Yo-Yo descends until the string is unwound. The string then begins to rewind, and the Yo-Yo climbs. Find the string tension and acceleration of the Yo-Yo in descent and in ascent. Neglect the mass of the shaft and assume the shaft radius is sufficiently small so that the string is essentially vertical.

20. A pencil of length l and mass m lying flat on a frictionless horizontal tabletop receives an impulse on one end at a right angle to the pencil. What is the orientation and position of the pencil at a time t after the impulse?

21. A rod of mass m and length l hangs vertically from a horizontal frictionless wire, as shown. Attached to the end of the rod is a small ball, also of mass m. The rod is free to move along the wire.

a. Find the location of the center of mass for rod plus ball, taking the hook as the origin of the coordinate system.

b. Find the moment of inertia for rod plus ball about the hook.

c. Use the parallel-axis theorem and the result in part *b* to find the moment of inertia about the center of mass.

d. The rod-ball system is struck by a hammer blow (impulse $\mathscr{I}$) a distance h from the hook. Set up equations for the linear and angular motion of the system.

e. Find h such that the hook does not move along the wire at the instant of blow.

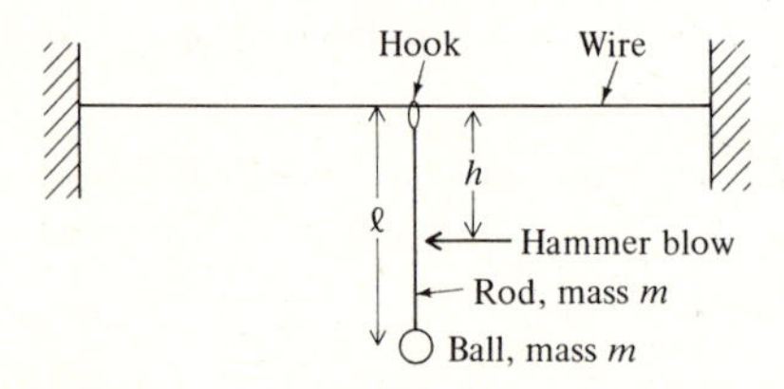

22. A ball of radius a rolling with velocity v on a level surface collides inelastically with a step of height $h < a$, as shown. Find the minimum velocity for which the ball will "trip" up over the step. Assume that no slipping occurs at the impact point.

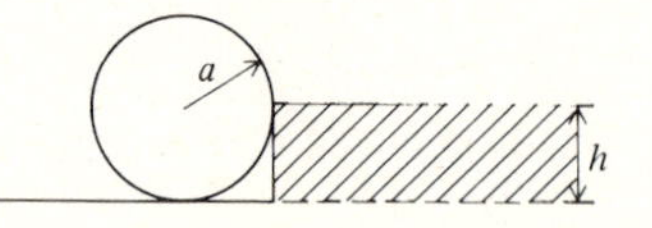

23. In the sport of bowling, if the ball is rolled straight down the middle of the alley, pins on the sides will often be left standing (wide splits). A good right-handed bowler will impart a spin to the ball on release, causing it to curve to the left as it goes down the alley and strike the pins somewhat to the side. Describe the required spin and show that the trajectory of the ball is parabolic in shape.

24. One of the Super-Ball examples discussed in the text concerned a ball dropped straight down with spin. Discuss the subsequent motion through several bounces.

25. Under what conditions will a Super-Ball bounce back and forth as illustrated? How does the spin change?

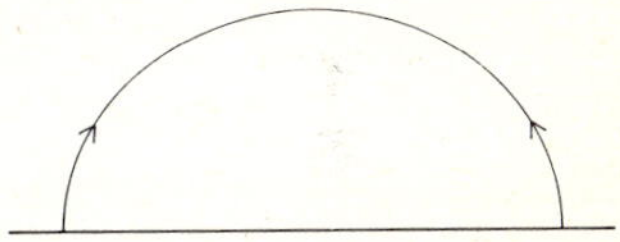

CHAPTER 6

Accelerated Coordinate Systems

The simple form of Newton's second law,

$$\mathbf{F} = m \frac{d^2\mathbf{r}_I}{dt^2} \tag{6-1}$$

for a particle of mass m holds only in inertial (unaccelerated) coordinate systems, as denoted by the I subscript above. On the other hand, physical events are sometimes more simply viewed with reference to an accelerated coordinate system. For example, observations of motion on the earth's surface are more simply expressed in terms of a coordinate system fixed on the rotating earth than in terms of an inertial coordinate system. For this reason it is useful to derive the form of the second law which directly applies in accelerated reference frames.

6-1 TRANSFORMATION TO MOVING COORDINATE FRAMES

To transform the law of motion to an accelerated reference frame, we first need to relate the time derivatives of vector quantities in moving and fixed coordinate systems. In a moving frame, an arbitrary vector quantity **A** can be written

$$\mathbf{A} = A_x\hat{\mathbf{i}} + A_y\hat{\mathbf{j}} + A_z\hat{\mathbf{k}} \tag{6-2}$$

where the directions of the unit vectors change in time. The time derivative of **A** is

$$\frac{d\mathbf{A}}{dt} = \left(\frac{dA_x}{dt}\hat{\mathbf{i}} + \frac{dA_y}{dt}\hat{\mathbf{j}} + \frac{dA_z}{dt}\hat{\mathbf{k}}\right) + \left(A_x\frac{d\hat{\mathbf{i}}}{dt} + A_y\frac{d\hat{\mathbf{j}}}{dt} + A_z\frac{d\hat{\mathbf{k}}}{dt}\right) \tag{6-3}$$

The first term on the right-hand side of this equation is the time rate of change of **A** with reference to the axes of the accelerated frame. We denote this time rate of change of **A** in the moving frame by

$$\frac{\delta\mathbf{A}}{\delta t} \equiv \frac{dA_x}{dt}\hat{\mathbf{i}} + \frac{dA_y}{dt}\hat{\mathbf{j}} + \frac{dA_z}{dt}\hat{\mathbf{k}} \tag{6-4}$$

The second term in Eq. (6-3) is due to the rotation of the coordinate system, which causes the direction of the unit vectors to change with time. Since the coordinate system is rigid, we can directly apply the result of Eq. (5-45),

$$\frac{d\mathbf{r}}{dt} = \boldsymbol{\omega} \times \mathbf{r} \tag{6-5}$$

to find the time derivatives of the unit vectors

$$\frac{d\hat{\mathbf{i}}}{dt} = \boldsymbol{\omega} \times \hat{\mathbf{i}} \qquad \frac{d\hat{\mathbf{j}}}{dt} = \boldsymbol{\omega} \times \hat{\mathbf{j}} \qquad \frac{d\hat{\mathbf{k}}}{dt} = \boldsymbol{\omega} \times \hat{\mathbf{k}} \tag{6-6}$$

Here $\boldsymbol{\omega}$ is the angular velocity of rotation of the accelerated frame relative to a fixed frame. Upon substitution of Eqs. (6-4) and (6-6) into Eq. (6-3), we have

$$\frac{d\mathbf{A}}{dt} = \frac{\delta\mathbf{A}}{\delta t} + \boldsymbol{\omega} \times (A_x\hat{\mathbf{i}} + A_y\hat{\mathbf{j}} + A_z\hat{\mathbf{k}})$$

or more simply,

$$\frac{d\mathbf{A}}{dt} = \frac{\delta \mathbf{A}}{\delta t} + \boldsymbol{\omega} \times \mathbf{A} \tag{6-7}$$

Accordingly, the time derivative $d\mathbf{A}/dt$ in a fixed reference frame consists of a part $\delta\mathbf{A}/\delta t$ from the time rate of change of **A** relative to the axes of the moving frame and a part $\boldsymbol{\omega} \times \mathbf{A}$ from the rotation of these axes relative to the fixed axes.

We next apply the result in Eq. (6-7) to the vector **r**, which specifies the location of a particle with respect to the moving axes. The first time derivative is

$$\frac{d\mathbf{r}}{dt} = \frac{\delta \mathbf{r}}{\delta t} + \boldsymbol{\omega} \times \mathbf{r} \tag{6-8}$$

The second time derivative can likewise be evaluated with the aid of Eq. (6-7).

$$\frac{d^2\mathbf{r}}{dt^2} = \frac{d}{dt}\left(\frac{\delta \mathbf{r}}{\delta t} + \boldsymbol{\omega} \times \mathbf{r}\right) = \frac{\delta}{\delta t}\left(\frac{\delta \mathbf{r}}{\delta t} + \boldsymbol{\omega} \times \mathbf{r}\right) + \boldsymbol{\omega} \times \left(\frac{\delta \mathbf{r}}{\delta t} + \boldsymbol{\omega} \times \mathbf{r}\right)$$

$$= \frac{\delta^2 \mathbf{r}}{\delta t^2} + \frac{\delta \boldsymbol{\omega}}{\delta t} \times \mathbf{r} + 2\boldsymbol{\omega} \times \frac{\delta \mathbf{r}}{\delta t} + \boldsymbol{\omega} \times (\boldsymbol{\omega} \times \mathbf{r}) \tag{6-9}$$

It also follows from Eq. (6-7) that the time derivative of $\boldsymbol{\omega}$ in Eq. (6-9) is independent of coordinate frame.

$$\frac{d\boldsymbol{\omega}}{dt} = \frac{\delta \boldsymbol{\omega}}{\delta t} \equiv \dot{\boldsymbol{\omega}} \tag{6-10}$$

To find the law of motion in the accelerated frame, we relate the location of the particle in the accelerated and inertial frames by the vector **R** connecting the origins of the two frames.

$$\mathbf{r}_I = \mathbf{r} + \mathbf{R} \tag{6-11}$$

as illustrated in Fig. 6-1. Then, substituting the result in Eq. (6-9) into

$$\frac{d^2\mathbf{r}_I}{dt^2} = \frac{d^2\mathbf{r}}{dt^2} + \frac{d^2\mathbf{R}}{dt^2}$$

we get

$$\frac{d^2\mathbf{r}_I}{dt^2} = \frac{\delta^2 \mathbf{r}}{\delta t^2} + \boldsymbol{\omega} \times (\boldsymbol{\omega} \times \mathbf{r}) + 2\boldsymbol{\omega} \times \frac{\delta \mathbf{r}}{\delta t} + \dot{\boldsymbol{\omega}} \times \mathbf{r} + \frac{d^2\mathbf{R}}{dt^2} \tag{6-12}$$

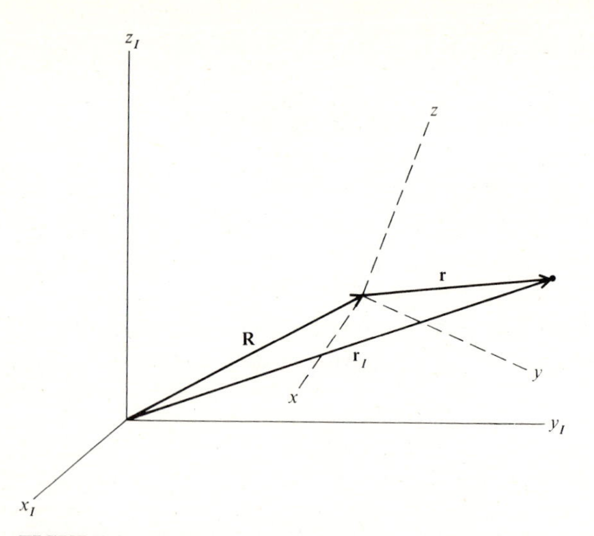

FIGURE 6-1 Inertial and accelerated coordinate frames.

The form of Newton's law in the accelerated frame now follows directly from Eqs. (6-1) and (6-12).

$$m\frac{\delta^2\mathbf{r}}{\delta t^2}=\mathbf{F}-m\left[\boldsymbol{\omega}\times(\boldsymbol{\omega}\times\mathbf{r})+2\boldsymbol{\omega}\times\mathbf{v}+\dot{\boldsymbol{\omega}}\times\mathbf{r}+\frac{d^2\mathbf{R}}{dt^2}\right] \qquad (6\text{-}13)$$

where $\mathbf{v}=\delta\mathbf{r}/\delta t$ is the velocity and $\delta^2\mathbf{r}/\delta t^2$ is the acceleration of a particle as observed in the moving coordinate system.

6-2 FICTITIOUS FORCES

We can write the equation of motion (6-13) for an accelerated frame in a form similar to Eq. (6-1) for an inertial frame.

$$m\frac{\delta^2\mathbf{r}}{\delta t^2}=\mathbf{F}_{\text{eff}} \qquad (6\text{-}14)$$

The acceleration $\delta^2\mathbf{r}/\delta t^2$ observed in the moving frame is generated by the *effective force*

$$\mathbf{F}_{\text{eff}}=\mathbf{F}-m\left[\boldsymbol{\omega}\times(\boldsymbol{\omega}\times\mathbf{r})+2\boldsymbol{\omega}\times\mathbf{v}+\dot{\boldsymbol{\omega}}\times\mathbf{r}+\frac{d^2\mathbf{R}}{dt^2}\right] \qquad (6\text{-}15)$$

The names associated with the so-called *fictitious-force* terms on the right-hand side of Eq. (6-15) are

Centrifugal force:

$$\mathbf{F}_{cf} = -m\boldsymbol{\omega} \times (\boldsymbol{\omega} \times \mathbf{r}) \tag{6-16}$$

Coriolis force:

$$\mathbf{F}_{\text{Cor}} = -2m\boldsymbol{\omega} \times \mathbf{v} \tag{6-17}$$

Azimuthal force:

$$F_{az} = -m\dot{\boldsymbol{\omega}} \times \mathbf{r} \tag{6-18}$$

Translational force:

$$\mathbf{F}_{tr} = -m\frac{d^2\mathbf{R}}{dt^2} \tag{6-19}$$

The centrifugal force of Eq. (6-16) is due to the rotational motion of the coordinate system. Since $\boldsymbol{\omega} \cdot \mathbf{F}_{cf} = 0$, the centrifugal force is perpendicular to the rotation axis $\hat{\boldsymbol{\omega}}$. If the angular velocity $\boldsymbol{\omega}$ is chosen to lie along the z axis of the moving frame, as in Fig. 6-2, then

$$\begin{aligned}\mathbf{F}_{cf} &= -m[\boldsymbol{\omega}(\boldsymbol{\omega} \cdot \mathbf{r}) - \mathbf{r}\omega^2] = m\omega^2(x\hat{\mathbf{i}} + y\hat{\mathbf{j}}) \\ &= m\omega^2\boldsymbol{\rho}\end{aligned} \tag{6-20}$$

where $\boldsymbol{\rho}$ is the cylindrical-radius vector to the particle from the z axis.

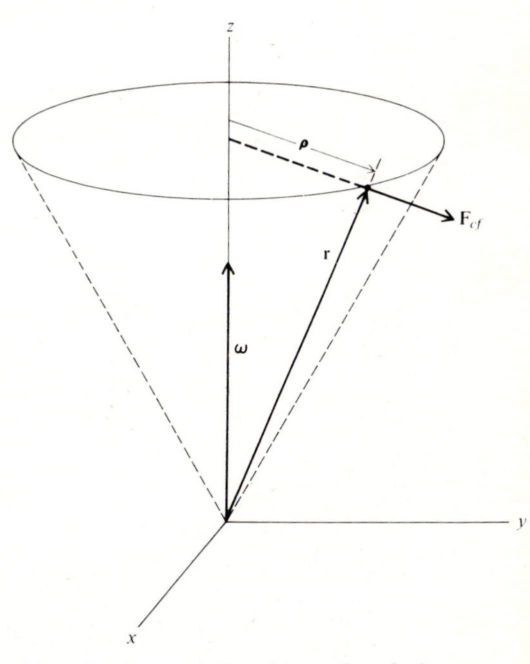

FIGURE 6-2 Centrifugal force $\mathbf{F}_{cf}$ due to rotation with angular velocity $\boldsymbol{\omega}$.

The centrifugal force is directed radially outward from the axis of rotation. The result in Eq. (6-20) is the same as Eq. (4-12).

The centrifugal force accounts for the parabolic shape of the surface of a spinning pail of water. Using Eq. (6-15), the equation of motion of a small mass of water m on the surface in a frame rotating with the pail is

$$m\frac{\delta^2\mathbf{r}}{\delta t^2} = \mathbf{F}' + m\mathbf{g} - m\boldsymbol{\omega} \times (\boldsymbol{\omega} \times \mathbf{r}) \tag{6-21}$$

where the force $\mathbf{F}'$, due to the pressure gradient, is normal to the surface. Since in equilibrium $\delta^2\mathbf{r}/\delta t^2 = 0$, the *effective-gravity* term

$$\mathbf{g}_{\text{eff}} \equiv \mathbf{g} - \boldsymbol{\omega} \times (\boldsymbol{\omega} \times \mathbf{r}) \tag{6-22}$$

must also be normal to the surface, by Eq. (6-21). In cylindrical coordinates, $\mathbf{g}_{\text{eff}}$ is given by

$$\mathbf{g}_{\text{eff}} = -g\hat{\mathbf{z}} + \omega^2\rho\hat{\boldsymbol{\rho}} \tag{6-23}$$

From the geometry of Fig. 6-3, the normal requirement on $\mathbf{g}_{\text{eff}}$ can be written

$$\tan\theta = \frac{dz}{d\rho} = \frac{\omega^2\rho}{g} \tag{6-24}$$

Integration of Eq. (6-24) gives

$$z = \frac{\omega^2}{2g}\rho^2 + \text{constant} \tag{6-25}$$

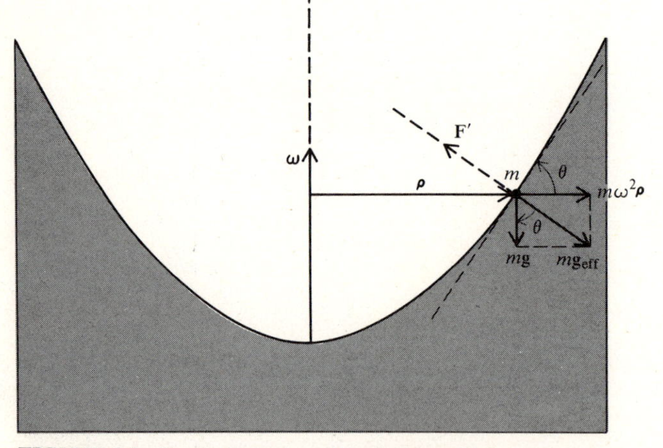

FIGURE 6-3 Parabolic surface of a spinning pail of water.

which is a parabolic shape. The solution in Eq. (6-25) can alternatively be found from Eq. (6-23) by potential-energy methods. The potential energy due to the force $m\mathbf{g}_{\text{eff}}$ is

$$V(z, \rho) = m(gz - \tfrac{1}{2}\omega^2\rho^2) \tag{6-26}$$

Since there can be no component of force tangential to the surface in equilibrium, the potential energy in Eq. (6-26) must be constant on the surface, and the result in Eq. (6-25) is thus obtained.

The Coriolis force in Eq. (6-17) is present when the particle is in motion relative to the accelerated coordinate system. Since $\boldsymbol{\omega} \cdot \mathbf{F}_{\text{Cor}} = 0$ and $\mathbf{v} \cdot \mathbf{F}_{\text{Cor}} = 0$, this force is perpendicular to both $\boldsymbol{\omega}$ and $\mathbf{v}$. The effects of the Coriolis force are important in such problems as calculations of long-range artillery trajectories and descriptions of large-scale atmospheric weather phenomena.

The azimuthal force in Eq. (6-18) occurs only when the angular-velocity vector changes with time. Inasmuch as $\mathbf{r} \cdot \mathbf{F}_{az} = 0$, this force always points in a direction perpendicular to $\mathbf{r}$. If $\boldsymbol{\omega}$ is decreasing in magnitude but constant in direction, the azimuthal force tends to maintain the rotational velocity of the particle.

The translational force in Eq. (6-19) is due to the acceleration of the origin of the moving frame relative to an inertial frame. In the special case when the motion of the accelerated coordinate system is purely translational (that is, $\boldsymbol{\omega} = 0$), the equation of motion in Eqs. (6-14) and (6-15) reduces to

$$m\frac{\delta^2\mathbf{r}}{\delta t^2} = \mathbf{F} - m\frac{d^2\mathbf{R}}{dt^2} \tag{6-27}$$

The problem of a pendulum with a moving support provides an interesting application of Eq. (6-27). We choose the origin of the moving coordinate system to coincide with the instantaneous location of the support, as illustrated in Fig. 6-4. We shall restrict our discussion to angular motion in the xy plane defined by $\mathbf{F}$ and $\mathbf{R}$. In terms of the tension $\mathbf{T}$ and the gravitational force $mg\hat{\mathbf{i}}$ acting on the pendulum bob, the equation of motion in the moving frame is

$$m\frac{\delta^2\mathbf{r}}{\delta t^2} = \mathbf{T} + m(g - A_x)\hat{\mathbf{i}} - mA_y\hat{\mathbf{j}} \tag{6-28}$$

where $\mathbf{A} = d^2\mathbf{R}/dt^2$ is the translational acceleration. Since physical

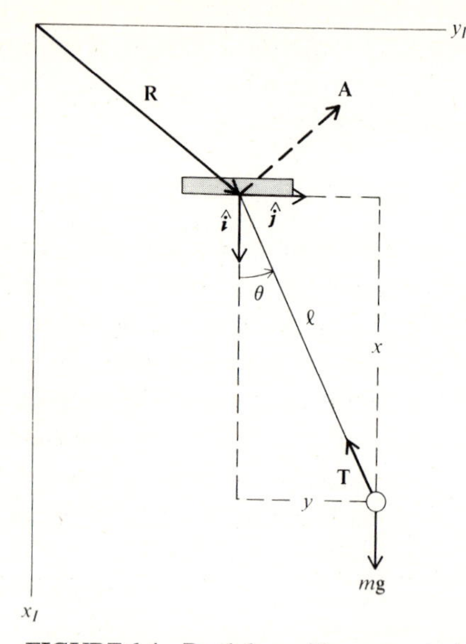

FIGURE 6-4 **Pendulum with a support which moves with acceleration A.**

motion occurs along the θ direction, it is advantageous to write Eq. (6-28) in polar coordinates using Eq. (2-71),

$$\frac{\delta^2 \mathbf{r}}{\delta t^2} = \hat{\mathbf{r}}(\ddot{r} - r\dot{\theta}^2) + \hat{\mathbf{l}}(r\ddot{\theta} + 2\dot{r}\dot{\theta})$$

and the relations

$$\begin{aligned} \hat{\mathbf{i}} &= \hat{\mathbf{r}} \cos\theta - \hat{\mathbf{l}} \sin\theta \\ \hat{\mathbf{j}} &= \hat{\mathbf{r}} \sin\theta + \hat{\mathbf{l}} \cos\theta \end{aligned} \qquad (6\text{-}29)$$

obtained from Eq. (2-70), Since $T_\theta = 0$, we find

$$\begin{aligned} ml\ddot{\theta} &= -m(g - A_x)\sin\theta - mA_y\cos\theta \\ -ml\dot{\theta}^2 &= T_r + m(g - A_x)\cos\theta - mA_y\sin\theta \end{aligned} \qquad (6\text{-}30)$$

The angular motion of the pendulum bob is therefore determined by

$$\ddot{\theta} + \left[\frac{g}{l} - \frac{A_x(t)}{l}\right]\sin\theta = -\frac{A_y(t)}{l}\cos\theta \qquad (6\text{-}31)$$

For $A_x(t) = 0$ and $A_y(t) = \ddot{y}_s$, this reduces to the result in Eq. (2-139) for horizontal motion of the pendulum support. For uniform vertical acceleration of the support (A_x = constant and $A_y = 0$), the natural frequency of small oscillations for $A_x < g$ is

$$\omega_0 = \sqrt{\frac{g - A_x}{l}} \tag{6-32}$$

When $A_x = g$, the pendulum undergoes free-fall motion and it behaves as if the gravity field has vanished. The fact that gravity can be made to disappear (or appear) locally by a coordinate transformation led Einstein to a theory of gravity, the general theory of relativity, in which gravity is linked closely to geometry.

6-3 MOTION OF THE EARTH

For motion of a particle on the earth, it is convenient to choose a coordinate system that is fixed on the earth's surface. In this reference frame which rotates with nearly constant angular velocity, the equation of motion from Eq. (6-13) is

$$m\frac{\delta^2 \mathbf{r}}{\delta t^2} = \mathbf{F} - m\left[\boldsymbol{\omega} \times (\boldsymbol{\omega} \times \mathbf{r}) + 2\boldsymbol{\omega} \times \mathbf{v} + \frac{d^2\mathbf{R}_e}{dt^2}\right] \tag{6-33}$$

where $\mathbf{R}_e$ is the vector connecting the center of the earth with the origin of the rotating coordinate system. Inasmuch as the time rate of change of $\mathbf{R}_e$ is entirely due to the earth's rotation, we can use Eq. (6-5) to obtain

$$\frac{d\mathbf{R}_e}{dt} = \boldsymbol{\omega} \times \mathbf{R}_e$$

and hence

$$\frac{d^2\mathbf{R}_e}{dt^2} = \frac{d}{dt}(\boldsymbol{\omega} \times \mathbf{R}_e) = \boldsymbol{\omega} \times \frac{d\mathbf{R}_e}{dt} = \boldsymbol{\omega} \times (\boldsymbol{\omega} \times \mathbf{R}_e) \tag{6-34}$$

Inserting this expression into Eq. (6-33), we have

$$m\frac{\delta^2 \mathbf{r}}{\delta t^2} = \mathbf{F}' + m\mathbf{g} - m\boldsymbol{\omega} \times (\boldsymbol{\omega} \times \mathbf{R}_e) - m\boldsymbol{\omega} \times (\boldsymbol{\omega} \times \mathbf{r}) - 2m\boldsymbol{\omega} \times \mathbf{v} \tag{6-35}$$

where the net external force has been separated into the gravitational force $m\mathbf{g}$ and other external forces $\mathbf{F}'$. If the earth were perfectly spherical and isotropic, $\mathbf{g}$ would be constant in magnitude and directed toward the center of the earth. In fact, local irregularities, distortions from sphericity, and deviations from uniform density cause slight variations in $\mathbf{g}$ at different points on the earth.

The condition for a particle at rest on the earth ($\mathbf{v} = 0$) to be in equilibrium ($\delta^2\mathbf{r}/\delta t^2 = 0$) from Eq. (6-35) is

$$\mathbf{F}' = -m\{\mathbf{g} - \boldsymbol{\omega} \times [\boldsymbol{\omega} \times (\mathbf{R}_e + \mathbf{r})]\} \tag{6-36}$$

For example, if m is the bob on a plumb line, the tension $\mathbf{F}'$ in the string is opposite to the direction determined by

$$\mathbf{g}_{\text{eff}} = \mathbf{g} - \boldsymbol{\omega} \times [\boldsymbol{\omega} \times (\mathbf{R}_e + \mathbf{r})] \tag{6-37}$$

The plumb bob thus points in the direction of $\mathbf{g}_{\text{eff}}$. We conclude that $\mathbf{g}_{\text{eff}}$ is the effective gravitational acceleration on the earth. The magnitude of the correction term to $\mathbf{g}$ for $r \ll R_e$ is

$$|\boldsymbol{\omega} \times [\boldsymbol{\omega} \times (\mathbf{R}_e + \mathbf{r})]| \approx \omega^2 R_e \sin\theta \tag{6-38}$$

where θ is the colatitude angle between $\boldsymbol{\omega}$ and $(\mathbf{R}_e + \mathbf{r})$. For the earth's angular velocity of rotation,

$$\omega = \frac{2\pi}{\tau} = \frac{2\pi}{24 \times 3{,}600} = 0.73 \times 10^{-4} \text{ rad/s} \tag{6-39}$$

we find that the correction term is quite small.

$$\omega^2 R_e \sin\theta = (0.73 \times 10^{-4})^2(6{,}371 \times 10^3) \sin\theta = 0.03 \sin\theta \text{ m/s}^2 \tag{6-40}$$

The direction of the correction term is radially outward from the rotation axis, as illustrated in Fig. 6-5.

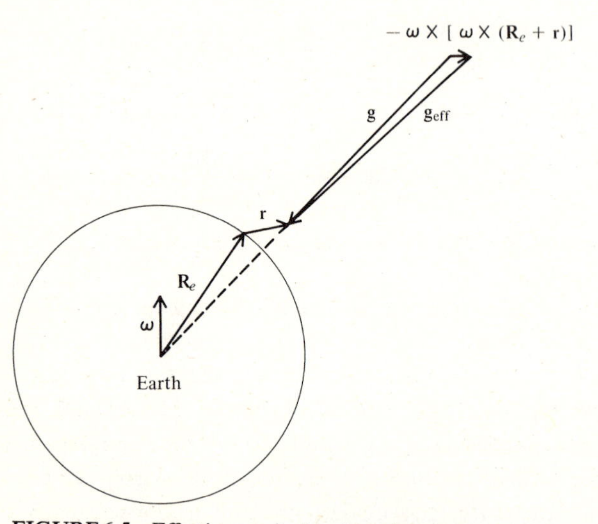

FIGURE 6-5 Effective gravitational acceleration $\mathbf{g}_{\text{eff}}$ (with the relative magnitude of the centrifugal acceleration exaggerated for clarity).

The differential equation (6-35) which describes the motion of a particle on the earth can be expressed in terms of $\mathbf{g}_{\text{eff}}$ in Eq. (6-37) as

$$m\frac{\delta^2\mathbf{r}}{\delta t^2} = \mathbf{F}' + m\mathbf{g}_{\text{eff}} - 2m\boldsymbol{\omega} \times \mathbf{v} \tag{6-41}$$

For convenience we choose the z axis of the coordinate system so that

$$\mathbf{g}_{\text{eff}} = -g_{\text{eff}}\hat{\mathbf{z}} \tag{6-42}$$

The y axis is taken to point north and the x axis east, as pictured in Fig. 6-6. The components of $\boldsymbol{\omega}$ along these axes are

$$\boldsymbol{\omega} = 0\hat{\mathbf{i}} + \omega \sin\theta\hat{\mathbf{j}} + \omega\cos\theta\hat{\mathbf{k}} \tag{6-43}$$

where θ is the colatitude angle. Substituting Eq. (6-43) into the Coriolis force term,

$$\mathbf{F}_{\text{Cor}} = -2m\boldsymbol{\omega} \times \mathbf{v}$$

of Eq. (6-41), we have

$$\mathbf{F}_{\text{Cor}} = 2m\omega[(v_y\cos\theta - v_z\sin\theta)\hat{\mathbf{i}} - v_x\cos\theta\hat{\mathbf{j}} + v_x\sin\theta\hat{\mathbf{k}}] \tag{6-44}$$

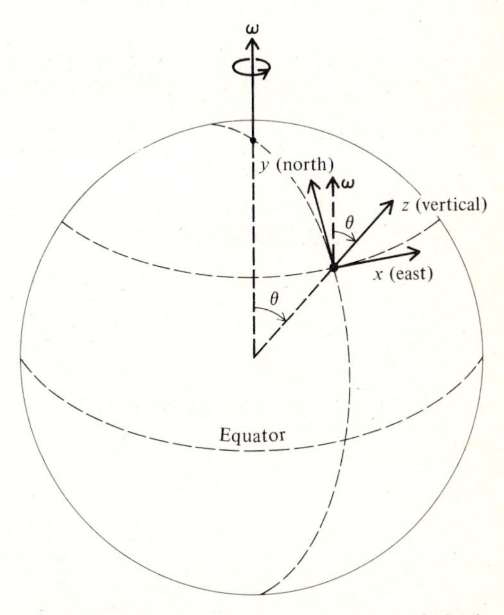

FIGURE 6-6 Coordinate frame fixed on the surface of the earth at colatitude angle θ.

The direction of deflection of the particle from its direction of motion due to the Coriolis force follows directly from Eq. (6-44). In the Northern Hemisphere, $0 \leq \theta \leq \pi/2$, we find

Velocity direction	Deflection direction
North ($v_y > 0$)	East
East ($v_x > 0$)	South and up
South ($v_y < 0$)	West
West ($v_x < 0$)	North and down
Up ($v_z > 0$)	West
Down ($v_z < 0$)	East

For motion parallel to the earth's surface ($v_z = 0$), the particle is always deflected to the right in the Northern Hemisphere and to the left in the Southern Hemisphere.

The equations of motion for a projectile ($\mathbf{F}' = 0$) on the earth from Eqs. (6-41), (6-42), and (6-44) are

$$\frac{\delta^2 x}{\delta t^2} = 2\omega\left(\frac{\delta y}{\delta t}\cos\theta - \frac{\delta z}{\delta t}\sin\theta\right)$$

$$\frac{\delta^2 y}{\delta t^2} = -2\omega\frac{\delta x}{\delta t}\cos\theta \tag{6-45}$$

$$\frac{\delta^2 z}{\delta t^2} = -g_{\text{eff}} + 2\omega\frac{\delta x}{\delta t}\sin\theta$$

A first integral over time can be immediately performed. For the initial conditions

$$x_0 = y_0 = z_0 = 0$$
$$v_x{}^0 = v_z{}^0 = 0$$
$$v_y{}^0 \neq 0$$

at time $t = 0$, we obtain

$$\frac{\delta x}{\delta t} = 2\omega(y\cos\theta - z\sin\theta)$$

$$\frac{\delta y}{\delta t} = v_y{}^0 - 2\omega x\cos\theta \tag{6-46}$$

$$\frac{\delta z}{\delta t} = -g_{\text{eff}}\,t + 2\omega x\sin\theta$$

When these results are substituted back into the right-hand side of Eqs. (6-45), we find

$$\frac{\delta^2 x}{\delta t^2} = 2\omega g_{\text{eff}}\, t \sin\theta + 2\omega v_y{}^0 \cos\theta - 4\omega^2 x$$

$$\frac{\delta^2 y}{\delta t^2} = -4\omega^2 (y\cos\theta - z\sin\theta)\cos\theta \tag{6-47}$$

$$\frac{\delta^2 z}{\delta t^2} = -g_{\text{eff}} + 4\omega^2 (y\cos\theta - z\sin\theta)\sin\theta$$

Inasmuch as $\omega \approx 0.7 \times 10^{-4}$ rad/s is a very small quantity, we can drop terms of order ω^2 in these equations. An approximate solution can then be directly found by two successive time integrations.

$$x = \tfrac{1}{3}\omega g_{\text{eff}}\, t^3 \sin\theta + \omega v_y{}^0 t^2 \cos\theta$$

$$y = v_y{}^0 t \tag{6-48}$$

$$z = -\tfrac{1}{2} g_{\text{eff}}\, t^2$$

The ratio of the easterly deflection to the northerly distance traveled is

$$\frac{x}{y} = \frac{2\pi t}{\tau}\left(\cos\theta + \frac{t g_{\text{eff}}}{3 v_y{}^0}\sin\theta\right) \tag{6-49}$$

where $\tau = \omega/2\pi = 24$ h. For $v_y{}^0 = 150$ km/h,

$$\frac{x}{y} \approx \frac{t}{4}(\cos\theta + 2.2 \times 10^{-5} t \sin\theta) \tag{6-50}$$

with t units of hours. For moderate velocities the deflection x becomes comparable with the distance y only for transit times of several hours.

The trade winds and weather circulations of high- and low-pressure areas are striking examples of Coriolis force effects. The equatorial region of the earth generally receives more heat from the sun. The warm air rises and is replaced by a flow of air from the temperate regions. The air moving south from the Northern Hemisphere is deflected westward by the Coriolis effect. This accounts for the steady prevailing winds to the west and south, known as the *trade winds*. On a smaller scale a low-pressure region of the order of 200 km across is associated with a counterclockwise circulation of the air because of the Coriolis force. The pressure gradient is largely balanced by the Coriolis force. Under certain circumstances this cyclonic motion builds up to great intensity and destructive power in the form of a

FIGURE 6-7 Foucault pendulum which hangs in the United Nations Building in New York City. *(Photo courtesy of United Nations.)*

hurricane, cyclone, or typhoon. High-pressure areas force air outward. This airflow deflects to the right and produces clockwise circulation in the Northern Hemisphere. Vortices on a still smaller scale such as tornados, dust devils, water spouts, and the bathtub vortex are not directly influenced by Coriolis effects to any great extent. Nevertheless, these vortices often have a counterclockwise motion because of the general counterclockwise movement which spawns them.

6-4 FOUCAULT'S PENDULUM

In 1851 the Coriolis effect due to the earth's rotation was dramatically demonstrated by Jean Foucault, using a simple pendulum which could oscillate for a long time without being appreciably damped by friction. Today Foucault pendulums are on exhibit in many public buildings, planetariums, and in churches in Russia. One of the most famous hangs in the United Nations Building in New York. The oscillation plane of a Foucault pendulum is observed to rotate slowly with time. This precession confirms the existence of a relative rotation of the earth about its axis with respect to the stars.

The motion of the Foucault pendulum can be determined from Eq. (6-41). We take $\mathbf{r}$ to represent the distance of the bob of mass m from its equilibrium position, as indicated in Fig. 6-8. At rest the pendulum hangs along the direction $\mathbf{g}_{\text{eff}}$, and the tension in the string is $\mathbf{F}' = -m\mathbf{g}_{\text{eff}}$. If the earth did not rotate, then $\mathbf{g}_{\text{eff}}$ would be constant and the Coriolis force term in Eq. (6-41) would not be present. In that case the motion would occur in a fixed plane. From Eqs. (2-109) and (2-115), the coordinate and velocity solutions for small oscillations in a vertical plane would be given by

$$\begin{aligned} \mathbf{r} &= \hat{\mathbf{i}}' r_0 \cos \omega_0 t \\ \mathbf{v} &= -\hat{\mathbf{i}}' r_0 \omega_0 \sin \omega_0 t \end{aligned} \tag{6-51}$$

where

$$\omega_0 = \sqrt{\frac{g}{l}} \tag{6-52}$$

is the angular frequency of simple harmonic motion. We can use this solution in making a crude estimate of the relative importance of the forces in Eq. (6-41).

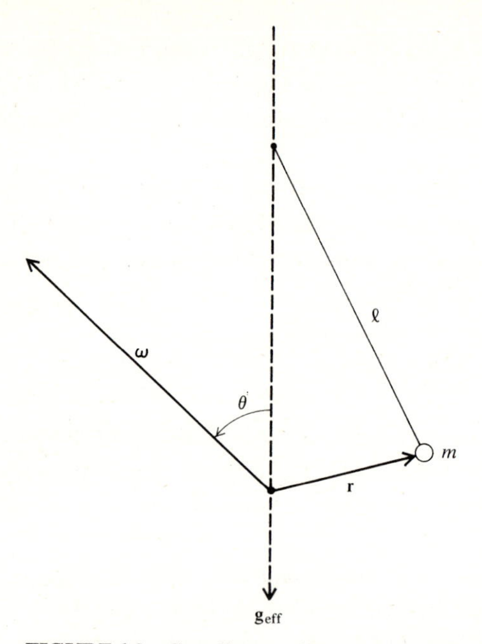

FIGURE 6-8 Coordinate **r** of Foucault pendulum bob.

For small displacements, the gravity force on the bob is

$$\mathbf{F}_G \approx -mg_{\text{eff}} \sin \phi \hat{\mathbf{r}} \approx -mg_{\text{eff}} \frac{r}{l} \hat{\mathbf{r}}$$

According to the discussion in Sec. 6-3, we can neglect the variation of g_{eff} with **r** and essentially take $g_{\text{eff}} \approx g$. This gravitational restoring force can be expressed in terms of the angular frequency in Eq. (6-52) as

$$\mathbf{F}_G = -m\omega_0^2 \mathbf{r} \tag{6-53}$$

From Eqs. (6-51) and (6-53) the order of magnitude of the gravity force is

$$|\mathbf{F}_G| \approx m\omega_0^2 r_0 \tag{6-54}$$

where we have used the fact that $\cos \omega_0 t$ is of order unity. A similar estimate for the magnitude of the Coriolis force term in Eq. (6-41), using Eq. (6-51), gives

$$|\mathbf{F}_{\text{Cor}}| = 2m|\boldsymbol{\omega} \times \mathbf{v}| \approx m\omega v \approx m\omega\omega_0 r_0 \tag{6-55}$$

The ratio of the estimates in Eqs. (6-54) and (6-55) is

$$\frac{|\mathbf{F}_{\text{Cor}}|}{|\mathbf{F}_G|} \approx \frac{\omega}{\omega_0} \tag{6-56}$$

The angular velocity of the earth from Eq. (6-39) is $\omega \approx 10^{-4}$ rad/s. Since any reasonable pendulum will have $\omega_0 \gg \omega$, the gravity force is much more important than the Coriolis force in determining the pendulum motion. The solution in Eq. (6-51) thus represents quite a good first approximation to the pendulum motion.

Despite its relative smallness, the Coriolis force has a significant effect on the motion of the Foucault pendulum because the direction of this force is out of the plane of oscillation. The plane of oscillation is determined by $\mathbf{g}_{\text{eff}}$ and $\mathbf{r}$. Since $\mathbf{v}$ and $\mathbf{r}$ are parallel, the Coriolis force

$$\mathbf{F}_{\text{Cor}} = -2m\boldsymbol{\omega} \times \mathbf{v}$$

has a component perpendicular to the plane of oscillation which causes a corresponding acceleration of the pendulum bob. By use of the small oscillation solution in Eq. (6-51) for motion in the $x'z$ plane, we find

$$\mathbf{F}_{\text{Cor}} \approx (2mr_0\,\omega_0\,\omega \sin \omega_0 t)\hat{\boldsymbol{\omega}} \times \hat{\mathbf{i}}' \tag{6-57}$$

The cross product in Eq. (6-57) is given by

$$\hat{\boldsymbol{\omega}} \times \hat{\mathbf{i}}' = \cos\theta \hat{\mathbf{j}}' + \text{component along } \hat{\mathbf{k}}$$

where θ is colatitude angle and $\hat{\mathbf{j}}'$ is the unit vector in the $\hat{\mathbf{k}} \times \hat{\mathbf{i}}'$ direction. Thus, in the Northern Hemisphere, the Coriolis force in the horizontal plane,

$$\mathbf{F}_{\text{Cor}} = (2mr_0\,\omega_0\,\omega \cos\theta \sin \omega_0 t)\hat{\mathbf{j}}' \tag{6-58}$$

causes the bob to deflect to the right as viewed from above, as shown in Fig. 6-9. The precession of the Foucault pendulum is therefore clockwise north of the equator.

With the approximations in Eqs. (6-53) and (6-58) for the gravitational and Coriolis forces, the equation of motion for small oscillations is

$$m\frac{\delta^2\mathbf{r}}{\delta t^2} = -m{\omega_0}^2\mathbf{r} + 2mr_0\,\omega_0\,\omega \cos\theta \sin \omega_0 t\hat{\mathbf{j}}' \tag{6-59a}$$

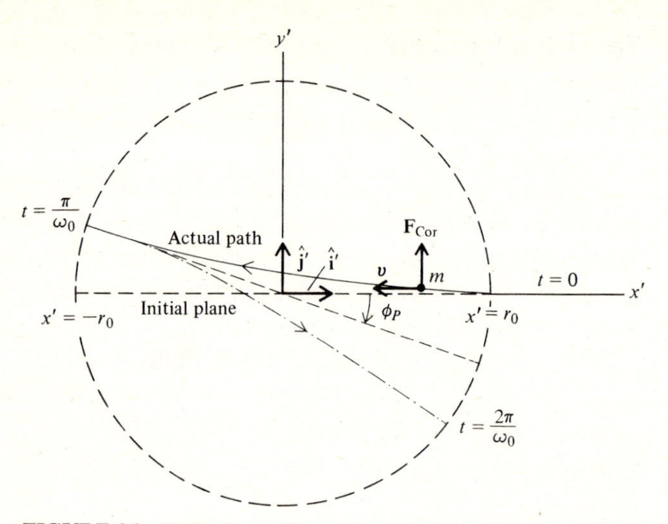

FIGURE 6-9 **Deflection of Foucault pendulum bob as viewed from above.**

where $\mathbf{r} = x'\hat{\mathbf{i}}' + y'\hat{\mathbf{j}}'$ with $y'/x' \ll 1$. The $\hat{\mathbf{j}}'$ component of Eq. (6-59*a*) is

$$\frac{\delta^2 y'}{\delta t^2} = -\omega_0{}^2 y' + 2r_0\,\omega_0\,\omega\cos\theta\sin\omega_0 t \tag{6-59b}$$

The y' motion is that of a harmonic oscillator driven at its resonance frequency ω_0; the solution of the y' equation is

$$y' = a\sin(\omega_0 t + \alpha) - (r_0\,\omega_0\cos\theta)(t\cos\omega_0 t) \tag{6-60}$$

as can be verified by substitution. Here a and α are arbitrary constants. For the initial conditions ($y' = 0$, $\dot{y}' = 0$; $x' = r_0$, $\dot{x}' = 0$) at $t = 0$, the motion is given by

$$x' = r_0\cos\omega_0 t$$

$$y' = \frac{r_0\,\omega\cos\theta}{\omega_0}(\sin\omega_0 t - \omega_0 t\cos\omega_0 t) \tag{6-61}$$

This trajectory is illustrated in Fig. 6-9. The deflection of the bob from the original oscillation plane is proportional to (ω/ω_0), as anticipated in Eq. (6-56). The deflection over one-half oscillation ($t = 0$ to $t = \pi/\omega_0$ and $x' = r_0$ to $x' = -r_0$) is

$$\Delta y' = \frac{\pi r_0\,\omega\cos\theta}{\omega_0} \tag{6-62}$$

The angle of precession is

$$\phi_P = \frac{\Delta y'}{r_0} = \frac{\pi\omega \cos\theta}{\omega_0} \tag{6-63}$$

From this we find the angular velocity of precession

$$\omega_P = \frac{\phi_P}{\Delta t} = \frac{\phi_P}{(\pi/\omega_0)} \tag{6-64}$$

to be

$$\omega_P = \omega \cos\theta$$

The precession vanishes at the equator and is a maximum at the north pole, where the pendulum precesses clockwise through a complete revolution every 24 h. From the viewpoint of an observer in space, the oscillation plane at the north pole remains fixed, while the earth turns counterclockwise beneath it.

6-5 DYNAMICAL BALANCE OF A RIGID BODY

The formulation of the equations of motion in a rotating reference system is also quite valuable in the description of rigid-body motion. As an introduction to the general treatment of rigid-body rotational motion, we discuss a simple example of a dumbbell formed by two point masses m at the ends of a massless rod of length l. The dumbbell rotates at a fixed inclination θ with constant angular velocity $\boldsymbol{\omega}$ about a pivot at the center of the rod, as shown in Fig. 6-10. The equation of motion (5-34) in a fixed reference frame for rotation about the pivot of the rod is

$$\mathbf{N} = \frac{d\mathbf{L}}{dt} \tag{6-65}$$

where $\mathbf{N}$ is the external torque on the rod applied at the pivot. The angular momentum is given by

$$\mathbf{L} = m(\mathbf{r}_1 \times \mathbf{v}_1 + \mathbf{r}_2 \times \mathbf{v}_2) = 2m\mathbf{r}_1 \times \mathbf{v}_1 \tag{6-66}$$

where $\mathbf{r}_2 = -\mathbf{r}_1$ and $\mathbf{v}_2 = -\mathbf{v}_1$ have been used in obtaining the last equality. From Eq. (6-5) the velocity $\mathbf{v}_1$ due to the rotation is

$$\mathbf{v}_1 = \boldsymbol{\omega} \times \mathbf{r}_1 \tag{6-67}$$

Thus $\mathbf{L}$ can be expressed in terms of $\boldsymbol{\omega}$ and $\mathbf{r}_1$ as

$$\mathbf{L} = 2m\mathbf{r}_1 \times (\boldsymbol{\omega} \times \mathbf{r}_1) = 2m[\boldsymbol{\omega} r_1^2 - \mathbf{r}_1(\boldsymbol{\omega} \cdot \mathbf{r}_1)] \tag{6-68}$$

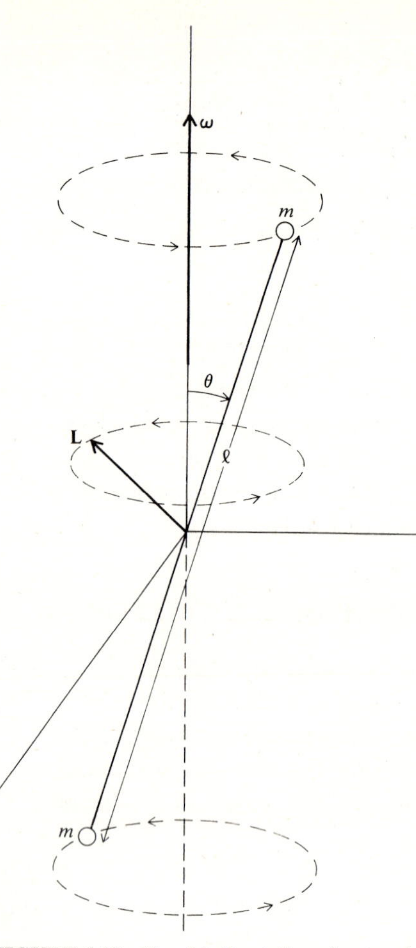

FIGURE 6-10 Dumbbell rotating about a pivot at center of the rod at a fixed inclination angle θ.

Since **L** is perpendicular to $\mathbf{r}_1$ and lies in the plane determined by $\mathbf{r}_1$ and $\boldsymbol{\omega}$, it also rotates with angular velocity $\boldsymbol{\omega}$. From Eq. (6-7), we then have

$$\frac{d\mathbf{L}}{dt} = \boldsymbol{\omega} \times \mathbf{L} \tag{6-69}$$

The external torque from Eqs. (6-65), (6-68), and (6-89) necessary to maintain the rotation is

$$\mathbf{N} = \boldsymbol{\omega} \times \mathbf{L} = 2m(\mathbf{r}_1 \times \boldsymbol{\omega})(\mathbf{r}_1 \cdot \boldsymbol{\omega}) \tag{6-70}$$

We can alternatively derive the result in Eq. (6-70) in a coordinate frame which rotates with the dumbbell. In a rotating reference frame the following rigid-body equation of motion can be derived from Eqs. (6-14) to (6-19):

$$\frac{\delta \mathbf{L}}{\delta t} = \mathbf{N} + \sum_i \mathbf{r}_i \times (\mathbf{F}^i_{cf} + \mathbf{F}^i_{\text{Cor}} + \mathbf{F}^i_{az} + \mathbf{F}^i_{tr}) \tag{6-71}$$

In a coordinate frame rotating with the dumbbell, $\delta \mathbf{L}/\delta t = 0$ and $\mathbf{F}^i_{\text{Cor}} = \mathbf{F}^i_{az} = \mathbf{F}^i_{tr} = 0$. Hence, to maintain the rotation, the torque applied at the pivot must balance the torque due to the centrifugal forces.

$$\mathbf{N} = -(\mathbf{r}_1 \times \mathbf{F}_{cf}{}^1 + \mathbf{r}_2 \times \mathbf{F}_{cf}{}^2) \tag{6-72}$$

From Eq. (6-16) the centrifugal forces are given by

$$\begin{aligned} \mathbf{F}_{cf}{}^1 &= -m\boldsymbol{\omega} \times (\boldsymbol{\omega} \times \mathbf{r}_1) \\ \mathbf{F}_{cf}{}^2 &= -m\boldsymbol{\omega} \times (\boldsymbol{\omega} \times \mathbf{r}_2) \end{aligned} \tag{6-73}$$

Using $\mathbf{r}_1 = -\mathbf{r}_2$ in Eqs. (6-72) and (6-73), the torque reduces to

$$\begin{aligned} \mathbf{N} &= 2m\mathbf{r}_1 \times [\boldsymbol{\omega} \times (\boldsymbol{\omega} \times \mathbf{r}_1)] \\ &= 2m\mathbf{r}_1 \times [\boldsymbol{\omega}(\boldsymbol{\omega} \cdot \mathbf{r}_1) - \mathbf{r}_1 \omega^2] \\ &= 2m(\mathbf{r}_1 \times \boldsymbol{\omega})(\mathbf{r}_1 \cdot \boldsymbol{\omega}) \end{aligned} \tag{6-74}$$

in agreement with the result in Eq. (6-70).

In terms of the angle θ between $\boldsymbol{\omega}$ and $\mathbf{r}_1$, the angular momentum and torque in Eqs. (6-68) and (6-70) of the rotating dumbbell can be written

$$\mathbf{L} = \tfrac{1}{2} m l^2 \omega (\hat{\boldsymbol{\omega}} - \cos\theta \hat{\mathbf{r}}_1) \tag{6-75}$$

$$\mathbf{N} = \tfrac{1}{2} m l^2 \omega \sin\theta \cos\theta \hat{\mathbf{n}} \tag{6-76}$$

where $\hat{\mathbf{n}} = (\mathbf{r}_1 \times \boldsymbol{\omega})/|\mathbf{r}_1 \times \boldsymbol{\omega}|$. For $\theta = \pi/2$, we find

$$\begin{aligned} \mathbf{L} &= (\tfrac{1}{2} m l^2)\boldsymbol{\omega} \\ \mathbf{N} &= 0 \end{aligned} \tag{6-77}$$

In this orientation the motion does not require an imposed torque.

For rotation of a dumbbell started at an angle θ with no external constraint on θ, the centrifugal forces cause the orientation of the rod to oscillate about $\theta = \pi/2$, as illustrated in Fig. 6-11.

From Eqs. (6-75) and (6-76), we see that torques on the rod are present whenever the angular momentum $\mathbf{L}$ does not lie along the axis of rotation $\boldsymbol{\omega}$. This result is generally true for rigid-body rotations.

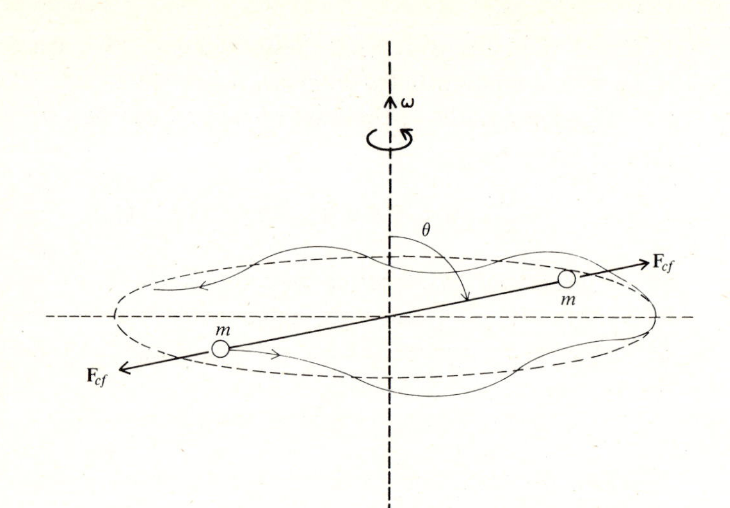

FIGURE 6-11 Motion of a rotating dumbbell which is free to change its θ orientation.

A practical application in which it is important that **L** and **ω** are parallel is the dynamic balancing of automobile tires. If a wheel is not balanced, noise and vibration result in the car and excessive wear occurs on the tire. There are two criteria for complete balance of a wheel.

(1) *Static balance:* Unless the CM of the wheel lies on the rotation axis, a centrifugal force is needed to maintain rotation. This tends to make the axle oscillate and imparts vibration to the car. A static balance consists of the application of weights on the rim of the wheel so that the wheel is in equilibrium at any fixed angle.

(2) *Dynamic balance:* Even when the CM lies on the wheel axis, it is possible that in rotation the angular momentum does not lie along the axis. If we specify the x axis as the rotation axis, $\boldsymbol{\omega} = \omega\hat{\mathbf{i}}$, the angular-momentum vector from Eq. (5-76) is

$$\mathbf{L} = (I_{xx}\hat{\mathbf{i}} + I_{yx}\hat{\mathbf{j}} + I_{zx}\hat{\mathbf{k}})\omega \tag{6-78}$$

Unless the products of inertia I_{yx} and I_{zx} vanish, **L** does not lie along **ω**. The time variation of **L** then leads to a torque, which tends to make the wheel wobble. A dynamic balance consists of the application of weights until the wheel spins smoothly with no wobble. Since tires are usually very nearly symmetrical, a static balance alone is usually sufficient to ensure good driving results.

6-6 PRINCIPAL AXES AND EULER'S EQUATIONS

For a rigid body of arbitrary shape, the rotational equation of motion (5-34) in a fixed coordinate system with origin at the center-of-mass point is

$$N_j = \frac{dL_j}{dt} = \frac{d}{dt}(I_{jk}\,\omega_k) \tag{6-79}$$

where a sum over the index k is implied. Since the moments and products of inertia I_{jk} relative to the fixed coordinate system change as a function of time as the body rotates, the description of the motion through Eq. (6-79) can be cumbersome and difficult. The analysis of the motion can often be greatly simplified by choosing instead a coordinate system that rotates with the body. In this reference frame the moments and products of inertia are time-independent. Using Eqs. (6-7) and (6-79), the equation of motion with respect to the moving-"body" axes is

$$N_j = \frac{\delta L_j}{\delta t} + (\boldsymbol{\omega} \times \mathbf{L})_j \tag{6-80}$$

A further simplification can be made by a judicious choice of the orientations of the rotating axes with respect to the rigid body. As we shall shortly prove, it is always possible to make a choice of axes in the body for which all the products of inertia vanish.

$$I_{ij} = 0 \qquad \text{for} \quad i \neq j \tag{6-81}$$

The axes for which Eq. (6-81) holds are called the *principal axes* of the rigid body. For these axes the angular-momentum components in Eq. (5-76) reduce to

$$\begin{aligned} L_1 &= I_{11}\omega_1 \equiv I_1\omega_1 \\ L_2 &= I_{22}\omega_2 \equiv I_2\omega_2 \\ L_3 &= I_{33}\omega_3 \equiv I_3\omega_3 \end{aligned} \tag{6-82}$$

where I_1, I_2, I_3 denote the principal moments of inertia. From Eq. (6-80) we obtain Euler's equations of motion for a rigid body in terms of the coordinate system aligned with the principal axes of the body.

$$N_1 = I_1\dot{\omega}_1 + (I_3 - I_2)\omega_3\omega_2$$
$$N_2 = I_2\dot{\omega}_2 + (I_1 - I_3)\omega_1\omega_3 \qquad (6\text{-}83)$$
$$N_3 = I_3\dot{\omega}_3 + (I_2 - I_1)\omega_2\omega_1$$

These equations are a convenient starting point for many discussions of rigid-body rotations.

To illustrate the application of Euler's equations, we return to the rotating rod of the preceding section. The principal axes of the body lie along and perpendicular to the rod, as illustrated in Fig. 6-12.

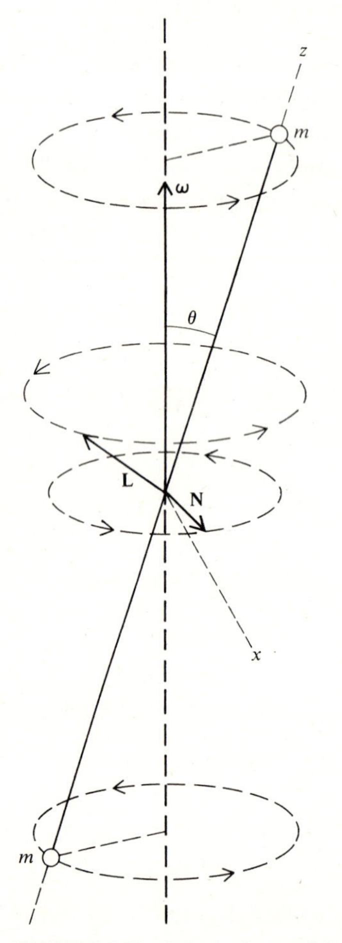

FIGURE 6-12 Principal axes of the dumbbell.

With the z axis along the rod and the x axis in the plane of the rod and $\boldsymbol{\omega}$, the components of $\boldsymbol{\omega}$ are

$$\begin{aligned} \omega_1 &= \omega \sin \theta \\ \omega_2 &= 0 \\ \omega_3 &= \omega \cos \theta \end{aligned} \tag{6-84}$$

where θ is the angle between $\boldsymbol{\omega}$ and the rod. The principal moments of inertia are

$$I_1 = I_2 = m\left(\frac{l}{2}\right)^2 + m\left(\frac{l}{2}\right)^2 = \tfrac{1}{2}ml^2 \tag{6-85}$$

$$I_3 = 0$$

Using Eqs. (6-84) and (6-85) in Eq. (6-83), we find

$$\begin{aligned} N_1 &= 0 \\ N_2 &= (\tfrac{1}{2}ml^2)\omega^2 \sin \theta \cos \theta \\ N_3 &= 0 \end{aligned} \tag{6-86}$$

where $\dot{\boldsymbol{\omega}} = 0$ has been used. This result obtained from Euler's equations is the same as Eq. (6-75).

In the derivation of Eq. (6-83) we have used the diagonal property in Eq. (6-81) of the inertia tensor in the principal-axes coordinate system. We will now establish this assertion. Suppose that the body rotates about one of its principal axes. For definiteness we take $\boldsymbol{\omega} = \omega_1^* \hat{\mathbf{i}}^*$. The asterisk is used here to distinguish principal axes from inertial axes. From Eq. (6-82) we find

$$\begin{aligned} L_1^* &= I_1 \omega_1^* \\ L_2^* &= 0 \\ L_3^* &= 0 \end{aligned} \tag{6-87}$$

or simply,

$$\mathbf{L} = I_1 \boldsymbol{\omega} \tag{6-88}$$

In the inertial coordinate system, $\boldsymbol{\omega}$ will in general have three components:

$$\boldsymbol{\omega} = \omega_1 \hat{\mathbf{i}} + \omega_2 \hat{\mathbf{j}} + \omega_3 \hat{\mathbf{k}}$$

From Eq. (6-88) the components of $\mathbf{L}$ along the inertial axes are

$$\begin{aligned} L_1 &= I_1 \omega_1 \\ L_2 &= I_1 \omega_2 \\ L_3 &= I_1 \omega_3 \end{aligned} \tag{6-89}$$

These components of **L** must be equivalent to the expression for **L** given in Eq. (5-76), namely,

$$\begin{aligned} L_1 &= I_{11}\omega_1 + I_{12}\omega_2 + I_{13}\omega_3 \\ L_2 &= I_{21}\omega_1 + I_{22}\omega_2 + I_{23}\omega_3 \\ L_3 &= I_{31}\omega_1 + I_{32}\omega_2 + I_{33}\omega_3 \end{aligned} \tag{6-90}$$

Equating the components in Eqs. (6-89) and (6-90), we find

$$\begin{aligned} (I_{11} - I_1)\omega_1 + I_{12}\omega_2 + I_{13}\omega_3 &= 0 \\ I_{21}\omega_1 + (I_{22} - I_1)\omega_2 + I_{23}\omega_3 &= 0 \\ I_{31}\omega_1 + I_{32}\omega_2 + (I_{33} - I_1)\omega_3 &= 0 \end{aligned} \tag{6-91}$$

For $\boldsymbol{\omega} \neq 0$, this system of homogeneous equations for $(\omega_1, \omega_2, \omega_3)$ has solutions only if the determinant of the coefficients of the $\boldsymbol{\omega}$ components vanishes.

$$\begin{vmatrix} (I_{11} - I_1) & I_{12} & I_{13} \\ I_{21} & (I_{22} - I_1) & I_{23} \\ I_{31} & I_{32} & (I_{33} - I_1) \end{vmatrix} = 0 \tag{6-92}$$

This leads to a cubic equation in I_1 of the form

$$I_1{}^3 + aI_1{}^2 + bI_1 + c = 0 \tag{6-93}$$

where a, b, and c are products of the inertia tensor elements I_{ij}. Of the three real solutions for I_1 from Eq. (6-93), any is appropriate to Eq. (6-87). The other two solutions refer to the principal moments about the other two principal axes, as can be shown by repeating the above argument for a rotation about $\hat{\mathbf{j}}^*$ or $\hat{\mathbf{k}}^*$ principal axis. By construction we have therefore shown that it is always possible to find a principal-axis system for any rigid body. It can be shown that $\hat{\mathbf{i}}^*$, $\hat{\mathbf{j}}^*$, $\hat{\mathbf{k}}^*$ are mutually orthogonal. In many applications the choice of principal axes is obvious from the symmetry of the body.

6-7 THE TENNIS RACKET THEOREM

The solution of Euler's equations for a rigid body with unequal principal moments of inertia can be beautifully illustrated with a tennis racket. The three principal axes of a tennis racket are readily identified to be (1) along the handle, (2) perpendicular to the handle in the plane of the strings, and (3) perpendicular to the handle and strings. When a tennis racket is tossed into the air with a spin about one of the principal axes, a curious phenomenon is observed. If the initial spin is about either the axis (1) or axis (3), the racket continues to spin uniformly about the initial axis and can easily be recaught. On the other hand, if the initial spin is about axis (2), the motion quickly becomes irregular, with spin developing about all three principal axes, which makes it difficult to catch the falling racket. The explanation of the observed behavior follows from Euler's equations. To apply Euler's equations to the tennis racket, we choose the origin of the principal-axes coordinate system at the CM of the racket, as illustrated in Fig. 6-13.

Since gravity is a uniform force in the vicinity of the earth's surface, there are no gravitational torques about the CM of the racket. If we neglect torques due to wind resistance, Euler's equations (6-83) simplify to

$$\begin{aligned} I_1\dot{\omega}_1 + (I_3 - I_2)\omega_3\omega_2 &= 0 \\ I_2\dot{\omega}_2 + (I_1 - I_3)\omega_1\omega_3 &= 0 \\ I_3\dot{\omega}_3 + (I_2 - I_1)\omega_2\omega_1 &= 0 \end{aligned} \tag{6-94}$$

To calculate the principal moments of inertia in Eq. (6-94), we use a grossly simplified model for the racket. We represent the mass distribution of the racket by a circular hoop of radius a and mass m_a

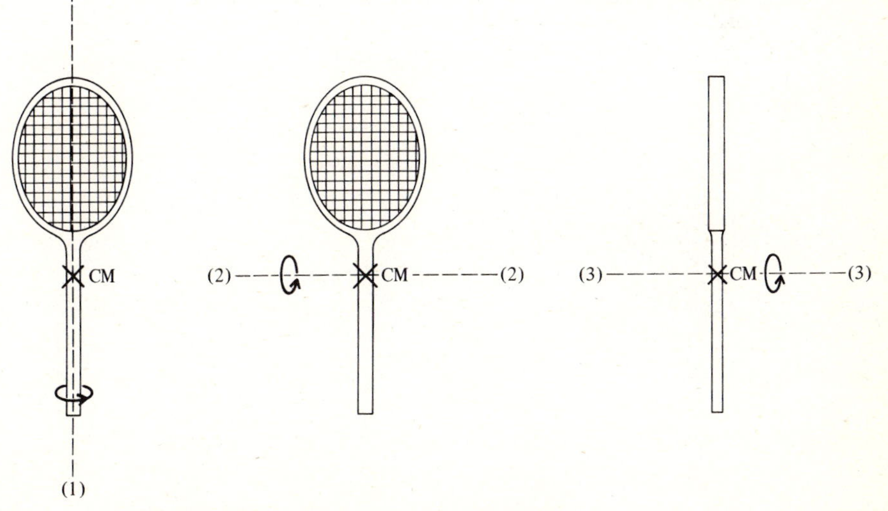

FIGURE 6-13 Principal axes of the tennis racket.

connected to a thin rod of length l and mass m_l. The total mass of the racket is $M = m_a + m_l$. The CM of the racket is located on principal axis (1) at a distance R from the center of the hoop, where

$$MR = m_a(0) + m_l \left(a + \frac{l}{2}\right)$$

or

$$R = \frac{m_l}{M}\left(a + \frac{l}{2}\right) \tag{6-95}$$

The moment of inertia of the racket about principal axis (1) comes entirely from the hoop. We use the perpendicular-axis rule in Eq. (5-97) to obtain

$$I_1 = \tfrac{1}{2} m_a a^2 \tag{6-96}$$

To compute the moment of inertia about principal axis (2), we will make use of the parallel-axis rule of Eq. (5-91). The moment of inertia of the hoop about an axis through its CM and parallel to principal axis (2) is $\frac{1}{2}m_a a^2$. By Eq. (5-91), the hoop makes a contribution to I_2 of

$$I_2^{\text{hoop}} = \tfrac{1}{2} m_a a^2 + m_a R^2 \tag{6-97}$$

since R is the perpendicular distance between the two parallel axes. The moment of inertia of the handle about an axis parallel to principal axis (2) passing through the CM of the handle is $\frac{1}{12}m_l l^2$. Again using Eq. (5-91), we find that the contribution of the handle to I_2 is

$$I_2^{\text{handle}} = \tfrac{1}{12} m_l l^2 + m_l \left(a + \frac{l}{2} - R\right)^2 \tag{6-98}$$

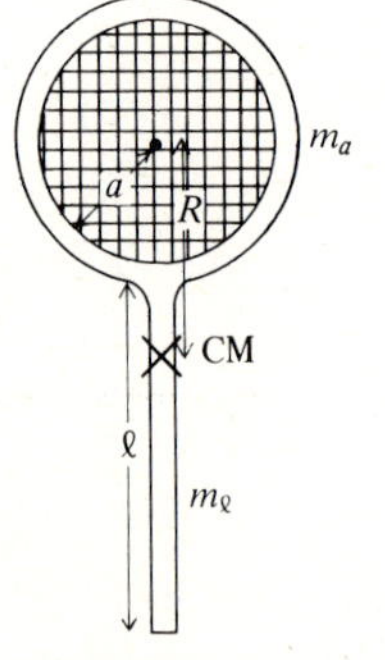

FIGURE 6-14 Dimensions of tennis racket model.

where $(a + l/2 - R)$ is the distance between these parallel axes. Combining Eqs. (6-97) and (6-98) and substituting for R from Eq. (6-95), we obtain

$$I_2 = \tfrac{1}{2}m_a a^2 + m_a \left(\frac{m_l}{M}\right)^2 \left(a + \frac{l}{2}\right)^2 + \tfrac{1}{12}m_l l^2 + m_l \left(\frac{m_a}{M}\right)^2 \left(a + \frac{l}{2}\right)^2$$

This can be further simplified to

$$I_2 = \tfrac{1}{2}m_a a^2 + \tfrac{1}{12}m_l l^2 + \frac{m_a m_l}{M}\left(a + \frac{l}{2}\right)^2 \tag{6-99}$$

Finally, for principal axis (3), the racket lies in a plane perpendicular to the axis, and we can use the perpendicular-axis rule of Eq. (5-96) to obtain

$$I_3 = I_1 + I_2 \tag{6-100}$$

By comparison of Eqs. (6-96), (6-99), and (6-100), we see that the principal moments of inertia are ordered as

$$I_1 < I_2 < I_3 \tag{6-101}$$

Characteristic parameters for our model of a wood tennis racket are

$$\begin{aligned} a &= 0.13 \text{ m} & l &= 0.38 \text{ m} \\ R &= 0.18 \text{ m} & M &= 0.34 \text{ kg} \\ m_l &= 0.18 \text{ kg} & m_a &= 0.15 \text{ kg} \end{aligned} \tag{6-102}$$

The principal moments of inertia from Eqs. (6-96), (6-99), (6-100), and (6-102) are

$$\begin{aligned} I_1 &= 0.1 \times 10^{-2} \text{ kg·m}^2 \\ I_2 &= 1.2 \times 10^{-2} \text{ kg·m}^2 \\ I_3 &= 1.3 \times 10^{-2} \text{ kg·m}^2 \end{aligned} \tag{6-103}$$

With this information, we return to the solution of Euler's equation in Eq. (6-94).

When we substitute the moment-of-inertia relation of Eq. (6-100) into Eq. (6-94), we get

$$\dot{\omega}_1 + \omega_3\omega_2 = 0 \tag{6-104}$$

$$\dot{\omega}_2 - \omega_1\omega_3 = 0 \tag{6-105}$$

$$\dot{\omega}_3 + r\omega_2\omega_1 = 0 \tag{6-106}$$

where $r = (I_2 - I_1)/(I_2 + I_1)$. Henceforth we shall use the approximation $r \approx 1$ as determined from the moments of inertia in Eq. (6-103). We are now in a position to discuss qualitatively the nature of the motion when the racket is initially spinning primarily about one of its principal axes.

If initially the spin is about axis (1), the product $\omega_2\,\omega_3$ in Eq. (6-104) can be neglected. We then find that ω_1 is constant.

$$\omega_1 = \omega_1(0) \tag{6-107}$$

To solve the remaining two equations, we introduce a complex variable,

$$\tilde{\omega} = \omega_3 + i\omega_2 \tag{6-108}$$

Since $\omega_2 = \text{Im}\,\tilde{\omega}$ and $\omega_3 = \text{Re}\,\tilde{\omega}$, Eqs. (6-105) and (6-106) become

$$\begin{aligned} \text{Im}\,\dot{\tilde{\omega}} - \omega_1\,\text{Re}\,\tilde{\omega} &= 0 \\ \text{Re}\,\dot{\tilde{\omega}} + \omega_1\,\text{Im}\,\tilde{\omega} &= 0 \end{aligned} \tag{6-109}$$

These two equations can be combined as a single equation for the complex variable $\tilde{\omega}$.

$$\dot{\tilde{\omega}} - i\omega_1\tilde{\omega} = 0 \tag{6-110}$$

This differential equation has the exponential solution

$$\tilde{\omega}(t) = a\,e^{i(\omega_1 t + \alpha)} \tag{6-111}$$

The corresponding results for ω_2 and ω_3 are

$$\begin{aligned} \omega_2(t) &= a \sin(\omega_1 t + \alpha) \\ \omega_3(t) &= a \cos(\omega_1 t + \alpha) \end{aligned} \tag{6-112}$$

By the initial conditions, the amplitude a is small, and we see that these components of the angular velocity remain small. In the approximate solution above, we find

$$\begin{aligned} |\tilde{\omega}| &= \sqrt{\omega_2(t)^2 + \omega_3(t)^2} = a \\ \omega &= \sqrt{\omega_1(t)^2 + \omega_2(t)^2 + \omega_3(t)^2} = \sqrt{\omega_1{}^2 + a^2} \end{aligned} \tag{6-113}$$

Thus the angular velocity vector $\boldsymbol{\omega} = \omega_1\hat{\mathbf{i}} + \omega_2\hat{\mathbf{j}} + \omega_3\hat{\mathbf{k}}$ precesses in a small cone about principal axis (1), as illustrated in Fig. 6-15. This explains the stability of the racket in rotational motion about axis (1).

For an initial spin of the racket primarily about axis (3), the solution to Euler's equations is similar to the case just treated. With $r = 1$, the mathematical structure of Eqs. (6-104) to (6-106) is preserved

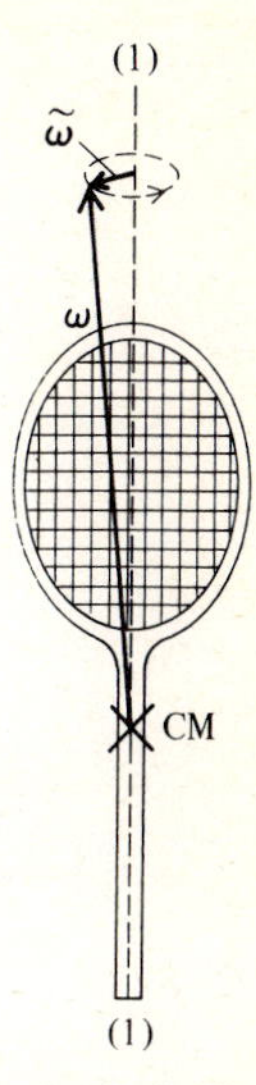

FIGURE 6-15 Stable precession of the angular velocity ω about principal axis (1) of the tennis racket.

under the interchange of ω_1 and ω_3. Thus, by analogy with Eqs. (6-107) and (6-112), the approximate solution is

$$\begin{aligned} \omega_3(t) &= \omega_3(0) \\ \omega_1(t) &= a \cos(\omega_3 t + \alpha) \\ \omega_2(t) &= a \sin(\omega_3 t + \alpha) \end{aligned} \tag{6-114}$$

Again the rotational motion is stable about this axis.

For initial spin about principal axis (2), the situation is different. In this case we initially neglect the product $\omega_1\omega_3$ in Eq. (6-105) to obtain

$$\omega_2(t) = \omega_2(0) \tag{6-115}$$

By separate addition and subtraction of Eqs. (6-104) and (6-106), we find

$$\begin{aligned} (\dot{\omega}_1 + \dot{\omega}_3) + \omega_2(\omega_1 + \omega_3) &= 0 \\ (\dot{\omega}_1 - \dot{\omega}_3) - \omega_2(\omega_1 - \omega_3) &= 0 \end{aligned} \tag{6-116}$$

The solutions for these linear combinations are

$$\begin{aligned} (\omega_1 + \omega_3) &= ae^{-\omega_2 t} \\ (\omega_1 - \omega_3) &= be^{+\omega_2 t} \end{aligned} \tag{6-117}$$

Upon solving for ω_1 and ω_3, we get

$$\begin{aligned} \omega_1(t) &= \tfrac{1}{2}(ae^{-\omega_2 t} + be^{+\omega_2 t}) \\ \omega_3(t) &= \tfrac{1}{2}(ae^{-\omega_2 t} - be^{+\omega_2 t}) \end{aligned} \tag{6-118}$$

In this motion the angular velocities about axes (1) and (3) grow rapidly with time and racket tumbles. The explicit form of the solution in Eqs. (6-115) and (6-118) is valid only for small times, only so long as the product $\omega_1\omega_3$ in Eq. (6-105) is negligible.

The condition of stability of the motion about a principal axis which has either the largest or the smallest moment of inertia and the instability about the other principal axis is often called the "tennis racket theorem." The conclusions of this theorem can be readily demonstrated by throwing a book or other oblong object into the air with a spin about one of the principal axes.

6-8 THE EARTH AS A FREE SYMMETRIC TOP

Since the earth is nearly spherical in shape, the gravitational torques exerted on the earth by the sun and the moon are quite small. To a good approximation the rotational motion can therefore be described by Euler's equations with no external torques. Since the earth is nearly axially symmetric, the principal moments of inertia for the two axes in the equatorial plane are equal.

$$I_1 = I_2 \equiv I \tag{6-119}$$

The third principal axis with moment of inertia I_3 is along the polar symmetry axis. From Eqs. (6-94) the differential equations for the earth's motion are

$$\begin{aligned} \dot{\omega}_1 + \frac{I_3 - I}{I}\,\omega_3\omega_2 &= 0 \\ \dot{\omega}_2 - \frac{I_3 - I}{I}\,\omega_1\omega_3 &= 0 \\ \dot{\omega}_3 &= 0 \end{aligned} \tag{6-120}$$

Any rigid body which obeys this set of torque-free equations is called a *free axially symmetric top*. The exact solution to this coupled set of equations is easily obtained by the method discussed in Sec. 6-7. The last equation above implies that ω_3 is constant.

$$\omega_3(t) = \omega_3(0) = \omega_3 \tag{6-121}$$

When we introduce the complex variable

$$\tilde{\omega} = \omega_1 + i\omega_2 \tag{6-122}$$

the remaining two differential equations can be written

$$\dot{\tilde{\omega}} - i\Omega\tilde{\omega} = 0 \tag{6-123}$$

where

$$\Omega = \omega_3 \left(\frac{I_3 - I}{I} \right) \tag{6-124}$$

The solution for $\tilde{\omega}$ is

$$\tilde{\omega}(t) = ae^{i(\Omega t + \alpha)} \tag{6-125}$$

which gives

$$\begin{aligned} \omega_1(t) &= a \cos (\Omega t + \alpha) \\ \omega_2(t) &= a \sin (\Omega t + \alpha) \end{aligned} \tag{6-126}$$

The magnitude of the angular-velocity vector ω is

$$\omega = \sqrt{\omega_1{}^2 + \omega_2{}^2 + \omega_3{}^2} = \sqrt{a^2 + \omega_3{}^2} \tag{6-127}$$

Since the components ω_1 and ω_2 in Eq. (6-126) trace out a circle of radius a while ω_3 and ω remain constant, the angular-velocity vector precesses uniformly in the body frame about the symmetry axis with angular velocity Ω.

The period of precession of $\boldsymbol{\omega}$ about the earth's symmetry axis is

$$\tau = \frac{2\pi}{\Omega} = \left(\frac{I}{I_3 - I} \right) \frac{2\pi}{\omega_3} \tag{6-128}$$

Since $2\pi/\omega_3 = 1$ day, the period of precession in days is determined by the moment-of-inertia ratio. For an earth of uniform density and oblate spheroidal shape, the value of this ratio, calculated from the measured radii of the earth, is

$$\frac{I}{I_3 - I} \approx 300 \tag{6-129}$$

Although the earth becomes more dense toward its center, the moment-of-inertia ratio is not appreciably changed from the uniform-density result. Thus the expected precessional period is about 300 days.

The direction of the earth's axis of rotation (i.e., the direction of $\boldsymbol{\omega}$) can be experimentally determined by location of the point in the night sky which appears to remain stationary as the earth rotates, as

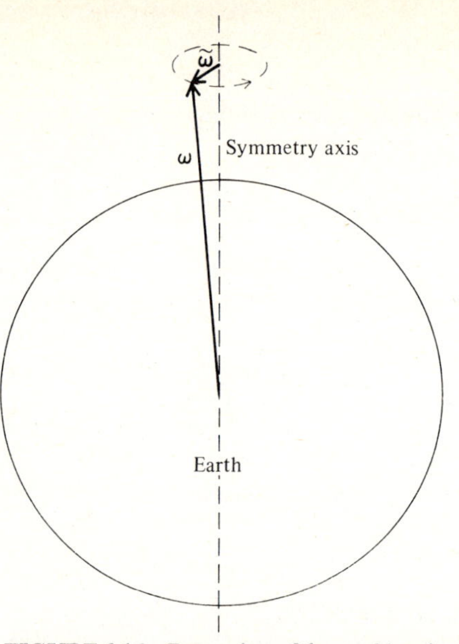

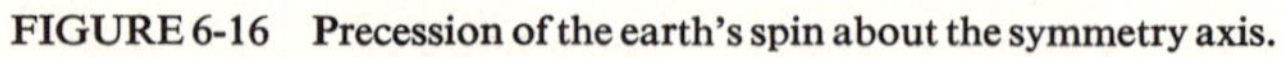

FIGURE 6-16 **Precession of the earth's spin about the symmetry axis.**

FIGURE 6-17 **Star trails in the night sky photographed with an 8-h exposure by a camera fixed on the earth.** *(Photo courtesy of Lick Observatory.)*

illustrated in Fig. 6-17. The direction of the earth's rotational axis is observed to precess about the symmetry axis with a period of about 440 days. The angle between $\boldsymbol{\omega}$ and the symmetry axis is quite small. In fact, at the north pole, $\boldsymbol{\omega}$ never moves more than 5 m from the symmetry axis. The actual motion of $\boldsymbol{\omega}$ is rather irregular, being strongly affected by earthquakes and seasonal changes. In fact, it is only due to these effects that the motion has a nonvanishing amplitude. On a quiet earth, viscous effects would rapidly damp out such a motion, and $\boldsymbol{\omega}$ would soon lie along the symmetry axis (this minimizes E for fixed $\mathbf{L}$). The discrepancy between the expected period of 300 days and the observed value of about 440 days is presumably due to the nonrigidity of the earth.

6-9 THE FREE SYMMETRIC TOP: EXTERNAL OBSERVER

The description of the earth's rotational motion as a free symmetric top in Sec. 6-8 was appropriate for an observer at rest in the rotating reference frame. In this section we concentrate on the motion of a free symmetric top as viewed by an external observer in an inertial frame. Since any object tossed into the air is basically a free top, the inertial description has a wide range of applications.

For a symmetric top the angular momentum and angular velocity with respect to the principal axes $(\hat{\mathbf{i}}, \hat{\mathbf{j}}, \hat{\mathbf{k}})$ in the top are

$$\begin{aligned} \mathbf{L} &= I(\omega_1\hat{\mathbf{i}} + \omega_2\hat{\mathbf{j}}) + I_3\omega_3\hat{\mathbf{k}} \\ \boldsymbol{\omega} &= (\omega_1\hat{\mathbf{i}} + \omega_2\hat{\mathbf{j}}) + \omega_3\hat{\mathbf{k}} \end{aligned} \tag{6-130}$$

where $\hat{\mathbf{k}}$ is in the direction of the symmetry axis. By eliminating $(\omega_1\hat{\mathbf{i}} + \omega_2\hat{\mathbf{j}})$ in these equations, the angular-velocity vector $\boldsymbol{\omega}$ can be expressed in terms of $\hat{\mathbf{L}}$ and $\hat{\mathbf{k}}$ as

$$\boldsymbol{\omega} = \frac{L}{I}\hat{\mathbf{L}} - \Omega\hat{\mathbf{k}} \tag{6-131}$$

where

$$\Omega \equiv \left(\frac{I_3 - I}{I}\right)\omega_3$$

as before, in Eq. (6-124). Since there is a linear relation Eq. (6-131) among $\boldsymbol{\omega}$, $\mathbf{L}$, and $\hat{\mathbf{k}}$, these three vectors must lie in a plane. The absence of torques on the top implies that $\mathbf{L}$ is constant in the inertial system. Thus the $\boldsymbol{\omega}\hat{\mathbf{k}}$ plane rotates around the direction of $\mathbf{L}$.

According to Eq. (6-131), the motion of the top as viewed from the inertial frame can be resolved into components ω_l along $\hat{\mathbf{L}}$ and ω_k along $\hat{\mathbf{k}}$ as

$$\omega_l = \frac{L}{I} \qquad \omega_k = -\Omega \tag{6-132}$$

Since $\hat{\mathbf{k}}$ is a vector fixed in the body (i.e., it rotates with the body), we have from Eqs. (6-6) and (6-131)

$$\dot{\hat{\mathbf{k}}} = \boldsymbol{\omega} \times \hat{\mathbf{k}} = \omega_l \hat{\mathbf{L}} \times \hat{\mathbf{k}} \tag{6-133}$$

Hence $\hat{\mathbf{k}}$ rotates with fixed angular velocity $\omega_l \hat{\mathbf{L}}$. The angular velocity $\boldsymbol{\omega}^*$ of the top as observed from the "precessing frame" which rotates with angular velocity $\omega_l \hat{\mathbf{L}}$ is

$$\boldsymbol{\omega}^* = \boldsymbol{\omega} - \omega_l \hat{\mathbf{L}} = -\Omega \hat{\mathbf{k}}$$

The motion of the top as seen from this frame is a rotation about the symmetry axis $\hat{\mathbf{k}}$ at the angular rate $-\Omega$. Since this is the rate that the top rotates with respect to $\boldsymbol{\omega}$ (which is a fixed vector in the precessing frame), we conclude that $+\Omega$ is the rate that $\boldsymbol{\omega}$ rotates with respect to the body, in agreement with the result Eq. (6-124) found from Euler's equations.

In the motion of the top, the angles that the symmetry axis $\hat{\mathbf{k}}$ makes with the vectors $\mathbf{L}$ and $\boldsymbol{\omega}$ remain constant, as can be shown from Eqs. (6-131) and (6-133) or from energy- and angular-momentum conservation. Since there are no torques, both the angular momentum $\mathbf{L}$ and the rotational kinetic energy T are constant. From Eq. (6-130) we can write $\mathbf{L}$ as

$$\mathbf{L} = I\omega_n \hat{\mathbf{n}} + I_3 \omega_3 \hat{\mathbf{k}} \tag{6-134}$$

where

$$\omega_n \hat{\mathbf{n}} \equiv \omega_1 \hat{\mathbf{i}} + \omega_2 \hat{\mathbf{j}}$$

is orthogonal to $\hat{\mathbf{k}}$. In terms of the components ω_n and ω_3, we have

$$L^2 = I^2 \omega_n{}^2 + I_3{}^2 \omega_3{}^2 \qquad 2T = \mathbf{L} \cdot \boldsymbol{\omega} = I\omega_n{}^2 + I_3 \omega_3{}^2 \tag{6-135}$$

where we have used the expression (5-82) for the rotational kinetic energy. The constancy of L^2 and T in Eq. (6-135) requires in turn that ω_n and ω_3 be constant. The magnitude of $\boldsymbol{\omega}$ in Eq. (6-130),

$$\omega = \sqrt{\omega_n{}^2 + \omega_3{}^3}$$

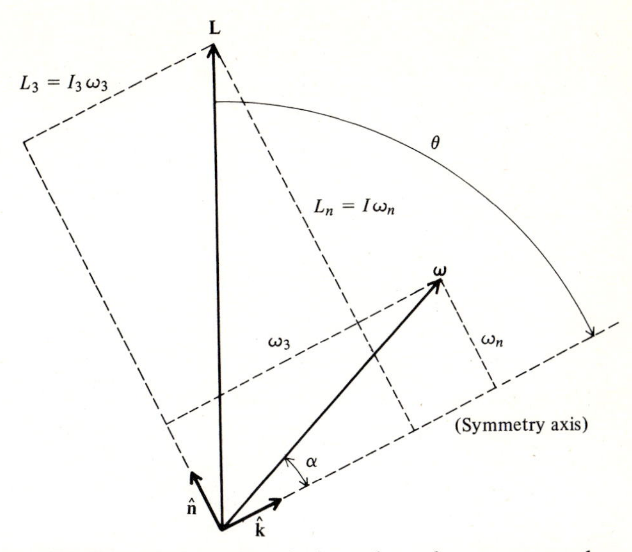

FIGURE 6-18 Components of angular velocity and angular momentum along the symmetry axis $\hat{\mathbf{k}}$ of the top and an axis $\hat{\mathbf{n}}$ perpendicular to the symmetry axis in the plane of $\boldsymbol{\omega}$ and **L**.

is then also constant. From the geometry of Fig. 6-18, the angles of interest are determined by

$$\tan\alpha = \frac{\omega_n}{\omega_3}$$
$$\tan\theta = \frac{L_n}{L_3} = \frac{I\omega_n}{I_3\omega_3} \tag{6-136}$$

The fixed relative orientation of **L**, $\boldsymbol{\omega}$, and $\hat{\mathbf{k}}$ follows immediately from these results. If we eliminate ω_n/ω_3 in Eqs. (6-136), we obtain

$$\tan\alpha = \frac{I_3}{I}\tan\theta \tag{6-137}$$

For an oblate top (pancake or coinlike), $I_3 > I$, and the angle α is larger than θ. For a prolate top (football- or cigar-shape), $\alpha < \theta$, which is the case illustrated in Fig. 6-18.

A simple geometric construction can be made to illustrate symmetrical free-top motion in an inertial reference frame. This construction is based on the constancy of the angles θ and α. As the plane containing the vectors $\boldsymbol{\omega}$ and $\hat{\mathbf{k}}$ precesses about **L**, the vector $\boldsymbol{\omega}$ sweeps out a cone (the space cone) of half-angle $(\theta - \alpha)$ about the fixed

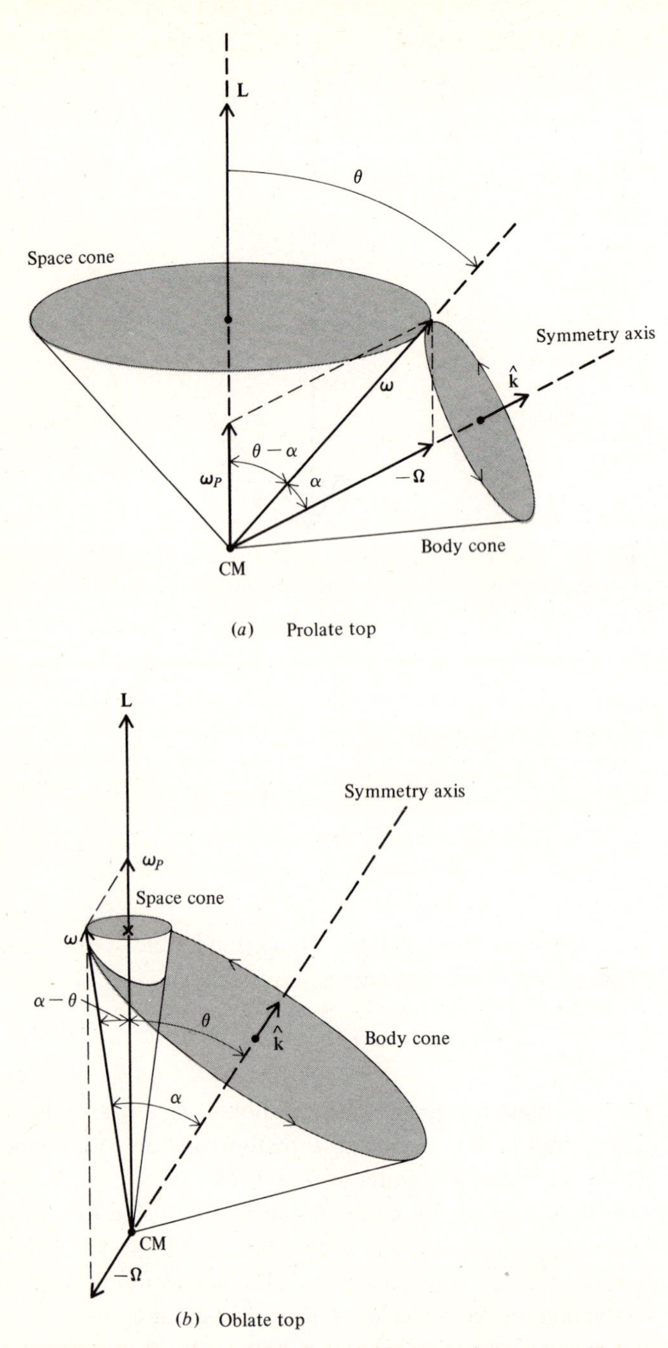

FIGURE 6-19 Space and body cones for (*a*) prolate top, (*b*) oblate top.

direction **L**. In the coordinate system fixed in the top, the vector $\boldsymbol{\omega}$ sweeps out a cone (the body cone) of half-angle α. Since $\boldsymbol{\omega}$ sweeps out both the space and body cones, the line of contact between the two cones is simply the vector $\boldsymbol{\omega}$. The points on the body cone which lie on the vector $\boldsymbol{\omega}$ are instantaneously at rest with respect to the fixed-space cone because $\boldsymbol{\omega}$ is the instantaneous axis of rotation of the top. As a consequence, the body cone must roll on the fixed-space cone without slipping. Thus we have a qualitative picture of the top's motion as the body cone rolling on the space cone. This is illustrated in Fig. 6-19*a* for a prolate top, and in Fig. 6-19*b* for an oblate top.

6-10 SPINNING TOP, INCLUDING GRAVITY

Untold generations of children have been fascinated by the precessing, rising, sleeping, and dying of spinning tops. The theory of spinning tops plays an important role in a wide variety of disciplines ranging from astronomy to applied mechanics to nuclear physics. In this section we discuss the motion of a symmetric top in a gravity field for a special case in which the point of contact of the top with supporting surface, the pivot, is fixed.

For the analysis of the motion, we choose the inertial axes with the origin at the fixed point of contact and the negative z axis in the direction of the gravity force. We calculate the angular momentum of the top and the gravitational torque on the top about the origin of the inertial system. The axis of symmetry $\hat{\mathbf{k}}$ of the top can be specified by spherical angles (θ, ϕ) with respect to the inertial axes, as shown in Fig. 6-20. The orientation of the $(\hat{\mathbf{i}}, \hat{\mathbf{j}})$ principal axes of the top about the symmetry axis $\hat{\mathbf{k}}$ can be specified by a further angle, ψ. Since angular velocities can be added vectorially, the angular velocity $\boldsymbol{\omega}$ of the top is then given in terms of the angles (θ, ϕ, ψ) by

$$\boldsymbol{\omega} = \dot{\phi}\hat{\mathbf{z}} - \dot{\theta}\hat{\mathbf{l}} + \dot{\psi}\hat{\mathbf{k}} \tag{6-138}$$

where $\hat{\mathbf{l}}$ is a unit vector perpendicular to both $\hat{\mathbf{k}}$ and $\hat{\mathbf{z}}$ (in the direction $\hat{\mathbf{k}} \times \hat{\mathbf{z}}$). We can resolve $\boldsymbol{\omega}$ along the orthogonal set of axes $(\hat{\mathbf{m}}, \hat{\mathbf{l}}, \hat{\mathbf{k}})$, where $\hat{\mathbf{m}} = \hat{\mathbf{l}} \times \hat{\mathbf{k}}$ and lies in the plane of $\hat{\mathbf{z}}$ and $\hat{\mathbf{k}}$, by using the relation

$$\hat{\mathbf{z}} = \sin\theta\hat{\mathbf{m}} + \cos\theta\hat{\mathbf{k}}$$

We obtain

$$\boldsymbol{\omega} = \dot{\phi}\sin\theta\hat{\mathbf{m}} - \dot{\theta}\hat{\mathbf{l}} + (\dot{\psi} + \dot{\phi}\cos\theta)\hat{\mathbf{k}} \tag{6-139}$$

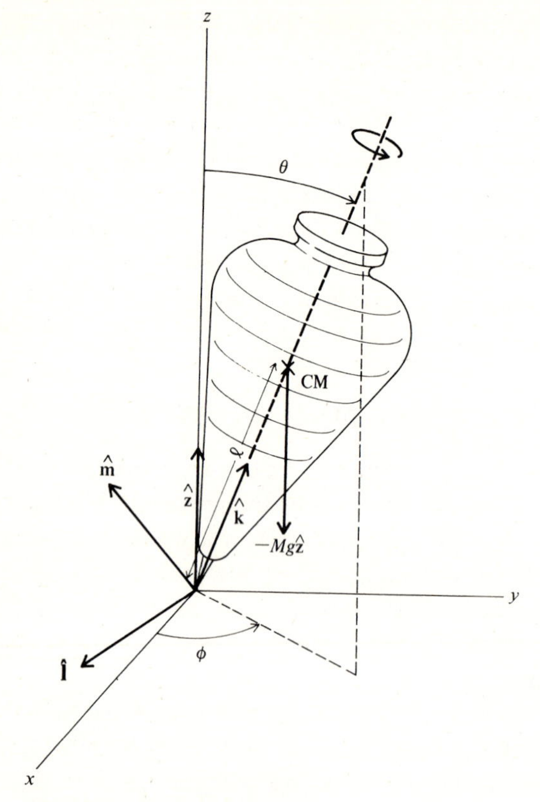

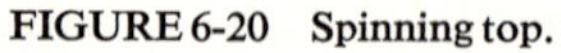

FIGURE 6-20 Spinning top.

This result can be cast in the form

$$\boldsymbol{\omega} = \omega_n \hat{\mathbf{n}} + \omega_3 \hat{\mathbf{k}} \tag{6-140}$$

with

$$\begin{aligned} \omega_n \hat{\mathbf{n}} &= \dot{\phi} \sin\theta \hat{\mathbf{m}} - \dot{\theta} \hat{\mathbf{l}} \\ \omega_3 &= \dot{\psi} + \dot{\phi} \cos\theta \end{aligned} \tag{6-141}$$

Since $\hat{\mathbf{n}} \cdot \hat{\mathbf{k}} = 0$ by inspection of Eq. (6-141), the $\hat{\mathbf{n}}$ and $\hat{\mathbf{k}}$ axes are also orthogonal. In Eq. (6-134) we have already established that the angular momentum $\mathbf{L}$ can be resolved along the axes $\hat{\mathbf{n}}$ and $\hat{\mathbf{k}}$ for any symmetric top as

$$\mathbf{L} = I\omega_n \hat{\mathbf{n}} + I_3 \omega_3 \hat{\mathbf{k}} \tag{6-142}$$

where in the present case the moments of inertia are to be calculated about the pivot of the top. From Eqs. (6-141) and (6-142), we have

$$\mathbf{L} = I(\dot{\phi} \sin \theta \hat{\mathbf{m}} - \dot{\theta} \hat{\mathbf{l}}) + I_3 \omega_3 \hat{\mathbf{k}} \tag{6-143}$$

We can now use Eqs. (6-140) to (6-143) to treat the motion of the top in the gravity field.

The gravitational torque on the top about the origin is

$$\mathbf{N} = -Mgl \sin \theta \hat{\mathbf{l}} \tag{6-144}$$

Since the $\hat{\mathbf{z}}$ axis of the inertial frame lies in the $\hat{\mathbf{k}}\hat{\mathbf{m}}$ plane, there is no N_z component of torque. The angular-momentum component L_z along the $\hat{\mathbf{z}}$ axis is therefore conserved. From Eq. (6-143), L_z is given by

$$L_z = I\dot{\phi} \sin \theta \hat{\mathbf{m}} \cdot \hat{\mathbf{z}} + I_3 \omega_3 \hat{\mathbf{k}} \cdot \hat{\mathbf{z}} \tag{6-145}$$

By reference to the geometry of Fig. 6-20, this simplifies to

$$L_z = I\dot{\phi} \sin^2 \theta + I_3 \omega_3 \cos \theta \tag{6-146}$$

The component of the torque N_3 along the $\hat{\mathbf{k}}$ axis also vanishes. From the equation of motion (6-80), we have

$$N_3 = \frac{dL_3}{dt} = 0 = \hat{\mathbf{k}} \cdot \frac{\delta \mathbf{L}}{\delta t} + \hat{\mathbf{k}} \cdot (\boldsymbol{\omega} \times \mathbf{L}) \tag{6-147}$$

Since $\boldsymbol{\omega}$, $\mathbf{L}$, and $\hat{\mathbf{k}}$ are coplanar for any symmetrical top [Eq. (6-131)], the term $\hat{\mathbf{k}} \cdot (\boldsymbol{\omega} \times \mathbf{L})$ vanishes. The unit vector $\hat{\mathbf{k}}$ is fixed in the body system, and so

$$\hat{\mathbf{k}} \cdot \frac{\delta \mathbf{L}}{\delta t} = \frac{\delta L_3}{\delta t} = I_3 \dot{\omega}_3 \tag{6-148}$$

We conclude from Eqs. (6-147) and (6-148) that L_3 and ω_3 are also constants of the motion.

$$L_3 = I_3 \omega_3 = \text{constant} \tag{6-149}$$

In addition to the conservation of the angular-momentum components L_z and L_3, the total energy of the frictionless top is conserved. The energy of the top with respect to the inertial frame is

$$E = \tfrac{1}{2}\boldsymbol{\omega} \cdot \mathbf{L} + Mgl \cos \theta \tag{6-150}$$

Using Eqs. (6-140) to (6-142), the energy can be written in terms of the spherical angles as

$$\begin{aligned} E &= \tfrac{1}{2}I\omega_n{}^2 + \tfrac{1}{2}I_3\omega_3{}^2 + Mgl \cos \theta \\ &= \tfrac{1}{2}I(\dot{\phi}^2 \sin^2 \theta + \dot{\theta}^2) + \tfrac{1}{2}I_3\omega_3{}^2 + Mgl \cos \theta \end{aligned} \tag{6-151}$$

In Eqs. (6-146), (6-149), and (6-151), we now have the information needed to construct a complete solution to the top problem. We can eliminate $\dot{\phi}$ from Eqs. (6-151) and (6-146) to obtain an equation for $\dot{\theta}$ in terms of θ. Unfortunately, this differential equation for $[\dot{\theta}(t)]^2$ is cubic in $\cos\theta$, and the solution cannot be given in terms of elementary functions. We shall therefore concentrate on approximate solutions which illustrate the basic features of top motion.

Approximate solutions to the motion can most readily be obtained by first differentiating Eqs. (6-146) and (6-151) with respect to time. This leads to

$$I\ddot{\phi}\sin^2\theta + 2I\dot{\phi}\dot{\theta}\cos\theta\sin\theta - I_3\omega_3\dot{\theta}\sin\theta = 0 \tag{6-152}$$

$$I\dot{\phi}\ddot{\phi}\sin^2\theta + I\dot{\phi}^2\dot{\theta}\cos\theta\sin\theta + I\dot{\theta}\ddot{\theta} - Mgl\dot{\theta}\sin\theta = 0 \tag{6-153}$$

Multiplying the first equation by $\dot{\phi}$ and subtracting the second equation from the first gives

$$I\ddot{\theta} = (Mgl - I_3\omega_3\dot{\phi} + I\dot{\phi}^2\cos\theta)\sin\theta \tag{6-154}$$

We now use this differential equation to investigate the conditions under which the motion of the top is pure precession. Motion of the top in ϕ corresponds to precession, and variation with θ is known as nutation. For pure precession the angle θ is constant. Since a constant value for θ requires that the right-hand side of Eq. (6-154) vanish, we can solve for $\dot{\phi}$ in terms of θ.

$$\dot{\phi} = \frac{I_3\omega_3}{2I\cos\theta}\left(1 \pm \sqrt{1 - \frac{4MglI\cos\theta}{I_3{}^2\omega_3{}^2}}\right) \tag{6-155}$$

For physical solutions, the quantity under the radical sign must not be negative. Since $\cos\theta > 0$ for a top on a table, the spin ω_3 must satisfy

$$\omega_3 \geq \sqrt{\frac{4MglI\cos\theta}{I_3{}^2}} \tag{6-156}$$

Only if the top has at least this minimum value of spin is pure precession possible. For a spin which is much greater than this minimum value, we can approximate the square root in Eq. (6-155) by the first two terms in a binomial expansion. We then find two possible approximate solutions for the precessional rate $\dot{\phi}$.

Slow precession:

$$\dot{\phi} = \frac{Mgl}{I_3 \omega_3} \equiv \omega_P \tag{6-157}$$

Fast precession:

$$\dot{\phi} = \frac{I_3 \omega_3}{I \cos \theta} \tag{6-158}$$

For the first solution, $\dot{\phi} \ll \omega_3$, and the angular momentum vector $\mathbf{L}$ lies nearly along the $\hat{\mathbf{k}}$ axis. This solution corresponds to slow gyroscopic precession, as discussed in Sec. 5-5. For the second solution, $\dot{\phi} \approx \omega_3$, and $\mathbf{L}$ lies nearly along the z axis. In this case, $L \cos \theta \approx I_3 \omega_3$ and $\dot{\phi} \approx L/I$, which is just the angular frequency ω_l in the force-free-top limit of Eq. (6-132). This solution with rapid precession about the vertical direction is independent of gravity in the limit $\omega_3 \gg (\omega_3)_{\min}$.

For a rapidly spinning top, slow precession and small nutation are frequently observed. To find an approximate solution for the motion with this condition, the quadratic terms in $\dot{\phi}$ and $\dot{\theta}$ in the differential equations (6-152) and (6-154) can be neglected. We then have

$$\begin{aligned} \ddot{\phi} \sin \theta &= \frac{I_3 \omega_3}{I} \dot{\theta} \\ \ddot{\theta} &= \left(\frac{Mgl}{I} - \frac{I_3 \omega_3}{I} \dot{\phi} \right) \sin \theta \end{aligned} \tag{6-159}$$

In terms of ω_P from Eq. (6-157) and ω_l defined as

$$\omega_l \equiv \frac{I_3 \omega_3}{I} \tag{6-160}$$

these equations can be written

$$\ddot{\phi} \sin \theta = \omega_l \dot{\theta} \tag{6-161}$$

$$\ddot{\theta} = \omega_l(\omega_P - \dot{\phi}) \sin \theta \tag{6-162}$$

If we time-differentiate Eq. (6-161) and substitute Eq. (6-162), we find

$$\frac{d^2 \dot{\phi}}{dt^2} + \omega_l^2 \dot{\phi} = \omega_l^2 \omega_P \tag{6-163}$$

where again we have dropped a quadratic term (of order $\ddot{\phi}\dot{\theta}$). We can immediately write down the solution to this equation for $\dot{\phi}$.

$$\dot{\phi}(t) = \omega_P + a \cos (\omega_l + \alpha) \tag{6-164}$$

For the initial conditions $\dot{\phi} = \omega_0$, $\phi = 0$, $\dot{\theta} = 0$, $\theta = \theta_0$ at $t = 0$, we find $\ddot{\phi}_0 = 0$ from Eq. (6-161) and

$$\dot{\phi}(t) = \omega_P - (\omega_P - \omega_0) \cos \omega_l t \qquad (6\text{-}165)$$

The solution for $\phi(t)$ follows by integration.

$$\phi(t) = \omega_P t - \left(\frac{\omega_P - \omega_0}{\omega_l}\right) \sin \omega_l t \qquad (6\text{-}166)$$

To solve for θ, we plug Eq. (6-165) into Eq. (6-161). This gives

$$\dot{\theta} = (\omega_P - \omega_0) \sin \omega_l t \sin \theta \qquad (6\text{-}167)$$

Since ω_P and ω_0 are small quantities, we can make the approximation $\sin \theta \approx \sin \theta_0$ on the right-hand side of Eq. (6-167). The solution for θ is then found by integration.

$$\theta(t) = \theta_0 + \left(\frac{\omega_P - \omega_0}{\omega_l}\right) \sin \theta_0 (1 - \cos \omega_l t) \qquad (6\text{-}168)$$

This completes the formal solution of the equations of motion in the approximation of slow precession and small nutation.

The solution for $\theta(t)$ in Eq. (6-168) exhibits nutation of the top between the angular limits θ_0 and $\theta_0 + 2[(\omega_P - \omega_0)/\omega_l] \sin \theta_0$. The sign of $(\omega_P - \omega_0)$ determines which is the upper and which is the lower bound on θ. The precession $\phi(t)$ in Eq. (6-166) has a sinusoidal motion associated with the nutation which is superimposed on the steady precession. When the initial precession ω_0 equals ω_P, the top undergoes steady precessional motion with no nutation. In Fig. 6-21 the curves traced out by the symmetry axis of the top are shown for various initial values of ω_0.

The nutation frequency of the top from Eq. (6-168) is ω_l. We see from Eqs. (6-157) and (6-160) that as the spin ω_3 of the top increases, the nutation frequency ω_l increases, while the precession frequency ω_P decreases. Furthermore, the nutation amplitude is inversely proportional to ω_l, so that nutation of a fast top is not so visible. When a fast top is spun on a hollow surface, however, a buzzing tone can often be heard with a frequency corresponding to the nutation frequency.

The phenomenon of nutation exhibited by our formal solution above can be understood from a more elementary viewpoint. For a top which is spinning rapidly, the angular momentum $\mathbf{L}$ is nearly along the symmetry axis $\hat{\mathbf{k}}$ of the top. The gravitational torque is

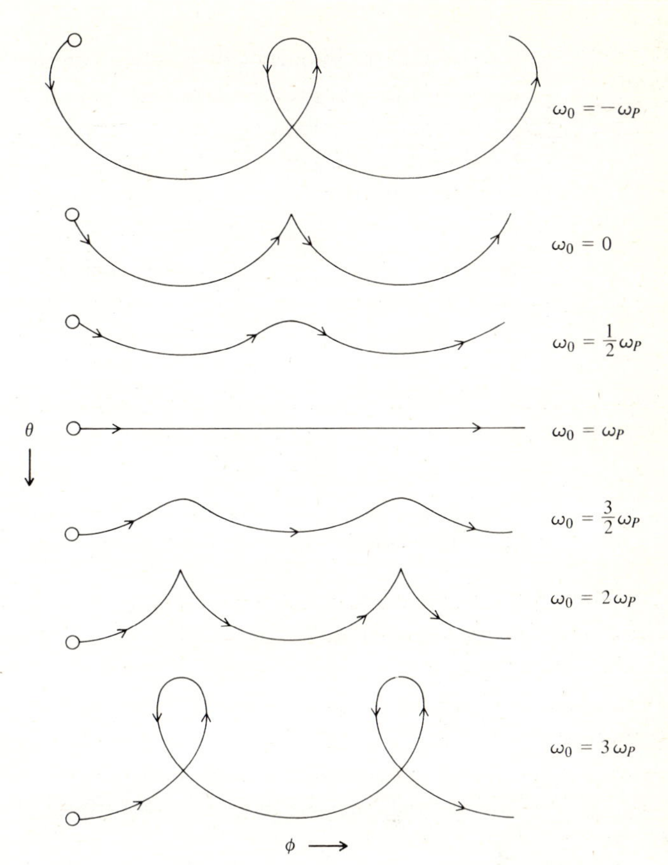

FIGURE 6-21 Nutation curves traced out by the symmetry axis of the top for various initial conditions. The top is started at the same value of θ in each case.

perpendicular to this axis and causes a gyroscopic precession about the vertical direction. The angular velocity of precession $\boldsymbol{\omega}_P = \omega_P \hat{\mathbf{z}}$ can be found by equating

$$\frac{d\mathbf{L}}{dt} = \mathbf{N} = -Mgl \sin \theta \hat{\mathbf{l}}$$

to

$$\frac{d\mathbf{L}}{dt} = \boldsymbol{\omega}_P \times \mathbf{L} = \omega_P L(\hat{\mathbf{z}} \times \hat{\mathbf{k}})$$

Since $\hat{\mathbf{z}} \times \hat{\mathbf{k}} = -\hat{\mathbf{l}} \sin \theta$ and $L \approx I_3 \omega_3$, we obtain

$$\omega_P \approx \frac{Mgl}{I_3 \omega_3} \tag{6-169}$$

for the angular frequency of steady precession about the z axis. For ω_3 very large, the precession rate ω_p is quite slow and the symmetry axis is nearly stationary. If the top is now given a slight push, it instantaneously acquires a small angular-momentum component $\Delta\mathbf{L}$ perpendicular to $\hat{\mathbf{k}}$. The resulting total angular momentum $\mathbf{L} + \Delta\mathbf{L}$ points in a direction slightly different from the symmetry axis $\hat{\mathbf{k}}$. The ensuing motion is like that of a free top with the symmetry axis precessing around $\mathbf{L} + \Delta\mathbf{L}$ in a small circle. The angular frequency of this circular motion is found from Eq. (6-132) to be

$$\omega_l = \frac{|\mathbf{L} + \Delta\mathbf{L}|}{I} \approx \frac{I_3\omega_3}{I} \tag{6-170}$$

The complete motion of the symmetry axis is a superposition of this rapid free-top circular motion about the direction $\mathbf{L} + \Delta\mathbf{L}$ on the slow precession of $\mathbf{L} + \Delta\mathbf{L}$ about the vertical direction.

6-11 ROLLING MOTION OF A TOP

The possible motions of a top with the point of contact fixed are confined to precession and nutation. For most actual tops, the point of contact rolls or skids on the supporting surface. We discuss in this section the rolling motion of a top with a blunt end, as pictured in Fig. 6-22.

If the top's peg rolls without slipping on a horizontal supporting surface, the only possible frictional force is perpendicular to the velocity of the CM, as discussed in Sec. 5-10. Consequently, the CM of the top moves in a circle of radius ρ with angular velocity Ω, as shown in Fig. 6-23. In uniform rolling motion the point of contact also traces out a circle, with angular velocity Ω and radius $\rho + l \sin\theta$, where θ is the angle of inclination with the vertical and l is the distance from the CM to the peg end. The direction of motion around the circle is determined by the direction of the top's spin.

The frictional force f at the point of contact balances the centrifugal force acting at the CM.

$$f = M\Omega^2\rho \tag{6-171}$$

Furthermore, this frictional force must be bounded by

$$|f| \le \mu Mg \tag{6-172}$$

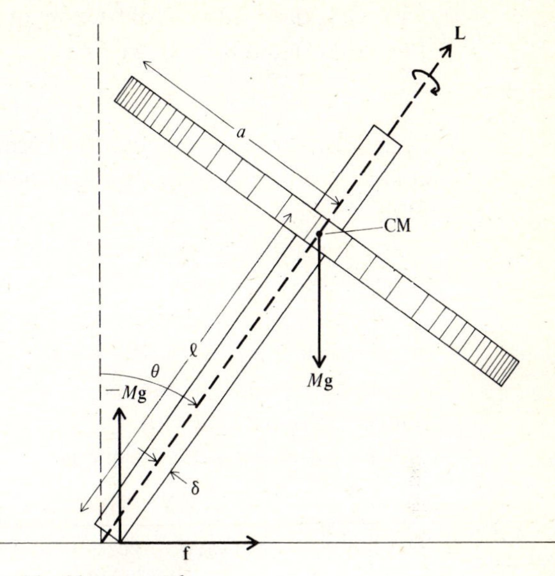

FIGURE 6-22 Top with a blunt peg end.

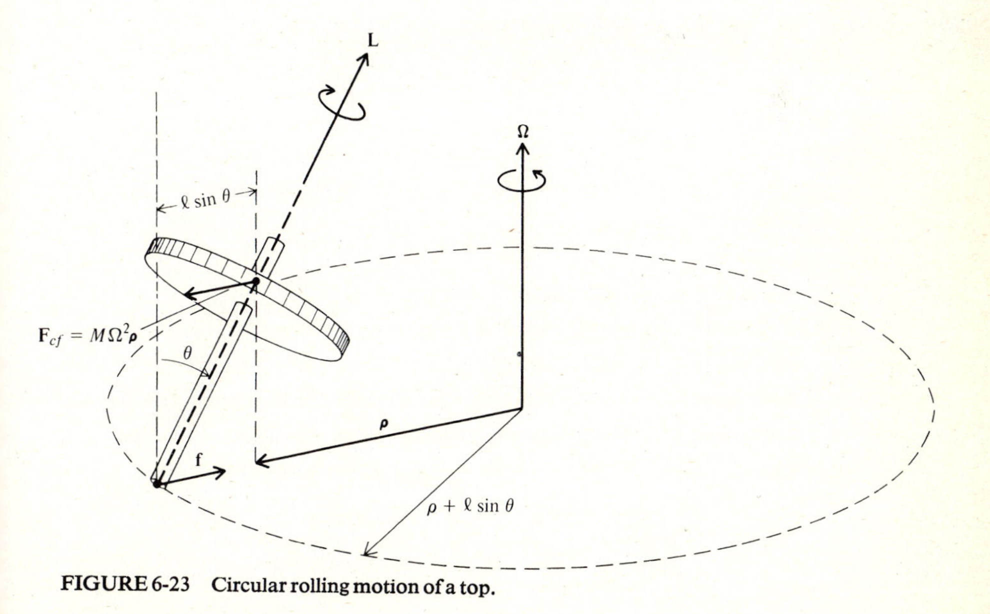

FIGURE 6-23 Circular rolling motion of a top.

where μ is the coefficient of friction, or else slipping will occur. From Eqs. (6-171) and (6-172), we find

$$\Omega^2|\rho| \leq \mu g \tag{6-173}$$

The frictional force f and the normal force Mg at the point of contact produce a torque **N** about the CM of the top. The direction of the torque is tangent to the circular path. Since for a fast top **L** lies nearly along the symmetry axis, **L** and **N** are perpendicular. Thus the effect of the torque is to cause the angular momentum to precess. For circular rolling motion, the rate of precession must be the same as the angular velocity Ω. In order that the precession of **L** be in the same direction as the rolling motion, the top must be inclined toward the center of the circle ($\theta > 0$).

The magnitude of the torque is

$$N = Mgl \sin\theta - fl\cos\theta \tag{6-174}$$

The requirement that **L** precess with angular velocity Ω gives

$$\frac{dL}{dt} = |\boldsymbol{\Omega} \times \mathbf{L}| \approx \Omega I_3 \omega_3 \sin\theta \tag{6-175}$$

for a fast top. The rotational equation of motion thereby leads to

$$Mgl \sin\theta - fl\cos\theta = \Omega I_3 \omega_3 \sin\theta \tag{6-176}$$

When we substitute the value of f from Eq. (6-171), we obtain

$$\frac{Mgl}{I_3 \omega_3}\left(1 - \Omega^2 \frac{\rho \cot\theta}{g}\right) = \Omega \tag{6-177}$$

For rolling motion the instantaneous velocity of the point of contact of the top's peg must be zero. From the geometry of Fig. 6-23, this implies that

$$\omega_3 \delta = \Omega(\rho + l \sin\theta) \tag{6-178}$$

where ω_3 is the top's spin and δ is the radius of the peg.

The radius ρ and angular velocity Ω of the top's rolling motion for a given spin ω_3 and inclination angle θ are given in terms of the top parameters δ, l, I_3 by Eqs. (6-177) and (6-178). As an example, we consider a small top formed of a light peg of radius δ inserted in a disk of radius a and mass M. For the dimensions we choose

$$\delta = 0.3 \text{ cm}$$
$$a = 3 \text{ cm}$$
$$l = 1.5 \text{ cm}$$

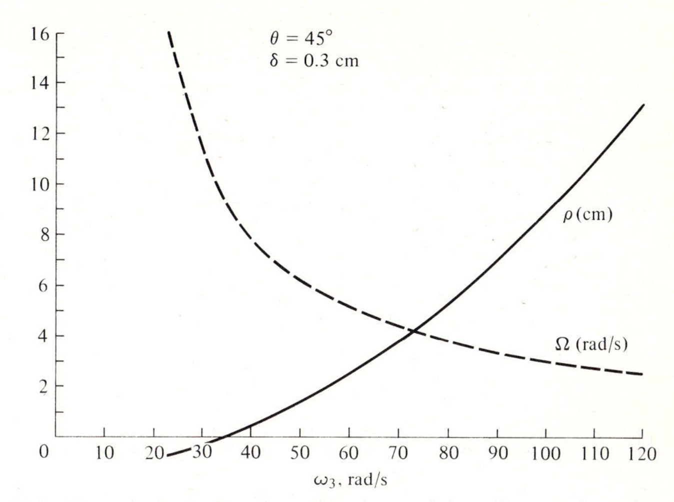

FIGURE 6-24 Numerical values vs. the top's spin ω_3 of the radius ρ and angular velocity Ω of rolling motion at an inclination angle $\theta = 45°$.

The moment of inertia about the top axis is approximately given by

$$I_3 \approx \tfrac{1}{2}Ma^2$$

Numerical computations of ρ and Ω from Eqs. (6-177) and (6-178) are plotted in Fig. 6-24 vs. the top spin ω_3 for the top parameters above and an inclination angle of $\theta = 45°$.

When $\rho = 0$, the CM of the top remains at a fixed location and the peg end traces out a circle on the horizontal surface, as sketched in Fig. 6-25. From Eqs. (6-177) and (6-178), the value of the spin ω_3 necessary for rolling motion with $\rho = 0$ is

$$(\omega_3)_{\rho=0} = \sqrt{\frac{Mgl^2 \sin\theta}{I_3\,\delta}} \approx 34 \text{ rad/s} \tag{6-179}$$

For a spin $\omega_3 > (\omega_3)_{\rho=0}$, the top moves in a larger circle at a lower angular rate Ω. For $\omega_3 < (\omega_3)_{\rho=0}$ the value of ρ is negative, which simply indicates that the CM is on the opposite side of the center of the circle from the peg end. In Fig. 6-26 the radius ρ and angular velocity Ω values are given as a function of the inclination angle θ for a top spin of $\omega_3 = 60$ rad/s.

If the peg radius δ is small, a high value of ω_3 is required, according to Eq. (6-179), to set the top in rolling motion. On the other hand, if the peg radius is large, the upper bound on $\Omega^2\rho$ in Eq. (6-173) places

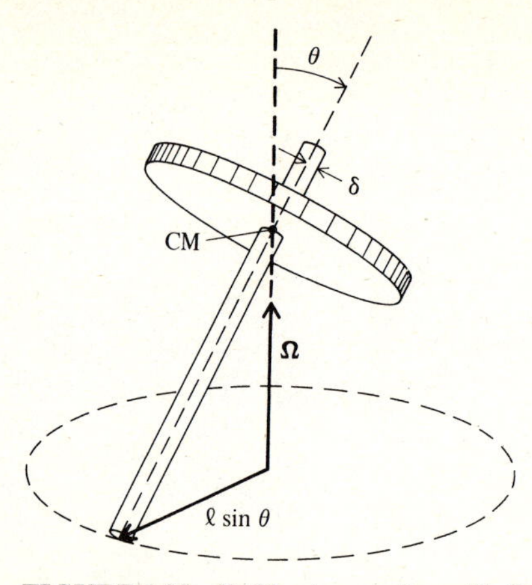

FIGURE 6-25 Rolling top motion with the CM at a fixed location.

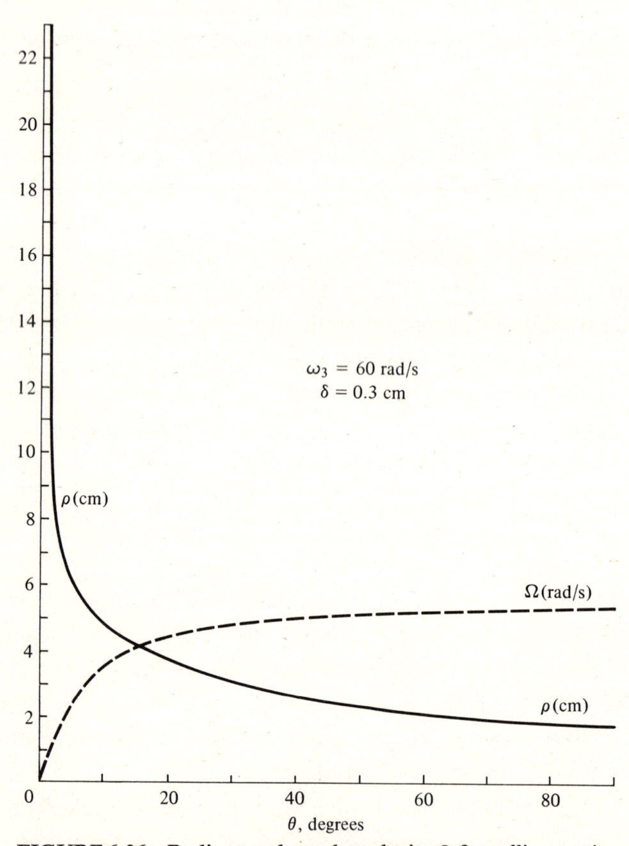

FIGURE 6-26 Radius ρ and angular velocity Ω for rolling motion of a top versus the inclination θ for a top spin $\omega_3 = 60$ rad/s.

a more severe restriction on the maximum value of ω_3 allowed for rolling motion, as can be seen from Eq. (6-178). In the next section we consider the motion of tops whose point of contact slips. A different phenomenon is thereby encountered—the rising motion of a top.

6-12 SLIPPING TOPS: RISING AND SLEEPING

When a spinning top similar to that in Fig. 6-22 is set down on a rough surface, the top usually slips initially. A frictional force directed opposite to the instantaneous skidding velocity acts to accelerate the CM of the top until the velocity of slipping is reduced to zero and pure rolling motion sets in. If the top is spinning rapidly when it is set down, it tends to maintain a fixed angle θ with the vertical, since the nutation is small. The normal force at the point of contact is then essentially the entire weight of the top, and so the frictional force is

$$|\mathbf{f}| \approx \mu Mg \tag{6-180}$$

where μ is the coefficient of friction. The resulting frictional torque

$$|\mathbf{N}| \approx \mu Mgl$$

about the CM of the top is roughly perpendicular to the peg for a thin peg, as illustrated in Fig. 6-27. For a rapidly spinning top, **L** lies nearly along the symmetry axis. Since the frictional torque is perpendicular to the symmetry axis, **L** precesses toward the vertical. The angular velocity of this precession is

$$\dot{\theta} = -\frac{|\mathbf{N}|}{|\mathbf{L}|} \approx -\frac{\mu Mgl}{I_3 \omega_3}$$

or

$$\dot{\theta} \approx -\mu\omega_P \tag{6-181}$$

by Eq. (6-169). The angular velocity of the rising motion is just the product of the precessional angular velocity and the coefficient of skidding friction.

As the top rises, kinetic energy is converted into potential energy and the spin of the top decreases. In addition, some of the energy is dissipated by friction. The rate of frictional dissipation of energy,

$$\frac{dE}{dt} = -fv \tag{6-182}$$

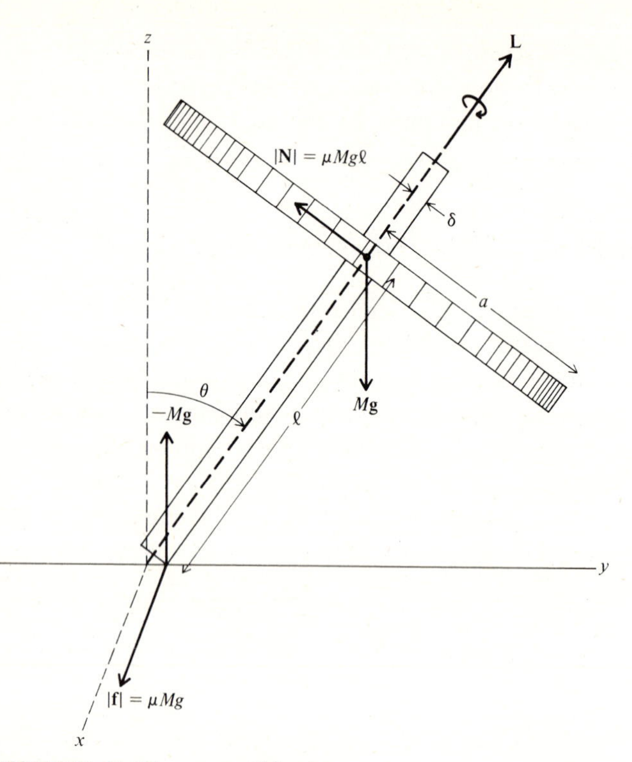

FIGURE 6-27 Forces on a rising top.

where f is the frictional force and v is the velocity at the point of contact, can be quite small for a thin peg. Nevertheless, the effects of friction in causing the top to rise are dramatic.

Once the top has risen to a vertical position, the point of contact is the symmetry axis, and the frictional force is much smaller. From Eq. (6-154), the equation of motion for very small θ is then approximately

$$I\ddot{\theta} - (Mgl - I_3\,\omega_3\,\dot{\phi} + I\dot{\phi}^2)\theta = 0 \tag{6-183}$$

where we have neglected the dissipation of energy by friction. In terms of the quantities ω_P and ω_l defined in Eqs. (6-157) and (6-160), this equation can be written

$$\ddot{\theta} + \omega_l\left(-\,\omega_P + \dot{\phi} - \frac{1}{\omega_l}\,\dot{\phi}^2\right)\theta = 0 \tag{6-184}$$

For a given $\dot{\phi}$, the motion in θ will be stable about $\theta = 0$ if

$$-\omega_P + \dot{\phi} - \frac{1}{\omega_l}\dot{\phi}^2 > 0 \tag{6-185}$$

The corresponding requirement on $\dot{\phi}$ is

$$\frac{\omega_l}{2}\left(1 - \sqrt{1 - \frac{4\omega_P}{\omega_l}}\right) < \dot{\phi} < \frac{\omega_l}{2}\left(1 + \sqrt{1 - \frac{4\omega_P}{\omega_l}}\right) \tag{6-186}$$

For a high value of the spin ω_3,

$$\frac{\omega_P}{\omega_l} = \frac{Mgl/I_3\,\omega_3}{I_3\,\omega_3/I} \ll 1 \tag{6-187}$$

and the condition in Eq. (6-186) is satisfied. The spinning top "sleeps" in the vertical position until friction slows down the spin to the value

$$\omega_3 = \sqrt{\frac{4MglI}{I_3^{\,2}}} \tag{6-188}$$

for which

$$\omega_l = 4\omega_P$$

and Eq. (6-186) is violated. At this point, the θ motion of the top becomes unstable and the top goes down as θ increases from zero.

To develop a feeling for the motion of a typical top, we consider as an example a top made of a thin disk of radius a and mass M which is supported by a narrow peg of length $l = a/2$ and negligible mass, as illustrated in Fig. 6-22. The moments of inertia about the point of contact of the peg with the table are

$$\begin{aligned} I_3 &= \tfrac{1}{2}Ma^2 \\ I &= \tfrac{1}{4}Ma^2 + Ml^2 = \tfrac{1}{2}Ma^2 \end{aligned} \tag{6-189}$$

If the top has a radius $a = 3$ cm and is set down with an initial spin of $\omega_3 = 60$ rad/s (about 10 r/s), the angular velocity of precession from Eq. (6-157) is

$$\omega_P = \frac{Mg(a/2)}{\frac{1}{2}Ma^2\omega_3} = \frac{g}{a\omega_3} = \frac{980}{3(60)} = 5.4 \text{ rad/s}$$

For a coefficient of friction $\mu = 1/10$, the angular velocity from Eq. (6-181) of the top's rise toward the vertical is

$$\dot{\theta} = -\mu\omega_P = -0.54 \text{ rad/s}$$

If the top is started at its maximum angle of inclination,

$$\theta = \arctan \frac{a/2}{a} = 0.46 \text{ rad}$$

the time to rise to the vertical is

$$t = \frac{\theta}{|\dot{\theta}|} = \frac{0.46}{0.54} = 0.86 \text{ s}$$

In this length of time the axis of the top has made

$$\frac{\omega_P t}{2\pi} = \frac{5.4(0.86)}{6.28} = 0.74 \text{ revolution}$$

about the vertical and

$$\frac{\omega_3 t}{2\pi} = \frac{60(0.86)}{6.28} = 8.2 \text{ revolutions}$$

about the symmetry axis. From Eq. (6-188), the condition for the motion at $\theta = 0$ to be stable is

$$\omega_3 > \sqrt{\frac{4MglI}{I_3^2}}$$

$$> \sqrt{\frac{4Mg(a/2)(\frac{1}{4}Ma^2)}{(\frac{1}{2}Ma^2)^2}} = \sqrt{\frac{4g}{a}}$$

From the parameters of our top, we find

$$\sqrt{\frac{4g}{a}} = \sqrt{\frac{4(980)}{3}} = 36 \text{ rad/s}$$

Since the inequality $\omega_3 > \sqrt{4g/a}$ is satisfied for the initial spin $\omega_3 =$ 60 rad/s, the top will "sleep" in its vertical position.

6-13 THE TIPPIE-TOP

When a tippie-top is spun on a smooth table, it turns itself upside-down, as pictured in Fig. 6-28. The usual high school and college rings likewise flip over to spin with their heavy ends upward. This fascinating behavior is due to a small frictional force at the point of contact with the table.

The frictional force is parallel to the table, opposing the velocity of slipping, as illustrated in Fig. 6-29. Since the horizontal direction

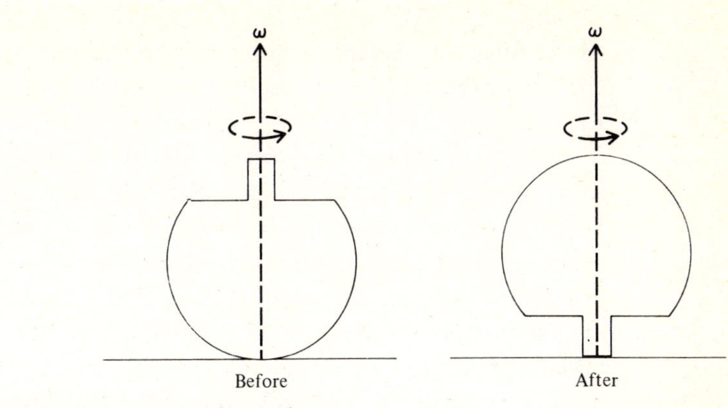

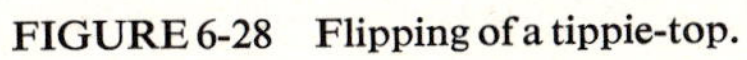
FIGURE 6-28 Flipping of a tippie-top.

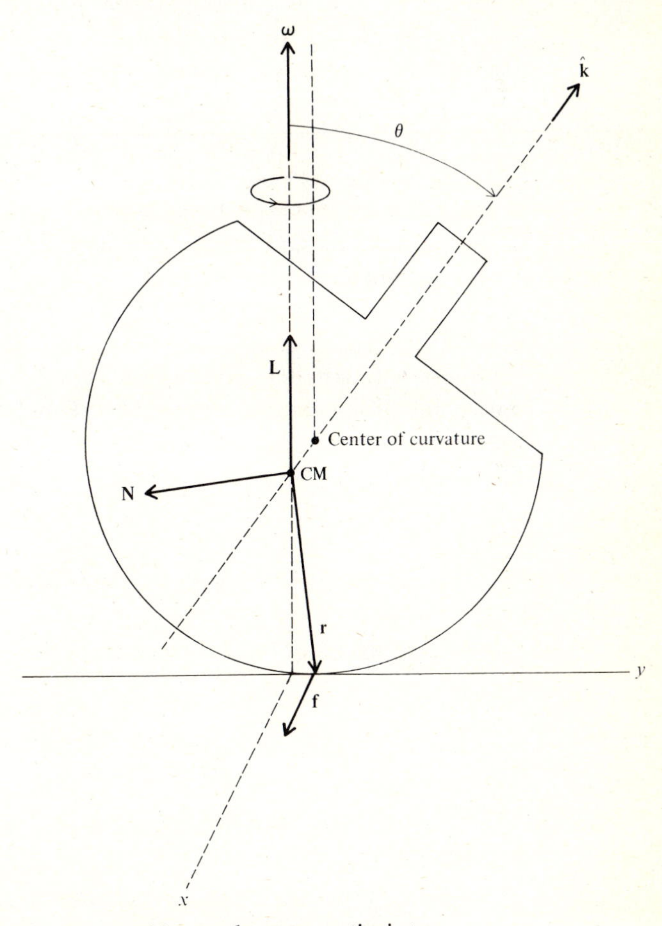

FIGURE 6-29 Frictional force and torque on a tippie-top.

of this force rotates rapidly with the angular frequency ω of the top, the time average of the force is zero, resulting in little effect on the motion of the CM.

The CM of a tippie-top is located close to the center of curvature, as indicated in Fig. 6-29. The gravitational torque can therefore be neglected. Furthermore, the frictional torque is nearly horizontal. This horizontal torque also rotates with angular frequency ω and time averages to zero. As a result, $d\mathbf{L}/dt \approx 0$, on the average, and the angular momentum of the top is nearly conserved. If the tippie-top is initially spun with the spin upward, as in Fig. 6-28, the approximately fixed direction of **L** is vertical with respect to the table.

The tipping motion is readily analyzed in a coordinate system which rotates with the top. In this reference frame the time average of the torque **N** is nonzero. The equation of motion (6-80) is

$$\mathbf{N} = \frac{\delta \mathbf{L}}{\delta t} + \boldsymbol{\omega} \times \mathbf{L} \tag{6-190}$$

As a simplifying approximation, we take the three principal moments of inertia about the CM as equal (the shape of the tippie-top is nearly spherical). Then

$$\mathbf{L} \approx I\boldsymbol{\omega} \tag{6-191}$$

and the $\boldsymbol{\omega} \times \mathbf{L}$ term in Eq. (6-190) vanishes. **L** remains vertical and **N** horizontal throughout the motion. Since **L** and **N** are perpendicular, the torque causes the angular momentum to precess uniformly in the body frame, according to Eq. (6-190). Taking the component of Eq. (6-190) along the symmetry axis $\hat{\mathbf{k}}$, we find

$$\hat{\mathbf{k}} \cdot \mathbf{N} = \hat{\mathbf{k}} \cdot \frac{\delta \mathbf{L}}{\delta t} = \frac{\delta(\hat{\mathbf{k}} \cdot \mathbf{L})}{\delta t} \tag{6-192}$$

Using the geometry of Fig. 6-29, which approximates **N** as horizontal, we have

$$\begin{aligned} \hat{\mathbf{k}} \cdot \mathbf{N} &= -N \sin \theta \\ \hat{\mathbf{k}} \cdot \mathbf{L} &= L \cos \theta \end{aligned} \tag{6-193}$$

where θ is the angle between **L** and $\hat{\mathbf{k}}$. From Eqs. (6-192) and (6-193), we obtain

$$\dot{\theta} = \frac{N}{L} \approx \frac{\mu M g R}{I\omega} \tag{6-194}$$

where R is the radius of the top. Thus θ increases with time, and the tippie-top flips over. Once the stem scrapes the table, the subsequent rise to the vertical is almost the same as an ordinary rising top, as treated in Sec. 6-12. We can estimate the time required for the tippie-top to flip over by use of Eq. (6-194). A spin of $\omega = 100$ rad/s is easily imparted to a tippie-top of radius $R = 1.5$ cm and moment of inertia $I \approx \frac{2}{3}MR^2$ (hollow sphere). For a coefficient of friction $\mu = 1/10$, we obtain

$$\dot{\theta} = \frac{3\mu g}{2R\omega} = \frac{3(0.1)(980)}{2(1.5)(100)} = 0.9 \text{ rad/s}$$

The flip time is, roughly,

$$t = \frac{\theta}{\dot{\theta}} \approx \frac{\pi}{0.9} \approx 3.5 \text{ s}$$

PROBLEMS

1. A particle of mass m moves in a smooth straight horizontal tube which rotates with constant angular velocity ω about a vertical axis which intersects the tube. Set up the equations of motion in polar coordinates and derive an expression for the distance of the particle from the rotation axis. If the particle is at $r = r_0$ at $t = 0$, what velocity must it have along the tube in order that it will be very close to the rotation axis after a very long interval of time?

2. A spherical planet of radius R rotates with a constant angular velocity ω. The effective gravitational acceleration g_{eff} is some constant, g, at the poles and $0.8g$ at the equator. Find g_{eff} as a function of the polar angle θ and g. With what velocity must a rocket be fired vertically upward from the equator to escape completely from the planet?

3. A particle of mass m is constrained to move in a vertical plane which rotates with constant angular velocity ω. Find the equations of motion of the particle, including the force of gravity.

4. A particle has velocity v on a smooth horizontal plane. Show that the particle will move in a circle, due to the rotation of the earth, and find the radius of the circle.

5. A ball is thrown vertically upward with velocity v_0 on the earth's surface. If air resistance is neglected, show that the ball lands a distance $(4\omega \sin \theta v_0{}^3/3g^2)$ to the west, where ω is the angular velocity of the earth's rotation and θ is the colatitude angle.

6. A bug which weighs 1 g crawls out along a radius of a phonograph record turning at $33\frac{1}{3}$ r/min. If the bug is 6 cm from the center and traveling at the rate of 1 cm/s, what are the forces on the bug? What added torque must the motor supply because of the bug?

7. A ring of mass m slides along a frictionless rod. The rod is rotated in a horizontal plane about a vertical axis, as shown, with angular velocity ω. Use lagrangian methods to find the equations of motion and the force on the rod. Solve the equations for the initial conditions $r = r_0$, $\dot{r} = 0$ at $t = 0$.

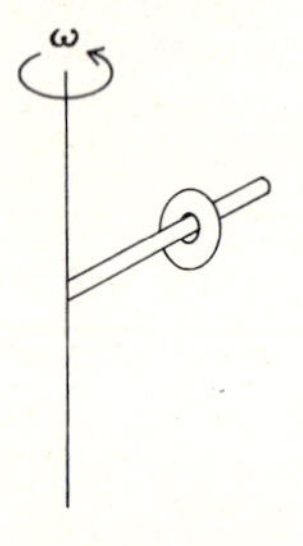

8. A bead of mass m is constrained to move on a hoop of radius R. The hoop rotates with constant angular velocity ω around a vertical axis which coincides with a diameter of the hoop.

a. Set up the lagrangian and obtain the equation of motion on the bead.

b. Find the critical angular velocity Ω below which the bottom of the hoop provides a stable-equilibrium position for the bead.

c. Find the stable-equilibrium position for $\omega > \Omega$.

9. A flat rectangular plate of mass M and sides a and $2a$ rotates with angular velocity ω about an axle through two diagonal corners, as shown. The bearings supporting the plate are mounted just at the corners. Find the force on each bearing.

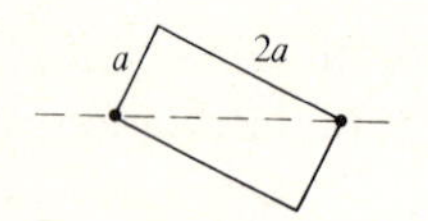

10. Show that the tennis racket in Sec. 6-7 is properly designed so that a hard stroke to the ball at the center of the racket head does not jar the player's hand.

11. A tennis racket is swung underhand and released so that it rises vertically with an initial spin about the unstable principal axis. At the instant of release the end of the handle is at rest. The racket subsequently rises to a height of 5 m.

a. Determine the time of rise to maximum height.

b. Find the initial angular velocity $\omega_2(0)$ about the CM.

c. For an initial spin about axis (3) that is 1 percent of $\omega_2(0)$, compute the time at which the racket begins to tumble.

12. Write $2E$ and $|\mathbf{L}|^2$ for a general rigid body, in terms of the principal-axis components of $\boldsymbol{\omega}$. From this, demonstrate the "tennis-racket theorem" for a free rigid body using conservation of E and $|\mathbf{L}|^2$.

13. A coin in a horizontal plane is tossed into the air with angular velocity components ω_1 about a diameter through the coin and ω_3 about the principal axis perpendicular to the coin. If ω_3 were equal to zero, the coin would simply spin around its diameter. For ω_3 nonzero, the coin will precess. What is the minimum value of ω_3/ω_1 for which the wobble is such that the same face of the coin is always exposed to an observer looking from above? This is a clever way to cheat at flipping coins!

14. Why is it difficult to spin a pencil on its point? Illustrate with a pencil of length 10 cm and diameter 0.5 cm.

15. For a top with a spherical peg end, show that the effective peg radius is $\delta = \delta_0 \sin \theta$, where θ is the inclination angle. Would you expect this top to rise higher than a similar top with a cut-off peg?

16. Analyze the motion of a tippie-top using an inertial frame on the table. (*Hint:* Find the implications for $\boldsymbol{\omega}$ of the precession of $\mathbf{L}$ about the vertical direction.)

CHAPTER 7

Gravitation

According to Newton's law of universal gravitation, each particle in the universe is attracted to every other particle with a force proportional to their masses, inversely proportional to the square of the distance between them, and directed along the line joining them. This law provides a nearly complete understanding of the motions of the planets, satellites, and stars. Indeed, it has only been in this century that a few tiny discrepancies have been uncovered whose explanation requires the more complete theory of gravity provided by Einstein's general relativity.

7-1 ATTRACTION OF A SPHERICAL BODY

The statement of Newton's law of gravity applies to the attraction between two point masses, whereas celestial bodies are roughly spherical collections of particles. The proof that a spherically symmetric body acts as if its mass is concentrated at its center is an essential step in the application of the law of gravitation to celestial mechanics. We give a proof of this familiar result, using the concept of potential energy.

The gravitational potential energy between two point particles separated by a distance r is

$$V(r) = -\frac{GMm}{r} \tag{7-1}$$

The corresponding force on m due to M is given by

$$\mathbf{F} = -\nabla V(r) = GMm \frac{d}{dr}\left(\frac{1}{r}\right)\hat{\mathbf{r}} = -\frac{GMm}{r^2}\hat{\mathbf{r}} = -\frac{GMm}{r^3}\mathbf{r} \tag{7-2}$$

where $\mathbf{r} = \mathbf{r}_m - \mathbf{r}_M$. We first calculate the gravitational potential energy due to a spherical shell of mass M and a point mass m located a distance R from the center of the shell, as shown in Fig. 7-1. If the radius of the shell is a, the surface mass density is

$$\sigma = \frac{M}{4\pi a^2} \tag{7-3}$$

The circular ring element in Fig. 7-1 has differential area $dA = 2\pi(a \sin\theta)(a\, d\theta)$ and mass.

$$\begin{aligned} dM &= (2\pi a^2 \sin\theta\, d\theta)\sigma \\ &= \frac{M}{2}\sin\theta\, d\theta \end{aligned} \tag{7-4}$$

Using the law of cosines,

$$r^2 = a^2 + R^2 - 2aR\cos\theta \tag{7-5}$$

we can express dM in terms of $r\,dr$. By differentiation of Eq. (7-5), we obtain

$$r\,dr = aR\sin\theta\, d\theta = 2aR\frac{dM}{M}$$

or

$$dM = \frac{Mr\,dr}{2aR} \tag{7-6}$$

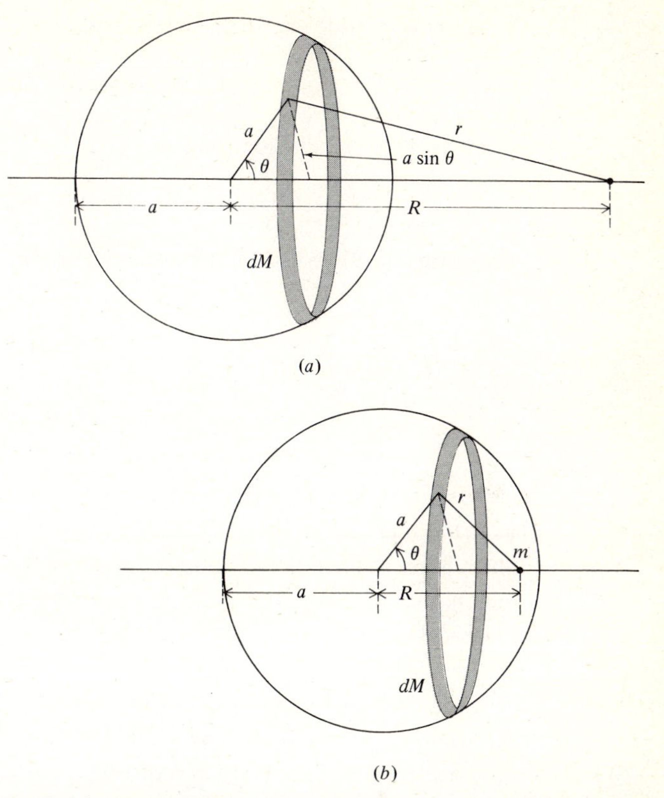

FIGURE 7-1 Gravitational attraction of a point mass m and a differential ring element dM on a spherical shell of mass M. (a) m outside the shell; (b) m inside the shell.

Since every particle in the ring is the same distance r from the mass m, the potential energy due to m and the ring mass dM is

$$dV(r) = -\frac{Gm\,dM}{r} \tag{7-7}$$

By substitution of dM from Eq. (7-6), we get

$$dV(r) = -\frac{GMm}{2aR}\,dr \tag{7-8}$$

The total potential energy due to m and the shell can be directly found by integration over r.

$$V(r) = \int_{r_{\min}}^{r_{\max}} dV(r) = -\frac{GMm}{2aR}(r_{\max} - r_{\min}) \tag{7-9}$$

When m is outside the shell, we see from Fig. 7-1 that

$$(r_{\max} - r_{\min}) = (R + a) - (R - a) = 2a \tag{7-10}$$

whereas when m is inside the shell,

$$(r_{\max} - r_{\min}) = (R + a) - (a - R) = 2R \tag{7-11}$$

From Eqs. (7-9) to (7-11), the potential energy is

$$V(R) = \begin{cases} -\dfrac{GMm}{R} & R > a \qquad (7\text{-}12a) \\[2ex] -\dfrac{GMm}{a} & R < a \qquad (7\text{-}12b) \end{cases}$$

When m is outside of the shell, the potential energy in Eq. (7-12a) is as if the mass M of the shell were concentrated at the center of the shell. Since a spherically symmetric solid body can be represented as a summation of concentric shells, the gravitational force on m due to a spherical body is as if the total mass M were concentrated at the center of the sphere.

If, on the other hand, m is inside the shell, $V(R)$ is constant and m feels no net force due to the shell. It follows that if m is inside a spherically symmetric body at a distance R from its center, m experiences a gravitational attraction only from the mass which lies inside of R. This part of the mass acts as if it were concentrated at the center of the sphere.

The total external gravitational force on an arbitrary extended body, composed of masses m_i at locations $\mathbf{r}_i$, is

$$\mathbf{F} = \sum_i m_i \mathbf{g}_i$$

where $\mathbf{g}_i$ is the total gravitational acceleration at $\mathbf{r}_i$ due to external masses. The total external gravitational torque on the body about the origin of the coordinate system is

$$\mathbf{N} = \sum_i m_i \mathbf{r}_i \times \mathbf{g}_i$$

We define a *center-of-gravity location* $\mathbf{R}_G$ by

$$\mathbf{N} \equiv \mathbf{R}_G \times \mathbf{F}$$

so that

$$\mathbf{R}_G \times \left(\sum_i m_i \mathbf{g}_i\right) = \sum_i m_i \mathbf{r}_i \times \mathbf{g}_i \tag{7-13}$$

Then the total force on the body can be considered to act at the center-of-gravity point. For a uniform gravitational acceleration $\mathbf{g}_i = \mathbf{g}$, we have

$$\mathbf{F} = \left(\sum_i m_i\right)\mathbf{g} = M\mathbf{g}$$

$$\mathbf{N} = \left(\sum_i m_i \mathbf{r}_i\right) \times \mathbf{g} = M\mathbf{R}_G \times \mathbf{g}$$

In this case the center of gravity $\mathbf{R}_G$ coincides with the center of mass $\mathbf{R}$. The center of gravity and center of mass also coincide for a spherical body, as has been demonstrated above. However, $\mathbf{R}_G$ is usually not the same as $\mathbf{R}$ for a nonspherical body in a nonuniform gravitational field. In general, the center-of-gravity location depends on the orientation of the body, as well as on the distance of the body from the external masses.

7-2 THE TIDES

When a body of finite extent is located in an external gravitational field, it is subjected to tide-generating forces. These tidal forces are due to the fact that the various parts of the body are at different distances from the external force center, and are thereby attracted more or less strongly. Tidal forces are disruptive and may even tear the body apart—this being a possible origin of the rings of Saturn.

The ocean tides on earth are caused by the variation from place to place of the gravitational attraction due to the moon and the sun. The atmosphere, the ocean, and the "solid" earth, all experience tidal forces, but only the effects on the ocean are commonly observed. To calculate the gross features of the midocean tides, we begin with a static theory in which the rotation of the earth about its own axis is initially neglected. The daily rotation of the earth will be invoked later to explain the propagation of the tides. In our simplified model we assume that the earth is entirely covered by ocean, with the surface always in a condition of equilibrium.

To calculate the tide-generating force, we consider the motion of a small mass m on the ocean's surface under the combined influence of the gravitational attraction of the earth and a distant mass M, as shown in Fig. 7-2. The tide-generating force due to several distant masses can be readily obtained from the result of this calculation for a single distant mass. The coordinates of the masses m, M_e, M in an inertial frame are represented by the vectors $\mathbf{r}_1$, $\mathbf{r}_2$, $\mathbf{r}_3$, respectively, as illustrated in Fig. 7-2. For convenience, we denote the relative coordinates of the masses by

$$\begin{aligned}\mathbf{r} &= \mathbf{r}_1 - \mathbf{r}_2 \\ \mathbf{d} &= \mathbf{r}_1 - \mathbf{r}_3 \\ \mathbf{R} &= \mathbf{r}_2 - \mathbf{r}_3\end{aligned} \tag{7-14}$$

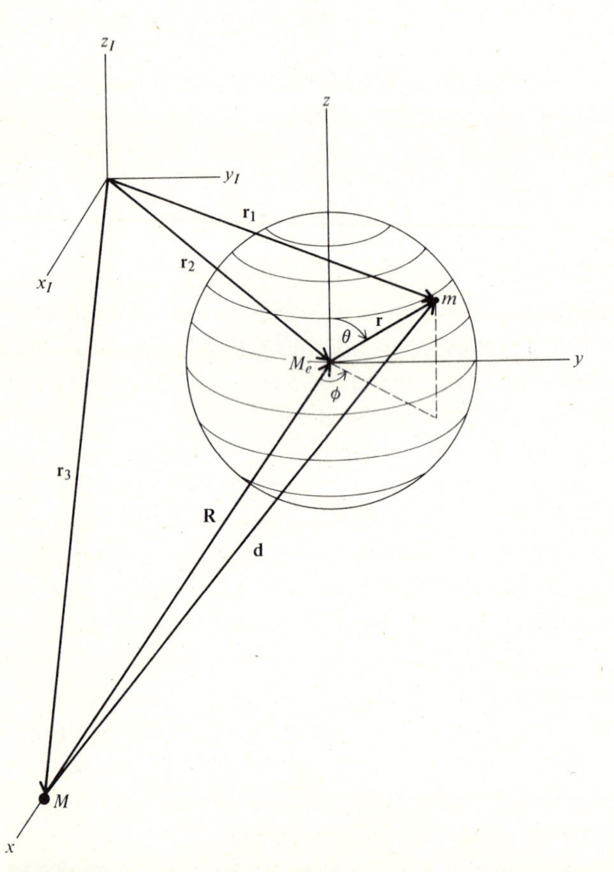

FIGURE 7-2 Location of a point on the earth's surface and a distant mass M in an inertial frame and an earth-centered frame.

With this notation, the motion of m and M_e due to gravitational forces is determined by

$$m\ddot{\mathbf{r}}_1 = -\frac{GmM_e\hat{\mathbf{r}}}{r^2} - \frac{GmM}{d^2}\hat{\mathbf{d}} \tag{7-15a}$$

$$M_e\ddot{r}_2 = -\frac{GM_eM}{R^2}\hat{\mathbf{R}} \tag{7-15b}$$

By dividing the first equation by m, the second equation by M_e, and then subtracting, we find the equation of motion for the relative coordinate **r**.

$$\ddot{\mathbf{r}} = -\frac{GM_e\hat{\mathbf{r}}}{r^2} - GM\left(\frac{\hat{\mathbf{d}}}{d^2} - \frac{\hat{\mathbf{R}}}{R^2}\right) \tag{7-16}$$

This result could have been directly obtained from Eq. (5-7). The first term on the right-hand side of Eq. (7-16) is the central gravity force of the earth on a particle of unit mass. The second term is the tide-generating force per unit mass due to the presence of the distant mass M. The tide-generating force is the difference between the gravitational force on the surface of the earth and at the center of the earth. The direction and relative magnitude of the tide-generating force due to M are plotted in Fig. 7-3 for points around the earth's equator. The effect of this force is to produce two tidal bulges which, as the earth rotates, are observed twice daily as high tides.

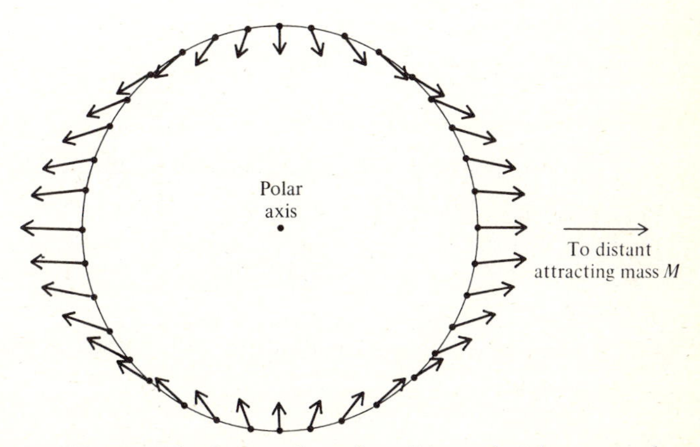

FIGURE 7-3 Tide-generating force on the surface of the earth at the equator due to a distant mass.

The gravitational forces can be expressed in terms of a potential energy

$$V(\mathbf{r}) = -\int_{\mathbf{r}_S}^{\mathbf{r}} \mathbf{F} \cdot d\mathbf{r} \tag{7-17}$$

Since only the force per unit mass enters in Eq. (7-16), we define a gravitational potential Φ as the potential energy per unit mass.

$$\Phi(\mathbf{r}) = \frac{V(\mathbf{r})}{m} = -\frac{1}{m}\int \mathbf{F}(\mathbf{r}) \cdot d\mathbf{r} \tag{7-18}$$

where by Eq. (7-16)

$$\frac{\mathbf{F}(\mathbf{r})}{m} = -\nabla\Phi(\mathbf{r}) = -\frac{GM_e\hat{\mathbf{r}}}{r^2} - \frac{GM\hat{\mathbf{d}}}{d^2} + \frac{GM\hat{\mathbf{R}}}{R^2} \tag{7-19}$$

The potential $\Phi(\mathbf{r})$ corresponding to the force of Eq. (7-19) can be computed by carrying out the integral in Eq. (7-18) using a convenient path. In this case, it is easier to guess $\Phi(\mathbf{r})$ and then to verify that it yields the correct force. We select, for simplicity, the position in the orbit for which $\hat{\mathbf{R}}$ is directed along the negative x axis, as in Fig. 7-2. The location of m from the earth's center is specified by the vector $\mathbf{r}$, whose components are x, y, z. The vector $\mathbf{d}$ by Eq. (7-14) is

$$\mathbf{d} = \mathbf{R} + \mathbf{r} \tag{7-20}$$

Since $\mathbf{R}$ does not change as the position of m is varied, we can regard $\mathbf{d}$ as a function of $\mathbf{r}$. The gravitational potential which gives the correct force is

$$\Phi(\mathbf{r}) = -\frac{GM_e}{r} - \frac{GM}{d} + \frac{GMx}{R^2} \tag{7-21}$$

The first and third terms correspond to the first and third terms of Eq. (7-19), since

$$\nabla_{\mathbf{r}}\left(\frac{1}{r}\right) = -\frac{\hat{\mathbf{r}}}{r^2} \qquad \nabla_{\mathbf{r}}(x) = \hat{\mathbf{i}} = -\hat{R} \tag{7-22}$$

where $\nabla_{\mathbf{r}}$ means the gradient with respect to the vector $\mathbf{r} = \hat{\mathbf{i}}x + \hat{\mathbf{j}}y + \hat{\mathbf{k}}z$

$$\nabla_{\mathbf{r}} \equiv \hat{\mathbf{i}}\frac{\partial}{\partial x} + \hat{\mathbf{j}}\frac{\partial}{\partial y} + \hat{\mathbf{k}}\frac{\partial}{\partial z} \tag{7-23}$$

Since $\mathbf{R}$ is held constant [by Eq. (7-20)], a change in a component of $\mathbf{r}$ gives an identical change in the component of $\mathbf{d}$; hence,

$$\nabla_{\mathbf{r}} = \nabla_{\mathbf{d}} \tag{7-24}$$

and we have for the second term of Eq. (7-21)

$$\nabla_{\mathbf{r}}\left(\frac{1}{d}\right) = \nabla_{\mathbf{d}}\left(\frac{1}{d}\right) = -\frac{\hat{\mathbf{d}}}{d^2} \tag{7-25}$$

In terms of spherical polar coordinates (r, θ, ϕ), the cartesian components of $\mathbf{r}$ are

$$\begin{aligned} x &= r \sin\theta \cos\phi \\ y &= r \sin\theta \sin\phi \\ z &= r \cos\theta \end{aligned} \tag{7-26}$$

The distance d between the distant mass M and the small mass m is

$$d^2 = (R - x)^2 + y^2 + z^2 = R^2 + r^2 - 2Rx \tag{7-27}$$

Since the distance R from M to M_e is large compared with the distance from the center of the earth to m, we can simplify Eq. (7-21) by expanding d in powers of the small quantity r/R. From Eq. (7-27), we have

$$\frac{1}{d} = \frac{1}{(R^2 + r^2 - 2Rx)^{1/2}} = \frac{1}{R\left(1 - \dfrac{2x}{R} + \dfrac{r^2}{R^2}\right)^{1/2}}$$

If we take

$$\beta = -\frac{2x}{R} + \frac{r^2}{R^2} \tag{7-28}$$

we can use the binomial expansion

$$(1 + \beta)^n = 1 + n\beta + \frac{n(n-1)}{2}\beta^2 + \cdots \tag{7-29}$$

for $n = -1/2$. This series converges for $\beta < 1$, and the first few terms are an accurate approximation when $\beta \ll 1$. Retaining powers of r/R (or x/R) up through $(r/R)^2$, we obtain

$$\frac{1}{d} \sim \frac{1}{R} + \frac{x}{R^2} + \frac{(3x^2 - r^2)}{2R^3} + \cdots \tag{7-30}$$

In this approximation, the potential in Eq. (7-21) is given by

$$\Phi(\mathbf{r}) = -\frac{GM_e}{r} - \frac{GMr^2}{R^3}\left(\tfrac{3}{2}\sin^2\theta\cos^2\phi - \tfrac{1}{2}\right) \tag{7-31}$$

where we have used Eq. (7-26) and dropped a constant term.

For equilibrium of the ocean surface, the net tangential force on m must vanish. Equivalently, the potential at any point on the ocean's surface must be constant. This constant can be chosen at our convenience, since a constant may be added to the potential without changing the force. For the equilibrium surface potential we choose $\Phi(\mathbf{r}) = -GM_e/R_e$, where R_e is undistorted spherical radius of the earth (i.e., when the distant M is absent). Using this condition in Eq. (7-31), we obtain

$$r - R_e = \frac{M}{M_e}\frac{r^3 R_e}{R^3}\left(\tfrac{3}{2}\sin^2\theta\cos^2\theta - \tfrac{1}{2}\right)$$

Since the height of the tidal displacement

$$h(\theta, \phi) \equiv r - R_e \tag{7-32}$$

is quite small compared with R_e, we obtain

$$h(\theta, \phi) \approx \frac{M}{M_e}\frac{R_e^4}{R^3}\left(\tfrac{3}{2}\sin^2\theta\cos^2\phi - \tfrac{1}{2}\right) \tag{7-33}$$

For a given colatitude angle θ in Eq. (7-33), the high tides occur at $\phi = 0$ and $\phi = \pi$, and low tides occur at $\phi = \pi/2$ and $\phi = 3\pi/2$. The difference in height between high and low tide is

$$\Delta h = \frac{3}{2}\frac{M}{M_e}\frac{R_e^4}{R^3}\sin^2\theta \tag{7-34}$$

The tidal displacement h is largest at $\theta = 90°$ (on the equator). The tidal distortion is illustrated in Fig. 7-4. The tide for an ocean devoid of continents has a football-like shape, with the major axis in the direction of the distant mass. The calculation of such an ideal tide was first made by Newton in 1687.

The preceding discussion applies to the tidal forces induced by one astronomic body. From Eq. (7-33) the ratio of the maximum heights of the lunar (L) and solar ($\odot$) tides on earth is

$$\frac{h_L}{h_\odot} = \left(\frac{M_L}{M_\odot}\right)\left(\frac{a_e}{a_L}\right)^3 \tag{7-35}$$

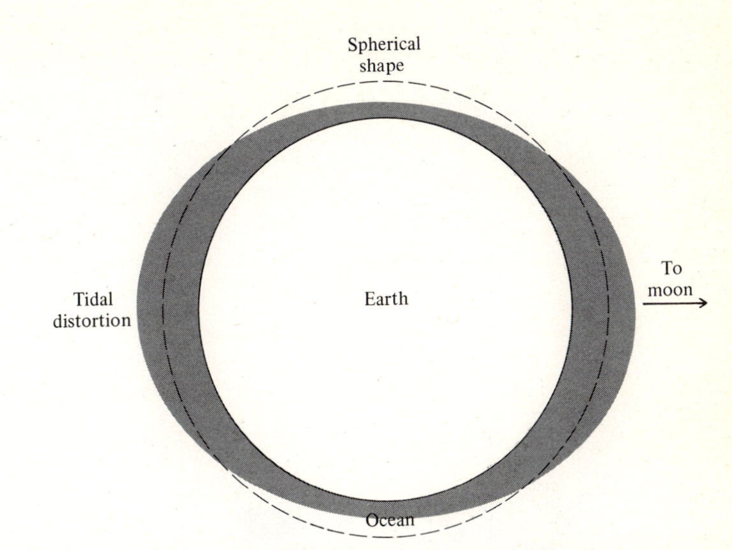

FIGURE 7-4 **Tidal distortion at the earth's equator on an exaggerated scale.**

where a_L is the earth-moon distance and a_e is the earth-sun distance. The numerical value of this ratio is

$$\frac{h_L}{h_\odot} = \frac{(1/81.5)M_e}{\frac{1}{3} \times 10^6\, M_e}\left(\frac{1.5 \times 10^8 \text{ km}}{3.8 \times 10^5 \text{ km}}\right)^3 = 2.2$$

The sun's tidal effect is smaller than the moon's, but it is not negligible. When the sun and moon are lined up (new or full moon), an especially large tide results (spring tide), and when they are at right angles (first or last quarter moon), their tidal effects partially cancel (neap tide). The diagram in Fig. 7-5 illustrates these four orientations of the moon relative to the earth and sun.

The tidal range due to the moon along the earth-moon axis can be calculated from Eq. (7-34). We get

$$\Delta h\left(\theta = \frac{\pi}{2}\right) = \frac{3}{2}\left(\frac{1}{81.5}\right)\left(\frac{6{,}371}{384{,}000}\right)^3 (6{,}371 \times 10^3) = 0.56 \text{ m}$$

This figure compares roughly with the measured tidal difference in midocean. As the earth rotates about its own axis, the tidal maxima, which lie on the earth-moon axis, will pass a given point on the earth's surface approximately two times a day. More precisely, since the orbital rotation of the moon about the earth (with period of $27\frac{1}{3}$ days) is in the same sense as the earth's own rotation (with period 24 h),

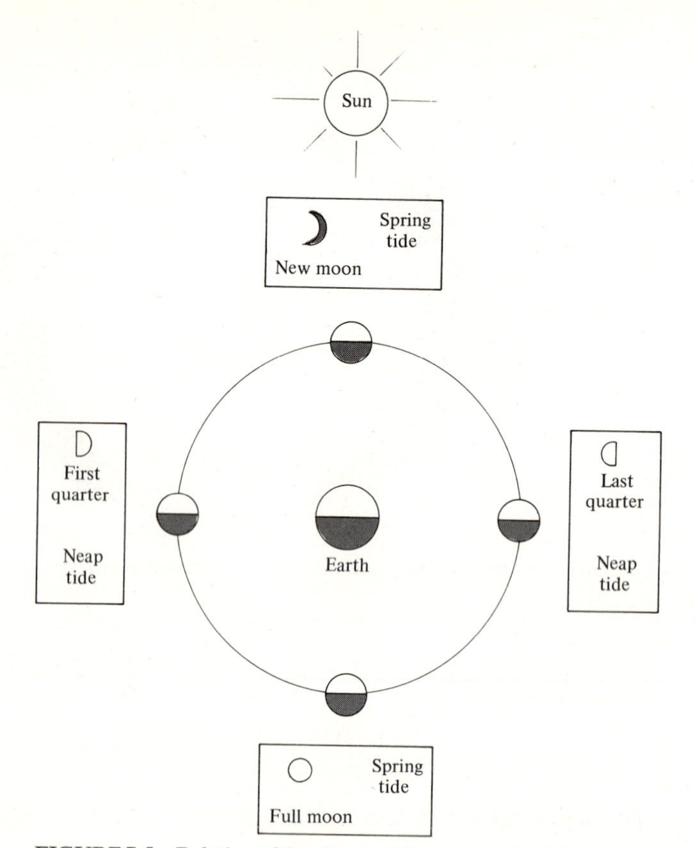

FIGURE 7-5 Relation of the phases of the moon to the tides on earth.

two tidal maxima pass a given spot on earth every $(24 + 24/27\frac{1}{3})$ h. Thus high tide occurs every 12 h and 26.5 min, and high tide is observed about 53 min later each day.

The two high tides are not of the same height as a result of the inclination of the earth's axis to the normal of the moon's orbital plane about the earth. In the Northern Hemisphere the high tide which occurs closest to the moon is higher, as illustrated in Fig. 7-6.

The tides are really more complicated than described above. Along coastal regions the configuration of the land masses and the ocean bottom cause considerable amplification or suppression of the tidal range. Over the world, tidal ranges vary by amounts over an order of magnitude 0 to 20 m.

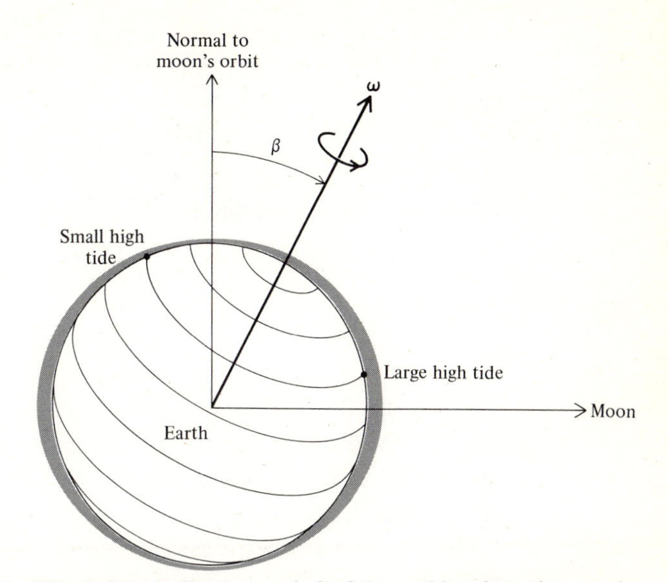

FIGURE 7-6 **Effect of the inclination angle β of the earth's axis to the moon's orbital plane on the heights of tides. β varies from 17 to 29° as the moon's elliptical orbit precesses slowly about the normal to the plane of the earth's heliocentric orbit.**

The friction of the moving tidal waves against ocean bottoms and the collisions of the tides with the continental shorelines dissipate energy at a rate estimated at 2 billion horsepower. To compensate for this energy loss, the earth's rotation about its axis must slow down at the rate of 4.4×10^{-8} s per rotation. The cumulative loss of time over a century is about 28 s. This gradual lengthening of the day is confirmed by the observation that various astronomical events such as eclipses and motions of planets seem to run systematically 28 s ahead of calculations based on observations a century ago.

Since the angular momentum of the earth-moon system is conserved, the slowing down of the earth's rotation about its axis must be accompanied by an equal increase of the net orbital angular momentum of the moon and earth about their center of mass. We can calculate the effect on the motion of the moon. The angular momentum due to the orbital motion is

$$L = \mu v r \tag{7-36}$$

where $\mu = M_e m_L/(M_e + m_L)$ is the reduced mass; $\mathbf{r} = \mathbf{r}_L - \mathbf{r}_e$ and

$\mathbf{v} = \mathbf{v}_L - \mathbf{v}_e$ are the relative coordinate and velocity. In a circular-orbit approximation, the gravitational and centrifugal forces must balance.

$$\frac{GM_e m_L}{r^2} = \frac{\mu v^2}{r} \tag{7-37}$$

Combining Eqs. (7-36) and (7-37), we find

$$r = \frac{L^2}{GM_e m_L \mu}$$
$$v = \frac{GM_e m_L}{L} \tag{7-38}$$

The increase in the orbital angular momentum L results in an increase of the distance r of the moon from the earth and a decrease in the moon's linear velocity $v_M = vM_e/(m_L + M_e)$. The recession of the moon calculated from the rate of slowing down of the earth's rotation is about $\Delta r = 1$ cm per lunar revolution about the earth. Thus the moon moves in a very slowly unwinding spiral trajectory. As a crude estimate, we divide the present distance to the moon by the present rate of recession and find that the moon and earth would have been very close neighbors several billion years ago, about the time when the sun and its planets condensed out of a great cloud of gas.

7-3 AUTOMATIC ATTITUDE STABILIZATION OF ORBITING SATELLITES

For some purposes it is desirable to maintain an orbiting satellite or space station so that its orientation with respect to the earth's surface is always the same (i.e., a fixed attitude). For example, the antenna of a communications satellite should always be directed toward the earth. The initial rotation of a spherical satellite can be set so that the satellite makes one rotation per revolution. Then the same side of the satellite always faces the earth. However, any small perturbation on the satellite or any inaccuracy in the original rotation rate will cause it to drift out of synchronous rotation. A more suitable method of attitude maintenance utilizes torque due to the earth's gravitational pull on a nonspherical orbiting satellite. This technique was successfully employed during the Gemini 11 and 12 flights in 1966. We shall discuss two idealized examples that illustrate the gravitational method of attitude maintenance.

For our first example we consider a slender uniform rod in a circular orbit of radius R about the center of the earth. The angular frequency ω_0 of the orbit is determined by the requirement that the centrifugal and gravitational forces balance.

$$m\omega_0{}^2 R = \frac{GmM_e}{R^2} \tag{7-39}$$

This gives

$$\omega_0 = \sqrt{\frac{GM_e}{R^3}} \tag{7-40}$$

While the center of mass follows the circular orbit, the rod is assumed to be free to rotate about its CM in the plane of the orbit. The orientation angle of the rod relative to the earth direction will be denoted by ϕ, as illustrated in Fig. 7-7. To determine the torques which govern the rotation of the rod, we choose a coordinate system that rotates with the rod. The origin of this moving coordinate system is taken to be the CM of the rod, and the x axis is directed along the radius vector from the earth's center. From Eq. (6-33) the effective force $\delta\mathbf{F}$ on a mass element dm in the rod in this rotating reference frame is

$$\delta\mathbf{F} = d\mathbf{F} - dm\left[\boldsymbol{\omega}_0 \times (\boldsymbol{\omega}_0 \times \mathbf{r}) \times 2\boldsymbol{\omega}_0 \times \mathbf{v} + \frac{d^2\mathbf{R}}{dt^2}\right] \tag{7-41}$$

Here $d\mathbf{F}$ is the gravitational force due to the earth, $\mathbf{r}$ is the location of dm in the rotating frame, and $\mathbf{R}$ is the location of the origin of the moving frame with respect to the earth's center.

We shall first show that the effective torque on the rod,

$$\mathbf{N} = \int \mathbf{r} \times \delta\mathbf{F} \tag{7-42}$$

arises solely from the $d\mathbf{F}$ term in Eq. (7-41). Since the motion of the rod is in the plane of rotation, we know that

$$\boldsymbol{\omega}_0 \cdot \mathbf{r} = 0 \tag{7-43}$$

for $\mathbf{r}$ along the rod. As a direct consequence, the torque due to the centrifugal-force term $\boldsymbol{\omega}_0 \times (\boldsymbol{\omega}_0 \times \mathbf{r})$ in Eq. (7-41) vanishes.

$$\mathbf{r} \times [\boldsymbol{\omega}_0 \times (\boldsymbol{\omega}_0 \times \mathbf{r})] = \mathbf{r} \times [\boldsymbol{\omega}_0(\boldsymbol{\omega}_0 \cdot \mathbf{r}) - \mathbf{r}\omega_0{}^2] = 0$$

The vanishing of the torque due to the Coriolis force,

$$\mathbf{r} \times (\boldsymbol{\omega}_0 \times \mathbf{v}) = \boldsymbol{\omega}_0(\mathbf{r} \cdot \mathbf{v}) - \mathbf{v}(\mathbf{r} \cdot \boldsymbol{\omega}_0)$$

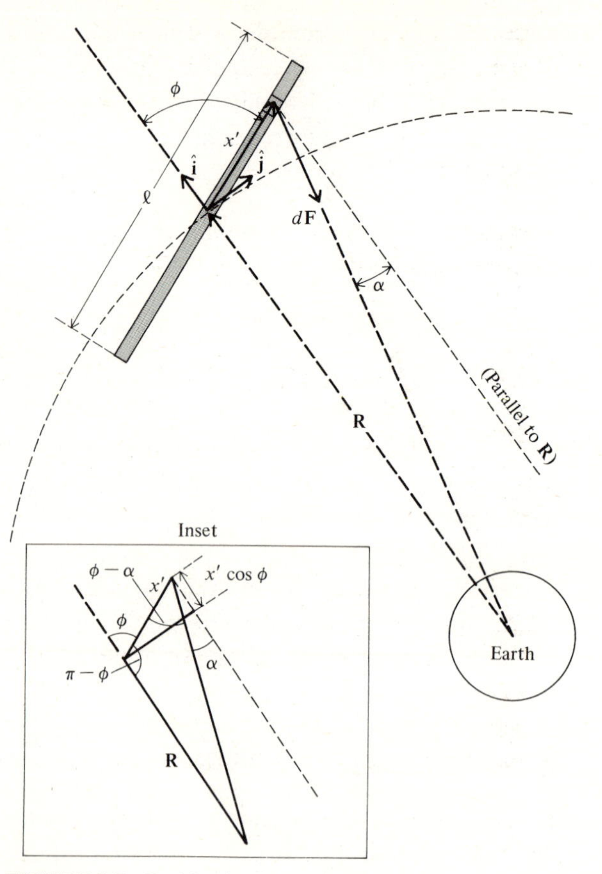

FIGURE 7-7 **Rod in circular orbit about the earth oscillating about its CM in the orbital plane.**

follows from Eq. (7-43) and the orthogonality of **r** and **v** for the rotating rod. Finally, in the last term of Eq. (7-41), the acceleration $d^2\mathbf{R}/dt^2$ is in the plane of the rotation, directed toward the center of the earth. When we take the cross product with **r**, the torque thus obtained points perpendicular to the plane of rotation. Furthermore, the torque is antisymmetric about the center of the rod. When integrated over *dm* along the rod, the torque due to this term vanishes.

Thus the net torque from Eqs. (7-41) and (7-42) is

$$\mathbf{N} = \int \mathbf{r} \times d\mathbf{F} \tag{7-44}$$

In terms of the mass density per unit length $\sigma = M/l$, the component of the gravitational force perpendicular to the rod gives rise to the differential torque.

$$\mathbf{r} \times d\mathbf{F} = \hat{\mathbf{z}} \frac{GM_E \sigma \, dx'}{(R + x' \cos \phi)^2} x' \sin (\phi - \alpha) \tag{7-45}$$

where x' is the coordinate measured along the rod and α is the angle between $\mathbf{R}$ and $d\mathbf{F}$, as illustrated in Fig. 7-7. The net torque is in turn given by

$$\mathbf{N} = \hat{\mathbf{z}} GM_E \sigma \int_{-l/2}^{l/2} \frac{dx' x' \sin (\phi - \alpha)}{(R + x' \cos \phi)^2} \tag{7-46}$$

From the geometry of Fig. 7-7 and the law of sines, we find

$$\frac{\sin(\phi - \alpha)}{R} = \frac{\sin(\pi - \phi)}{R + x' \cos \phi} = \frac{\sin \phi}{R + x' \cos \phi} \tag{7-47}$$

We use this relation to eliminate the angle α in Eq. (7-46).

$$\mathbf{N} = \hat{\mathbf{z}} \frac{GM_E \sigma}{R^2} \sin \phi \int_{-l/2}^{+l/2} \frac{dx' x'}{[1 + (x'/R) \cos \phi]^3} \tag{7-48}$$

Inasmuch as $l \ll R$, we can expand the denominator of the integrand in powers of x'/R by use of Eq. (7-29) with $n = -3$ and $\beta = (x'/R) \cos \phi$. Retaining only the leading terms, we obtain

$$\frac{1}{(1 + (x'/R) \cos \phi)^3} \approx 1 - \frac{3x'}{R} \cos \phi \tag{7-49}$$

In this approximation the torque becomes

$$\begin{aligned} \mathbf{N} &\approx \hat{\mathbf{z}} \frac{GM_E \sigma}{R^2} \sin \phi \int_{-l/2}^{l/2} dx' \left(x' - \frac{3x'^2}{R} \cos \phi \right) \\ &\approx \omega_0{}^2 \sigma \frac{l^3}{4} \sin \phi \cos \phi \hat{\mathbf{z}} \end{aligned} \tag{7-50}$$

The rigid-body equation of motion in the rotating frame about the CM of the rod is given by Eq. (6-80).

$$\frac{\delta \mathbf{L}}{\delta t} + \boldsymbol{\omega}_0 \times \mathbf{L} = \mathbf{N} \tag{7-51}$$

Since $\mathbf{L}$ and $\boldsymbol{\omega}_0$ are parallel, $\boldsymbol{\omega}_0 \times \mathbf{L} = 0$. Using

$$\mathbf{L} = I \boldsymbol{\omega} = I \frac{\delta \phi}{\delta t} \hat{\mathbf{z}} \tag{7-52}$$

with $I = \frac{1}{12}Ml^2 = \frac{1}{12}\sigma l^3$ for the moment of inertia of the rod about its CM, we obtain from Eqs. (7-50) and (7-52) the differential equation

$$\frac{\delta^2\phi}{\delta t^2} + 3\omega_0{}^2 \sin\phi \cos\phi = 0 \tag{7-53}$$

Thus the rod is in equilibrium for $\phi = 0, \pi/2, \pi$, or $3\pi/2$. The equilibrium is stable for $\phi = 0$ or π and unstable for $\phi = \pi/2$ or $3\pi/2$. For small oscillations about stable equilibrium, the equation of motion simplifies to

$$\frac{\delta^2\phi}{\delta t^2} + (\sqrt{3}\omega_0)^2 \phi = 0 \tag{7-54}$$

This equation is simple harmonic with frequency

$$\omega = \sqrt{3}\omega_0 \tag{7-55}$$

Once the rod is placed in orbit in a stable-equilibrium configuration, the gravitational torques automatically constrain it to oscillate about a constant attitude with respect to the surface of the earth.

We next turn to a somewhat more realistic model for an automatic-attitude-stabilizing satellite. We consider a dumbbell-shaped satellite consisting of two spheres of radius a and mass m connected by a rod of length l and negligible mass. As in the previous model, we assume that the satellite's center of mass is in a circular orbit at a distance R from the center of the earth and that the satellite is free to rotate about its CM in the orbital plane. The orbital angular frequency is again given by Eq. (7-40). To solve for the equation of motion for the dumbbell model, we shall use the lagrangian method. We choose a fixed coordinate system with origin at the earth's center. The polar coordinates of the satellite's CM are R and θ. The orientation of the satellite relative to the earth direction is measured by the angle ϕ. From the geometry of Fig. 7-8, we obtain the following relations (using $l \ll R$):

$$\begin{aligned} r_1 &\approx R + \frac{l}{2}\cos\phi \\ r_2 &\approx R - \frac{l}{2}\cos\phi \\ \theta_1 &\approx \theta - \frac{l}{2R}\sin\phi \\ \theta_2 &\approx \theta + \frac{l}{2R}\sin\phi \end{aligned} \tag{7-56}$$

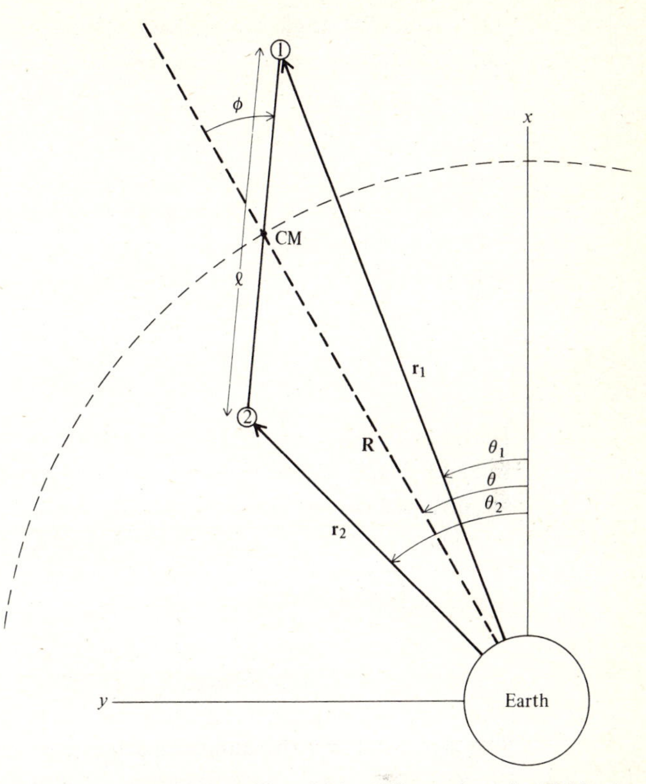

FIGURE 7-8 Orbiting dumbbell satellite which oscillates about its CM in the orbital plane.

The kinetic energy of the dumbbell satellite is

$$T = \tfrac{1}{2}m(\dot{r}_1^2 + r_1^2\dot{\theta}_1^2 + \dot{r}_2^2 + r_2^2\dot{\theta}_2^2) + (\tfrac{2}{5}ma^2)\dot{\phi}^2 \tag{7-57}$$

The last term in Eq. (7-57) takes into account the kinetic energy due to the rotational motion of the two spheres with angular velocity $\dot{\phi}$ with respect to their individual centers. The potential energy is given by

$$V = -GMm\left(\frac{1}{r_1} + \frac{1}{r_2}\right) = -m\omega_0{}^2R^3\left(\frac{1}{r_1} + \frac{1}{r_2}\right) \tag{7-58}$$

where ω_0 is defined by Eq. (7-40). The above expressions for T and V can now be expressed in terms of the orbit parameters R, $\dot{\theta} = \omega_0$, and the orientation angle ϕ by means of Eq. (7-56). Differentiation of Eq. (7-56) with respect to time leads to

$$\begin{aligned}\dot{r}_1 &= -\frac{l\dot{\phi}}{2}\sin\phi \\ \dot{r}_2 &= \frac{l\dot{\phi}}{2}\sin\phi \\ \dot{\theta}_1 &= \omega_0 - \frac{l\dot{\phi}}{2R}\cos\phi \\ \dot{\theta}_2 &= \omega_0 + \frac{l\dot{\phi}}{2R}\cos\phi\end{aligned} \tag{7-59}$$

When Eqs. (7-56) and (7-59) are substituted into the energy expressions of Eqs. (7-57) and (7-58) and terms of higher order than $(l/R)^2$ are dropped compared with unity, the kinetic and potential energies become

$$\begin{aligned}T &\approx \tfrac{1}{2}m\left[2R^2\omega_0{}^2 + \left(\frac{l^2}{2} + \frac{4}{5}a^2\right)\dot{\phi}^2 - 2\omega_0 l^2\dot{\phi}\cos^2\phi + \frac{\omega_0{}^2 l^2}{2}\cos^2\phi\right] \\ V &\approx -2m\omega_0{}^2R^2\left(1 + \frac{l^2}{4R^2}\cos^2\phi\right)\end{aligned} \tag{7-60}$$

The lagrangian for the dumbbell satellite is

$$L = T - V = m[3R^2\omega_0{}^2 + \tfrac{1}{4}(l^2 + \tfrac{8}{5}a^2)\dot{\phi}^2 - \omega_0 l^2\dot{\phi}\cos^2\phi + \tfrac{3}{4}\omega_0{}^2l^2\cos^2\phi] \tag{7-61}$$

The Lagrange equation for the angular variable is

$$\frac{d}{dt}\left(\frac{\partial L}{\partial\dot{\phi}}\right) - \frac{\partial L}{\partial\phi} = 0 \tag{7-62}$$

The appropriate derivatives of Eq. (7-61) are

$$\begin{aligned}\frac{d}{dt}\left(\frac{\partial L}{\partial\dot{\phi}}\right) &= \frac{m}{2}(l^2 + \tfrac{8}{5}a^2)\ddot{\phi} + m\omega_0 l^2\dot{\phi}\sin 2\phi \\ \frac{\partial L}{\partial\phi} &= m\omega_0 l^2\dot{\phi}\sin 2\phi - \tfrac{3}{4}m\omega_0{}^2l^2\sin 2\phi\end{aligned} \tag{7-63}$$

From Eqs. (7-62) and (7-63), we find the equation of motion

$$\ddot{\phi} + \tfrac{3}{2}\omega_0{}^2\left(\frac{l^2}{l^2 + \tfrac{8}{5}a^2}\right)\sin 2\phi = 0 \tag{7-64}$$

This equation is similar in form to Eq. (7-53). For small angles of oscillation, $\sin 2\phi \approx 2\phi$, and the motion is simple harmonic with angular frequency

$$\omega = \omega_0 \sqrt{\frac{3l^2}{l^2 + \frac{8}{5}a^2}} \tag{7-65}$$

It is interesting to note that for l much larger than a, the angular frequency ω does not depend on any property of the satellite, except being linear.

In the Gemini 11 and 12 flights in late 1966, this method of unattended station-keeping was shown to be a practical technique in space. The method used was to hook a tether to the Agena rocket stage orbiting near the Gemini satellite. The Agena-Gemini configuration then resembled the dumbbell model described above.

The moon provides another example of automatic attitude stabilization. The same face of the moon is always toward the earth's surface. Torques due to the earth's gravitational pull on the lopsided moon account for this attitude stabilization.

7-4 GRAVITY FIELD OF THE EARTH

A spherically symmetric body gravitates as if its mass were concentrated at its center. The spherical approximation for celestial bodies is often a good one. Nevertheless, small nonspherical distortions in the shapes are usually present which affect the gravity field in the vicinity of the bodies. The most common deviation from spherical symmetry is caused by rotation about an axis. Due to its rotation, our earth has an equatorial radius about 21.5 km in excess of the polar radius. The deviations from spherical symmetry in the shape of the earth can be mapped out through the effects on the gravity field. In this section we determine the modification in the earth's gravitational potential and gravity due to its rotational distortion.

The rotational distortion of the earth is axially symmetric about the north-south axis of rotation. The potential due to an element dM in the earth at (x, y, z) at a location (X, Y, Z) is

$$d\Phi = -\frac{G\,dM}{[(X-x)^2 + (Y-y)^2 + (Z-z)^2]^{1/2}} \tag{7-66}$$

The net potential is found by integration over all elements in the earth. If we set

$$
\begin{aligned}
r^2 &= x^2 + y^2 + z^2 \\
R^2 &= X^2 + Y^2 + Z^2
\end{aligned}
\tag{7-67}
$$

the potential is

$$
\Phi(X, Y, Z) = -\frac{G}{R} \int \frac{dM}{\left[1 - \frac{2(xX + yY + zZ)}{R^2} + \frac{r^2}{R^2}\right]^{1/2}} \tag{7-68}
$$

The square root in the integrand can be expanded in a power series in the small parameter.

$$
\beta = -\frac{2}{R}\left(\frac{xX + yY + zZ}{R}\right) + \frac{r^2}{R^2} \tag{7-69}
$$

Using the Taylor series expansion of Eq. (7-29) with $n = -1/2$, the potential to order $(r/R)^2$ is given by

$$
\begin{aligned}
\Phi = -\frac{G}{R}\Bigg\{\int dM + \frac{1}{R}\int dM\left(\frac{xX + yY + zZ}{R}\right) \\
+ \frac{1}{R^2}\int dM\left[\frac{3}{2}\left(\frac{xX + yY + zZ}{R}\right)^2 - \tfrac{1}{2}r^2\right]\Bigg\}
\end{aligned}
\tag{7-70}
$$

Since (x, y, z) are measured from the center of mass of the earth, the integrals over terms linear in these variables vanish. Then the potential expression simplifies to

$$
\Phi = -\frac{GM_e}{R} - \frac{G}{R^3}\int dM\left[\frac{3}{2}\left(\frac{x^2X^2 + y^2Y^2 + z^2Z^2}{R^2}\right) - \tfrac{1}{2}r^2\right] \tag{7-71}
$$

The first term is just the gravitational potential if the earth were spherical. Since the deviations from spherical shape are small, the correction terms in Eq. (7-71) are also small. The truncation of the expansion in Eq. (7-70) after a few terms is therein justified. Due to the axial symmetry of the earth, the integrals over x^2 and y^2 in Eq. (7-71) are equal.

$$
\int dMx^2 = \int dMy^2 \tag{7-72}
$$

Using Eq. (7-72) and the definitions of Eq. (7-67), the integral term in the potential can be simplified to

$$
\Phi = -\frac{GM_e}{R} - \frac{G}{R^3}\left(\frac{3Z^2}{2R^2} - \frac{1}{2}\right)\int dM(z^2 - x^2) \tag{7-73}
$$

The integral can be represented by the moments of inertia as defined in Eq. (5-75).

$$\int dM(z^2 - x^2) = I_{xx} - I_{zz} \equiv I - I_3 \tag{7-74}$$

Moreover, the ratio Z/R is simply the cosine of the polar angle θ of the mass m with respect to the rotation axis of the earth. With these substitutions, we get

$$\Phi(R, \theta) = -\frac{GM_e}{R} - \frac{G(I - I_3)}{R^2}\left(\tfrac{3}{2}\cos^2\theta - \tfrac{1}{2}\right) \tag{7-75}$$

At this stage it is convenient to introduce a dimensionless parameter

$$\varepsilon' = \frac{I_3 - I}{\tfrac{2}{5}M_e R_E^2} \tag{7-76}$$

as a measure of the earth's distortion. Here R_E is the equatorial radius of the earth. For a distorted earth of ellipsoidal shape and uniform density, the moments of inertia are

$$I_3 = \tfrac{2}{5}M_e R_E^2 \tag{7-77}$$

$$I = \tfrac{1}{5}M_e(R_E^2 + R_P^2) \tag{7-78}$$

where R_P is the polar radius. When Eqs. (7-77) and (7-78) are used in Eq. (7-76), we find

$$\varepsilon' = \frac{(R_E + R_P)(R_E - R_P)}{2R_E^2} \approx \frac{R_E - R_P}{R_E} \tag{7-79}$$

In reality, the earth consists of a molten core surrounded by a less dense mantle. As a result, we expect

$$\varepsilon' < \frac{R_E - R_P}{R_E} \tag{7-80}$$

An alternative measure of the distortion is the "flattening" parameter ε

$$\varepsilon = \frac{R_E - R_P}{R_E} \tag{7-81}$$

In terms of ε' defined by Eq. (7-76), the potential of Eq. (7-75) due to the distorted earth is

$$\Phi(R, \theta) = -\frac{GM_e}{R}\left[1 - \frac{2\varepsilon' R_E^2}{5R^2}\left(\tfrac{3}{2}\cos^2\theta - \tfrac{1}{2}\right)\right] \tag{7-82}$$

For $\varepsilon' \neq 0$ the potential energy is noncentral. The effects of the non-central term on the gravity field of the earth provide an experimental measure of the distortion ε'.

A small mass m rotating with the earth experiences a force due to the gravitational potential, given in Eq. (7-82), and a centrifugal force due to the rotation of the earth. The centrifugal force on m, from Eq. (6-20), is

$$F_{cf} = m\omega^2 \rho$$

where ρ is the perpendicular distance to m from the rotation axis. The centrifugal potential is

$$\Phi_{cf} = -\frac{1}{m}\int_0^\rho F_{cf}\, d\rho$$

or (7-83)

$$\Phi_{cf} = -\tfrac{1}{2}\omega^2 \rho^2$$

In terms of the distance from the earth's center R and the polar angle θ of m with respect to the earth's rotation axis, we have

$$\rho = R \sin\theta$$

and

$$\Phi_{cf}(R, \theta) = -\tfrac{1}{2}\omega^2 R^2 \sin^2\theta \tag{7-84}$$

The total potential is

$$\Phi(R, \theta) = -gR_E^2\left[\frac{1}{R} - \frac{2\varepsilon' R_E^2}{5R^3}\left(\tfrac{3}{2}\cos^2\theta - \tfrac{1}{2}\right) + \frac{\lambda R^2}{2R_E^3}\sin^2\theta\right] \tag{7-85}$$

where λ is the dimensionless parameter

$$\lambda = \omega^2 R_E / g \tag{7-86}$$

and g is the gravitational acceleration for a spherical earth,

$$g = \frac{GM_e}{R_E^2} \tag{7-87}$$

The numerical value of λ is

$$\lambda = \frac{(0.727 \times 10^{-4})^2(6{,}378 \times 10^3)}{9.78}$$

$$= \frac{1}{290} \tag{7-88}$$

The result in Eq. (7-85) can be used to determine the effective gravity force on m at any angle θ.

$$\mathbf{F} \equiv m\mathbf{g}_{\text{eff}} = -m\nabla\Phi(R, \theta) \tag{7-89}$$

In spherical coordinates the ∇ operator is given by (see Prob. 2-11)

$$\nabla = \hat{\mathbf{R}}\frac{\partial}{\partial R} + \hat{\mathbf{l}}\frac{1}{R}\frac{\theta}{\partial\theta} + \hat{\mathbf{m}}\frac{1}{R\sin\theta}\frac{\partial}{\partial\phi} \tag{7-90}$$

where $\hat{\mathbf{R}}$, $\hat{\mathbf{l}}$, and $\hat{\mathbf{m}}$ are the orthogonal unit vectors in the R, θ, and ϕ directions. Applying Eqs. (7-89) and (7-90) to Eq. (7-85), we obtain

$$\mathbf{g}_{\text{eff}} = -\hat{\mathbf{R}}g\left[\frac{R_E{}^2}{R^2} - \frac{6\varepsilon' R_E{}^4}{5R^4}\left(\tfrac{3}{2}\cos^2\theta - \tfrac{1}{2}\right) - \frac{\lambda R}{R_E}\sin^2\theta\right] + \hat{\mathbf{l}}g\sin\theta\cos\theta\left(\frac{6\varepsilon' R_E{}^4}{5R^4} + \frac{\lambda R}{R_E}\right) \tag{7-91}$$

For $\theta = 0$ or $\theta = \pi/2$, the acceleration is purely in the negative radial direction. At the pole ($\theta = 0$), the magnitude of the effective gravitational acceleration is

$$g_{\text{eff}}^P = g\left(\frac{R_E{}^2}{R_P{}^2} - \frac{6\varepsilon' R_E{}^4}{5R_P{}^4}\right) \tag{7-92}$$

while the effective value at the equator ($\theta = \pi/2$) is

$$g_{\text{eff}}^E = g\left(1 + \frac{3\varepsilon'}{5} - \lambda\right) \tag{7-93}$$

The difference in g_{eff} between poles and equator is

$$\Delta g_{\text{eff}} = g_{\text{eff}}^P - g_{\text{eff}}^E = g\left[\left(\frac{R_E{}^2 - R_P{}^2}{R_P{}^2}\right) - \tfrac{3}{5}\varepsilon'\left(1 + \frac{R_E{}^4}{R_P{}^4}\right) + \lambda\right] \tag{7-94}$$

With the definition in Eq. (7-81), we find

$$\frac{\Delta g_{\text{eff}}}{g} = 2\varepsilon - \tfrac{9}{5}\varepsilon' + \lambda \tag{7-95}$$

when we drop quadratic and higher-order terms in small quantities.

For a spherical earth, $\varepsilon = \varepsilon' = 0$, and from Eq. (7-88),

$$\frac{\Delta g_{\text{eff}}}{g} = \lambda = \frac{1}{290} \tag{7-96}$$

This correction to g_{eff} due to the centrifugal force alone is inadequate to account for the experimental value

$$\left(\frac{\Delta g_{\text{eff}}}{g}\right)_{\text{expr}} = \frac{1}{190} \tag{7-97}$$

For a distorted earth with uniform density, $\varepsilon' = \varepsilon$ from Eqs. (7-97) and (7-81), and from Eqs. (7-88), (7-95), and (7-97)

$$\varepsilon = 5\left[\left(\frac{\Delta g_{\text{eff}}}{g}\right)_{\text{expr}} - \lambda\right] = 5\left(\frac{1}{190} - \frac{1}{290}\right) = \frac{1}{110} \tag{7-98}$$

This gives a radius difference

$$R_E - R_P = \varepsilon R_E \approx \frac{6{,}371}{110} = 58 \text{ km} \tag{7-99}$$

The measured radius difference is 21.5 km. This indicates that $\varepsilon' < \varepsilon$, as would be expected for a nonuniform density. In the next section we will derive another relation between ε' and ε from the condition of rotational equilibrium for the earth's surface. This will allow us to make a precise determination of the shape parameters ε' and ε.

7-5 SHAPE OF THE EARTH

The centrifugal forces due to the rotation of the earth about its axis cause the distortion from a spherical shape. In the equilibrium state a particle of mass m on the earth's surface would experience no net force tangent to the surface. In such a condition of rotational equilibrium, the sum of the gravitational and centrifugal potentials must therefore be a constant on the surface. From Eq. (7-85) we have

$$\frac{1}{R} - \frac{2\varepsilon' R_E^2}{5R^3}\left(\tfrac{3}{2}\cos^2\theta - \tfrac{1}{2}\right) + \frac{\lambda R^2}{2R_E^3}\sin^2\theta = C \tag{7-100}$$

where C is a constant. If we evaluate Eq. (7-100) at the pole ($\theta = 0$), we get

$$\frac{1}{R_P} - \frac{2\varepsilon' R_E^2}{5R_P^3} = C \tag{7-101}$$

On the other hand, at the equator ($\theta = \pi/2$), we find

$$\frac{1}{R_E} + \frac{1}{5}\frac{\varepsilon'}{R_E} + \frac{1}{2}\frac{\lambda}{R_E} = C \tag{7-102}$$

The difference of Eqs. (7-101) and (7-102) is

$$\frac{R_E - R_P}{R_E R_P} - \frac{\varepsilon'}{5}\left(\frac{2R_E^2}{R_P^3} + \frac{1}{R_E}\right) - \frac{\lambda}{2R_E} = 0 \tag{7-103}$$

When we use Eq. (7-81) and retain only first-order terms in the small parameters ε and ε', we get

$$\varepsilon - \tfrac{3}{5}\varepsilon' - \tfrac{1}{2}\lambda = 0 \tag{7-104}$$

as the basic equation for the shape of the earth.

The first theoretical calculation of the earth's shape was made by Newton. His calculation omitted the contribution ε' due to the change in the gravitational potential from the distortion. In that limit

$$\varepsilon = \frac{\lambda}{2} = \frac{1}{580} \tag{7-105}$$

from Eqs. (7-104) and (7-88). This value is a factor of 2 smaller than the experimental value

$$\varepsilon_{\text{expr}} = \frac{R_E - R_P}{R_E} = \frac{21.5\text{ km}}{6{,}378\text{ km}} = \frac{1}{297} \tag{7-106}$$

In the approximation of a uniform density for the earth, $\varepsilon' = \varepsilon$ from Eqs. (7-79) and (7-88). Then, from Eqs. (7-104) and (7-88),

$$\varepsilon = \frac{5}{4}\lambda = \frac{5}{4}\left(\frac{1}{290}\right) = \frac{1}{232} \tag{7-107}$$

which is closer to the experimental value in Eq. (7-106).

In the absence of direct information on the earth's density (as could be obtained from seismographic studies), we can combine the calculated change in effective gravitational acceleration of Eq. (7-95) with the shape condition in Eq. (7-104) to solve for ε and ε' in terms of $\Delta g_{\text{eff}}/g$ and λ. We obtain

$$\begin{aligned} \varepsilon &= \tfrac{5}{2}\lambda - \frac{\Delta g_{\text{eff}}}{g} \\ \varepsilon' &= \tfrac{10}{3}\lambda - \frac{5}{3}\frac{\Delta g_{\text{eff}}}{g} \end{aligned} \tag{7-108}$$

From Eqs. (7-88) and (7-97) the numerical values of ε and ε' are

$$\begin{aligned} \varepsilon &= \frac{1}{298} \\ \varepsilon' &= \frac{1}{367} \end{aligned} \tag{7-109}$$

As anticipated in Eq. (7-80), the value of ε' is smaller than ε.

The value of ε' can be most accurately deduced from precession of satellite orbits induced by the noncentral term in the potential of the distorted earth in Eq. (7-82). A value similar to that in Eq. (7-109) is found.

PROBLEMS

1. The density of a planet of radius R with a molten core of radius $R/2$ is given by ρ for $r > R/2$ and 5ρ for $r < R/2$, where ρ is a constant. Calculate the gravitational potential at points inside and outside the planet.

2. Find the mass density $\rho(r)$ of a planet for which the gravitational force has constant magnitude throughout its interior.

3. If a narrow tunnel were dug through the earth along a diameter, show that the motion of a particle in the tunnel would be simple harmonic. Compare the period to the orbital period of a satellite in a circular orbit close to the earth. Assume that the density of the earth is uniform.

4. The gravitational attraction due to a nearby mountain range might be expected to cause a plumb bob to hang at an angle slightly different from vertical. If a mountain range could be represented by an infinite half-cylinder of radius a and density ρ_M lying on a flat plane, show that a plumb bob at a distance r_0 from the cylinder axis would be deflected by an angle $\theta \approx \pi a^2 G\rho_M/(r_0 g)$. In actual measurements of this effect, the observed deflection is much smaller. Next assume that the mountain range can be represented by a cylinder of radius a and density ρ_M which is flotating in a fluid of density $2\rho_M$, as illustrated. Show that the plumb-bob deflection due to the mountain range is zero in this model. Since the latter result is in much better agreement with observations, it is postulated that mountains, and also continents, are in isostatic equilibrium with the underlying mantle rock.

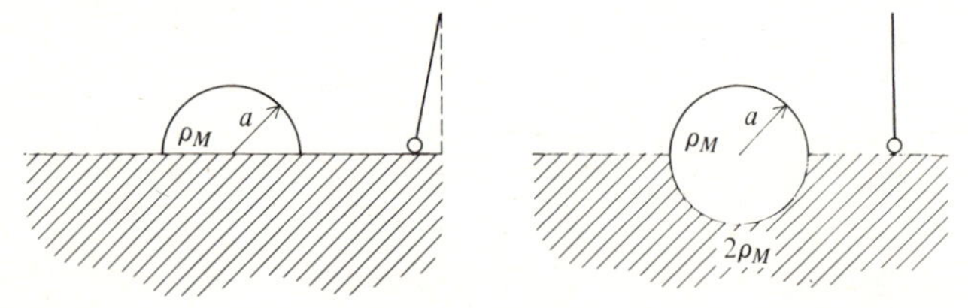

5. A dumbbell-shaped satellite consisting of two spheres of radius a and mass m connected by a rod of length l and of negligible mass is in a circular orbit about the earth. Find the frequency of its small oscillations in a plane passing through the center of the earth perpendicular to the orbit plane.

6. The mean radius of an oblate planet is defined as the radius of a sphere having the same volume as the planet. Show that $R_{\text{mean}} = R_E(1 - \varepsilon/3)$, when R_E is the equatorial radius and $\varepsilon = (R_E - R_P)/R_E$.

7. Assuming that Saturn is a planet of uniform density, calculate the flattening parameter $\varepsilon = (R_E - R_P)/R_E$. The mass of Saturn is 95.3 earth masses, its equatorial radius is 60,000 km, and its period of rotation about its axis is 10.4 h. Compare your answer with the measured distortion $\varepsilon = 1/9.2$. Is the assumption of uniform density justified?

8. For all spinning planets in which the density increases toward the center, the flattening parameter $\varepsilon = (R_E - R_P)/R_E$ is bounded by

$$0.5\lambda \leq \varepsilon \leq 1.25\lambda$$

where $\lambda = \omega^2 R_E/g$. Derive this result. Calculate ε/λ for the planets of our solar system for which the necessary measurements have been made.

9. A satellite will begin to break up when the maximum tidal force on the satellite exceeds the internal gravity force (neglecting the tensile strength of its material). In the following problem, assume that the planet (mass M, radius R) and satellite (mass $m \ll M$, radius r if M were absent) have the same uniform mass density ρ. Neglecting deviations from sphericity of the satellite's shape, show that breakup begins to occur when the center of the satellite is a distance $d = 1.26R$ from the center of the planet. Next allow for small distortions in the satellite's shape due to the tidal force. Using expressions in the text for the tidal potential and the gravitational potential of a slightly distorted sphere, show that the shape parameter of the satellite is given by $\varepsilon = -\frac{15}{4}(R/d)^3$, where d is the distances between the centers. Then show that breakup occurs at $d \approx 1.71R(\varepsilon = -0.75)$. Inasmuch as the distortion is large, the assumption of $\varepsilon \ll 1$ is unjustified. The result of an exact calculation is $d = 2.44R$; this is known as *Roche's limit*.

10. Pulsars are thought to be rapidly rotating neutron stars. The Crab nebula pulsar has a radius of about 10 km, a mass of about one solar mass, and revolves at a rate of 30 times per second.

a. Assuming a uniform mass density, compute the difference between the equatorial and polar radii of the Crab nebula pulsar.

b. Find the nearest distance that a man 2 m tall could approach the pulsar without being pulled apart. Assume that his body points toward the pulsar and that dismemberment begins to occur when the gravitational force between head and feet exceed five times his body weight. What is the period of revolution in a circular orbit about the pulsar at this minimum distance?

APPENDIX A

Tables of Data, Constants, and Units

TABLE A-1 SUN AND EARTH DATA

Mean distance from sun to earth	1.495×10^8 km
Mass of sun	$M_\odot = 1.987 \times 10^{30}$ kg
Mass of earth	$M_e = 5.97 \times 10^{24}$ kg
Sun-to-earth mass ratio	$M_\odot / M_e = 332{,}945$
Mean radius of earth	$R_e = 6{,}371$ km
Mean gravity on earth	$g = 9.8064$ m/s^2
Equatorial earth gravity	$g_E = 9.7805$ m/s^2
Polar earth gravity	$g_P = 9.8322$ m/s^2

TABLE A-2 MOON DATA

Semimajor axis of orbit	3.84×10^5 km
Eccentricity of orbit	0.055
Sidereal period about earth (days)	27.32
Inclination of orbit to ecliptic	5.15°
Radius	$R_L = 1{,}741$ km
Mean density, g/cm^3	3.33
Mass (earth = 1)	$M_L = 1/81.56$
Surface gravity (earth = 1)	0.165
Escape velocity, km/s	2.4
Orbital velocity about earth, km/s	1.0

TABLE A-3 PROPERTIES OF THE PLANETS

	Mercury	Venus	Earth	Mars	Jupiter	Saturn	Uranus	Neptune	Pluto
Mean distance from sun, AU*	0.4	0.72	1.00	1.52	5.2	9.5	19.2	30.1	39.5
Eccentricity of orbit	0.206	0.007	0.017	0.093	0.048	0.056	0.047	0.009	0.247
Sidereal period, years	0.24	0.61	1.00	1.9	11.9	29.5	86.0	164.8	248.4
Inclination of orbit to ecliptic	7.00°	3.39°	0.0°	1.85°	1.31°	2.50°	0.77°	1.78°	17.15°
Equatorial radius (earth = 1)†	0.39	0.97	1.0	0.52	11.0	9.4	4.0	3.8	0.45(?)
Mean density, g/cm^3	5.0	4.9	5.52	4.2	1.41	0.71	1.25	1.74	2.0(?)
Mass (earth = 1)‡	0.054	0.185	1.0	0.108	318.4	95.3	14.5	17.3	0.035(?)
Rotation period, days	59	−250	1.0	1.03	0.41	0.43	0.45	0.65	0.7(?)
$\varepsilon = (R_E - R_P)/R_E$			1/300	1/190	1/16.1	1/9.2	0.03		
Surface gravity (earth = 1)§	0.38	0.86	1.0	0.39	2.7	1.16	0.91	1.2	0.2(?)
Escape velocity, km/s	4.2	10.3	11.1	5.0	59.5	36.0	21.2	23.7	0.3(?)
Orbital velocity about sun, km/s	47.8	34.9	29.8	24.1	13.0	9.7	6.8	5.5	4.5

* Where earth = 1.495×10^8 km.
† Where earth = 6,378 km.
‡ Where earth = 5.975×10^{24} kg.
§ Where earth = $9.8\ m/s^2$.

TABLE A-4 SOME PHYSICAL CONSTANTS

Gravitational constant:
$G = 6.673 \times 10^{-11}$ N · m²/kg²

Electron charge:
$e = 1.602 \times 10^{-19}$ C

Proton mass:
$m_p = 1.6725 \times 10^{-27}$ kg $= 938.3$ MeV/c²

Neutron mass:
$m_n = 1.6748 \times 10^{-27}$ kg $= 939.6$ MeV/c²

Electron mass:
$m_e = 9.1096 \times 10^{-31}$ kg $= 0.511$ MeV/c²

α *particle* (He^{++}) *mass:*
$m_\alpha = 6.644 \times 10^{-27}$ kg $= 3727.4$ MeV/c²

Velocity of light:
$c = 2.998 \times 10^8$ m/s

TABLE A-5 SOME NUMERICAL CONSTANTS

$\pi = 3.1415927$
$e = 2.7182818$
$\ln 2 = 0.69314718$
$1 \text{ rad} = 57.2957795°$

TABLE A-6 CONVERSION FACTORS

	Multiply	By	To obtain
Distance:	feet	0.3048	meters
	meters	3.281	feet
	kilometers	3,281	feet
	feet	3.048×10^{-4}	kilometers
	kilometers	0.6214	miles
	miles	1.609	kilometers
	miles	5,280	feet
	meters	100	centimeters
	centimeters	10^{-2}	meters
	kilometers	1,000	meters
	centimeters	0.3937	inches
	inches	2.540	centimeters
	astronomical unit	1.495×10^{8}	kilometers
Velocity:	feet/second	0.3048	meters/second
	meters/second	3.281	feet/second
	meters/second	2.237	miles/hour
	miles/hour	0.4470	meters/second
	feet/second	0.6818	miles/hour
	miles/hour	1.609	kilometers/hour
	kilometers/hour	0.6214	miles/hour
	kilometers/second	2,237	miles/hour
	miles/hour	4.470×10^{-4}	kilometers/second
Mass, weight, and force:	pounds (weight)	0.4536	kilograms (mass)
	kilograms (mass)	2.205	pounds (weight)
	newtons	1	kg m/s^2
	newtons	10^{5}	dynes
	pounds	0.2248	newtons
	newtons	4.448	pounds
Liquid measure:	gallons	3.785	liters
	liters	0.2642	gallons
	liters	10^{-3}	cubic meters
Volume and pressure:	cubic feet	0.02832	cubic meters
	cubic meters	35.31	cubic feet
	pounds/square inch	68,950	dynes/square centimeter
	dynes/square centimeter	1.450×10^{-5}	pounds/square inch
Energy and power:	newton-meter	1	joule
	dyne-centimeter	1	erg
	joule	10^{7}	ergs
	electron volts	1.602×10^{-19}	joule
	joule	6.242×10^{18}	electron volts
	million electron volts	10^{6}	electron volts
	electron volts	10^{-6}	million electron volts
	joule	0.7376	foot-pound
	horsepower	746	watts
	watt	1	joule/second
Time:	mean solar		
	day	8.640×10^{4}	seconds
	year	3.156×10^{7}	seconds
	hour	3,600	seconds
	r/min	0.1047	radians/second
	radians/second	9.549	r/min

TABLE A-7 ABBREVIATIONS FOR UNITS

Length	centimeter	cm
	meter	m
Mass	gram	g
	kilogram	kg
Time	second	s
	hour	h
Velocity	meter per second	m/s
	kilometer per second	km/s
	kilometer per hour	km/h
	mile per hour	mi/h
Angular Velocity	revolutions per minute	r/min
	radians per second	rad/s
Energy	electron volt	eV
	million electron volt	MeV
Force	pound	lb
Charge	coulomb	C

APPENDIX B

Answers to Selected Problems

CHAPTER 1

2. $a = -\mu g$
3. $\mu \approx 2.25\ v_t \approx 137$ m/s
5. $x = \dfrac{2v_0 \sin\alpha(v_0 \cos\alpha + v_r)}{g} = 97.2$ m
6. $h - x = -\dfrac{v_t^2}{2g} \ln\left(1 - \dfrac{v^2}{v_t^2}\right)$; 87.4 m
8. $x = l_0 \cosh\left(\sqrt{\dfrac{g}{l}}\, t\right)$

10. $x = \dfrac{a}{mb}\left[t - \dfrac{1}{b}(1 - e^{-bt})\right];\ v = \dfrac{a}{mb}(1 - e^{-bt})$

11. $x = l + \dfrac{mg}{k}\left(1 - \cos\sqrt{\dfrac{g}{l}}\,t\right)$

14. 94.5 r/min

15. $440 + 1.5 \times 10^{-6}$ cps

16. $A\#/A = 0.0014;\ \nu\# = 466$ cps

18. $x = \dfrac{a}{M(\omega_0{}^2 + b^2)}\left(\dfrac{b}{\omega_0}\sin\omega_0 t - \cos\omega_0 t + e^{-bt}\right)$

CHAPTER 2

2. $\omega = \sqrt{\dfrac{aF_0}{m}}$

4. 10^4 man-years

6. 2.26 km/s

8. $V(\mathbf{r}) = \cos k$

9. $V(\mathbf{r}) = \dfrac{K(x - z)^3}{3}$

12. $x = \dfrac{mg}{k} + a\cos\left(\sqrt{\dfrac{k}{m}}\,t + \alpha\right) \qquad \theta = b\cos\left(\sqrt{\dfrac{g}{l}}\,t + \beta\right)$

13. $\theta \approx 47°$

15. $0 \le E \le mg,\ \frac{5}{2}mg \le E \le \infty$

17. $x = \dfrac{mg}{k}\left(1 - \cos\sqrt{\dfrac{k}{2m}}\,t\right)$

19. $\omega_- = \sqrt{\dfrac{2k}{m}},\ \omega_+ = \sqrt{\dfrac{k}{2m}}$

20. $\omega = \sqrt{\dfrac{g}{l}}\left[\dfrac{1}{1 \pm \sqrt{m_2/(m_1 + m_2)}}\right]^{1/2}$

21. $x_1 = \dfrac{(F/m)(\omega_2{}^2 - \omega^2)}{D};\ x_2 = \dfrac{F}{m}\dfrac{\omega_2{}^2}{D}$; where $F = F_0\cos\omega t$ and $D = (\omega_2{}^2 - \omega^2)(\omega_1{}^2 + \omega_2{}^2 - \omega^2) - \omega_2{}^4$

CHAPTER 3

2. $v = v_0 e^{-(\sigma/m_0)t}$

4. 145,000 N

6. $h = 330$ km, $h_\infty = 690$ km

8. 2.6 eV

11. Target mass 3 times the projectile mass

13. Compression $= \sqrt{\dfrac{mv^2}{6k}}$

16. $\dfrac{d\sigma}{dT} = \dfrac{A}{T_0}$

CHAPTER 4

1. $\mathbf{L} = \mathbf{L}_0 e^{-ct}$
2. 1/32 r/min
5. $F(r) = -\dfrac{L^2}{mr^3}\left(1 + \dfrac{6c}{r}\right)$
7. (*a*) 1.02 km/s, 27.4 days; (*b*) $\dfrac{v_e}{v_L} = 4.9$
11. 6.65 earth radii
14. Apogee = 1,080 km, period = 1.6 h
17. 13.8 km/s, 11.1 km/s
18. $\sigma = \pi R_L^2\left[1 + \left(\dfrac{v_{\text{esc}}^L}{v_0}\right)^2\right]$
19. $\sigma = \dfrac{2\pi}{v_0}\sqrt{\dfrac{2c}{m}}$

CHAPTER 5

1. 4,670 km from center of earth
3. 2.28 m
5. one-third full
7. $\dfrac{F_\odot}{F_e} = 2.18$
8. (*a*) $r_0 = \left(\dfrac{a}{b}\right)^2$, $\omega_0 = \dfrac{b^6}{a^5\sqrt{\mu}}$, energy required $= \dfrac{b^8}{12a^6}$

   (*b*) $L_{\max} = \dfrac{b^2\sqrt{\mu}}{2a}$, $r_B = 2r_0$

   (*c*) $v_1 = \dfrac{b^6}{8a^5}\dfrac{\sqrt{\mu}}{m_1}$, $v_2 = \dfrac{m_1}{m_2}v_1$
11. (*a*) $E = \frac{1}{2}\mu v_1^2$, $L = \mu b_0 v_1$

   (*b*) $r_0 = \dfrac{b_0}{\sqrt{2}}\left[1 + \sqrt{1 + \dfrac{8V_0}{\mu b_0^4 v_1^2}}\right]^{1/2}$

   (*c*) $\cos\beta = \dfrac{m_1 - m_2}{2\sqrt{m_1 m_2}}\dfrac{(v_1^2 - v_{1f}^2)^{1/2}}{v_{1f}}$
12. $v = \dfrac{R\omega}{3}$
15. $T = \frac{1}{4}g$
16. $l = \sqrt{\dfrac{I}{M}}$

19. $a_{up} = a_{down} = \dfrac{-g}{1 + R^2/2r^2}$, $T_{up} = T_{down} = \dfrac{MgR^2}{R^2 + 2r^2}$

21. (*a*) $\frac{3}{4}l$; (*b*) $I = \frac{4}{3}ml^2$
(*c*) $I_{cm} = \frac{5}{24}ml^2$; (*e*) $h = \frac{8}{9}l$

22. $v_{min} = \dfrac{\sqrt{10gb/7}}{1 - \frac{5}{7}b/a}$

25. $V_x{}^\circ = -\frac{2}{5}\omega_z{}^\circ a$, $\omega_z{}^1 = -\omega_z{}^\circ$

CHAPTER 6

2. $g_{eff} = g\sqrt{1 - 0.36 \sin^2 \theta}$, $v = 0.949v_{esc}$

4. $r = \dfrac{v}{2\omega \cos \theta}$

8. (*b*) $\Omega = \sqrt{\dfrac{g}{R}}$; (*c*) $\cos \theta = \dfrac{g}{R\omega^2}$

11. (*a*) 1 s; (*b*) 30 rad/s; (*c*) $\frac{1}{5}$ s

13. $\dfrac{\omega_3}{\omega_1} = \dfrac{1}{2}$

14. 6,460 rad/s is minimum spin for stable precession.

CHAPTER 7

1. $V(r) = G\pi\rho(\frac{10}{3}r^2 - 4R^2) \qquad 0 < r < R/2$

$V(r) = G\pi\rho\left(-\dfrac{2}{3}\dfrac{R^3}{r} + \frac{2}{3}r^2 - 2R^2\right); \qquad R/2 < r < R$

$V(r) = G\pi\rho\left(-\dfrac{2R^3}{r}\right) \qquad r > R$

2. $\rho(r) \propto \dfrac{1}{r}$

5. $\omega = \omega_0\sqrt{\dfrac{3l^2}{l^2 + \frac{8}{5}a^2}}$ same as in orbital plane

7. $\varepsilon = 1/7.3$

10. (*a*) $3m$; (*b*) 2,210 km, 1.79 s

Index